CHINA
SYNDROME

ALSO BY KARL TARO GREENFELD

Speed Tribes:
Days and Nights with Japan's Next Generation

Standard Deviations:
Growing Up and Coming Down in the New Asia

CHINA
SYNDROME

THE TRUE STORY OF THE 21ST CENTURY'S FIRST GREAT EPIDEMIC

KARL TARO GREENFELD

HarperCollins*Publishers*

HarperCollins books may be purchased for educational, business, or sales promotional use. For information, please write: Special Markets Department, HarperCollins Publishers, 10 East 53rd Street, New York, NY 10022.

FIRST EDITION

Designed by Nancy B. Field
Map designed by Dennis Wong

Printed on acid-free paper

Library of Congress Cataloging-in-Publication Data is available upon request.

ISBN-10: 0-06-058722-9
ISBN-13: 978-0-06-058722-2

06 07 08 09 10   ❖/RRD   10 9 8 7 6 5 4 3 2 1

For Esmee and Lola,

*may your immune systems always mount
the appropriate responses*

There are only four questions you need to ask about a virus:

What is it?
What does it do?
Where does it come from?
And how do you kill it?

—GUAN YI,
Virologist, University of Hong Kong

CONTENTS

Hohhot
INNER MONGOLIA
Beijing
Taiyuan
SHANXI
CHINA
Shanghai
Nanchang
JIANGXI
Foshan
GUANGDONG
Shantou
Zhongshan
Heyuan
Guangzhou
Hong Kong
Shenzhen

DRAMATIS PERSONAE

Henk Bekedam Country Director, China, World Health Organization (WHO)

Rob Breiman Epidemiologist, formerly of the United States Centers for Disease Control and Prevention (CDC) and now with the International Center for Diarrheal Disease Research in Dhaka, Bangladesh

Cao Hong Chief Respiratory Specialist, Zhongshan Number Three Hospital

Margaret Chan Director of Health, Hong Kong

Danny Yang Chin Hanoi index patient

Deng Zide Director of Infectious Diseases at Guangzhou Number Three Hospital (also known as Sun Yat-sen Hospital)

Trevor Ellis Chief Veterinarian, Hong Kong Department of Agriculture, Fisheries, and Conservation

Fang Lin Earliest suspected SARS case in Shenzhen

Matthew Forney Bejing Bureau Chief, *Time* magazine

Keiji Fukuda Chief, Epidemiology and Surveillance Section, United States Centers for Disease Control and Prevention

Julie Gerberding Director, U.S. CDC

Guan Yi Assistant Professor, Department of Microbiology, University of Hong Kong, co-head of Pandemic Preparedness in Asia; smuggled samples from China

David Heymann   Executive Director, Communicable Diseases, WHO

Hong Tao   Senior Microbiologist at the Chinese Center for Disease Control and Prevention (CDC)'s Institute of Virology

Hu Jintao   Chief Secretary of the Communist Party and President of China

Huang Wenjie   Chief of Respiratory Diseases at Guangzhou General Military Hospital

Huang Yong   *Time* magazine correspondent in Beijing

Susan Jakes   *Time* magazine correspondent in Beijing

Jiang Yanyong   Physician at Beijing 301 Hospital and whistle-blower

Jiang Zemin   Former Chief Secretary of the Communist Party and former President of China

Anna Kong   Amoy Gardens resident

Thomas G. Ksiazek   Senior Pathologist, U.S. CDC

Liu Jianlun   Guangzhou nephrologist and Hong Kong index patient

James Maguire   Epidemiologist, U.S. CDC

Hitoshi Oshitani   Regional Advisor in Communicable Infectious Disease Surveillance and Response, Western Pacific Region, WHO

Malik Peiris   Professor of Microbiology, University of Hong Kong, co-head of Pandemic Preparedness in Asia; led the team that isolated the virus

Mike Ryan   Geneva-based WHO epidemiologist

Alan Schnur   Coordinator for the Communicable Disease Surveillance and Response Department in China, for the WHO

Klaus Stöhr   Chief of the Global Influenza Programme at the WHO

Joseph Sung   Chief of Gastroenterology and Hepatology, Prince of Wales Hospital

John Tam   Chief of Pathology, Chinese University of Hong Kong

Carlo Urbani Hanoi-based epidemiologist and parasitologist, WHO

Rob Webster Pioneering expert in animal influenzas, and Guan Yi and Malik Peiris's mentor

Wen Jiabao Premier of China

Wu Yi Vice-Premier of China and highest-ranking woman in Chinese government

Xiao Zhenglun Deputy Director of the Guangzhou Institute of Respiratory Diseases and the leader of the Heyuan expedition

Xu Ruiheng Deputy Head, Guangdong CDC

E. K. Yeoh Secretary of Health, Welfare, and Food, Hong Kong

K. Y. Yuen Chairman, Department of Microbiology, University of Hong Kong

Sherif Zaki Senior Virologist, U.S. CDC

Zhang Wenkang Minister of Health, China

Zhong Nanshan Director, Guangzhou Institute of Respiratory Diseases and the most famous physican in China

PROLOGUE

YOU ARE HERE BECAUSE OF YOUR ANCESTORS' IMMUNE SYSTEMS. IF any of them— as tree-dwellers or hunter-gatherers on the plains of Africa, or as farmers or herders in Bronze Age villages, or during the great epidemics of civilized history—had succumbed to any of the many microbes that ruthlessly cull humanity, then you would not be reading this right now. Somehow, because of better nutrition or greater intelligence or geographic circumstance or, most likely, just plain dumb luck, whatever ailments, diseases, and infections your predecessors were stricken with weren't fatal, and those forebears successfully reproduced. The odds against that confluence of genetic good fortune are incalculable; statistically, a German Jew probably had a better chance of surviving the Holocaust.

But for those of us born into the antibiotic era, modern medicine and science have made infectious disease a remote threat. It seems like something that happens to very poor people very far away, in tropical villages or distant third-world cities. When there is an outbreak closer to home, like Lyme disease in Connecticut or mad cow in England, the media coverage and public reaction almost immediately verge on hysteria. We remain, on a basic, primordial level, terrified of disease. It is an unconscious fear, encoded into our DNA, and it surfaces whenever a nasty new microbe is alleged to be aloft, adrift, or, I suppose, afoot. Yet for all our vestigial fright, the vast majority of us have never lived through an infectious-disease outbreak. We are, of course, a historical aberration.

Those of us in the developed world today live remarkably disease-free lives, owing primarily to modern medicine, science, and better nutrition. Diseases thrive on starvation, and there are few going hungry in the lands of Carrefour, Park 'n' Shop, and Gristedes. But trace your own family tree back a generation, two at the most, and immediately the impact of disease is apparent in genealogical dead ends—great-uncles whose first wives died in childbirth, great-aunts whose tombstones read B.1920–D.1923. Whooping cough, measles, smallpox, plague, tuberculosis, dysentery, and influenza killed far more on the Atlantic, or Pacific, crossing or during covered-wagon journeys west than did storms, Indians, and frostbite combined. The immigrant's song is one of sickness. Every family's journey is a history of triumph over disease.

My own father had polio as a University of Michigan senior in 1948. He was hospitalized for five months and told he might never walk again. Fortunately, he made a full recovery. But if he had never had polio, how different would the course of his life have been? He would have finished college on schedule, returned to New York six months earlier, and then perhaps not have been in residence at the McDowell Colony fifteen years later, when my mother attended on a painting fellowship. My father was among the last generation to be afflicted by the disease. Today, the WHO has plans to make polio the second virus to be eradicated after smallpox. Yet for my grandparents, polio was a persistent fear. There were outbreaks every summer. Parents warned their children to stay out of the water and sought to keep them from playing in muddy or dirty environments. Every cough or fever was a source of tremendous anxiety, the start of a nervous vigil for the telltale sore neck and backache that were early symptoms of the disease. Today, this seems almost a parody of overprotective parenting, but it was a typical rational expression of the fear of disease.

Yet infectious disease is a capricious demon, and in my case, I would not be alive were it not for tuberculosis. My Japanese grandmother was afflicted with it as a teenager in Tottori, Japan, and, stigmatized by this disease of poverty, was deemed a bad matrimonial match. She was forced, instead, to marry a widower, my grandfather, whose first wife had given him two children but had died, along with the fetus, while laboring to deliver a subsequent child. Hanako Kometani, my grandmother, proved to be much stronger than anyone had estimated and would have three children of her own, two boys and one daughter—my mother. But had she not been damaged goods because of her bout with infectious disease, she would never have married a widower. And my mother would never have been born, and these words might never have been written.

In my own life, I've given hardly a thought to disease, save when getting poked a few times and swallowing some pills before trips to exotic places and making sure my children are fully vaccinated. The journalistic and fictional accounts of disease outbreaks I've read—ranging from Defoe's *Journal of the Plague Year* to Richard Preston's *Hot Zone*—seem like archival history or terrifying science fiction. The health matters that have concerned me and my family have had more to do with chronic illness, my father's clogged arteries, my own periodic intestinal problems, my wife's scoliosis, and, most vexing for my family, my brother's autism. Infectious diseases were a matter I had given very little thought to, before accepting a position at *Time Asia*, *Time* magazine's regional offshoot, and moving to Hong Kong in 2000.

TIME ASIA STRIVES TO COVER ASIA IN A MANNER SOMEWHAT SIMILAR to that in which *Time* reports on the United States. It is a difficult task, considering the diversity and breadth of Asia, the specious notion that Asia is one market, and the fact that the magazine has a much smaller staff than its U.S. parent publication. The circulation of the magazine is about three hundred thousand, versus the American edition's four million, and most years it ekes out a slight profit.

The magazine, earlier this decade, published a more eclectic mix of stories than did the U.S. edition, though that eclecticism came at the cost of some professionalism. Our editors, writers, and reporters tended not to be the fast-trackers you find in the Time-Life Building on Sixth Avenue in New York. If you ended up at TIME Asia, then it was likely that at some point you had screwed up, either by making a bad career choice or, say, getting arrested in Alaska with a quarter-pound of cocaine. We had reporters who hadn't finished high school, editors who had been in and out of rehab, such as myself, and staffers who were still wrestling with nasty substance habits. Certainly, there were a few A-student types with an earnest interest in Asia, who had struck out for the Far East, eager to make their mark in journalism; most of them would eventually head back to New York to join our mother publication or return to graduate school. But they were outnumbered by the Asia hands gone slightly to seed and the young ravers who found in our offices a place to recharge and reload between binges. Very few of our staff, as I say, arrived here on the heels of a winning streak.

After a year as deputy editor, I became the editor of the magazine in early 2002. My office, on the thirty-seventh floor of a high-rise in Quarry Bay, had a panoramic view of Victoria Harbor, Kowloon beyond that, and, in the distance, the lush mountains dividing Kowloon from the New Territories. Just past that, out of sight, was China. If I had requisitioned sleeker office furniture and kept my bookshelves organized and my desk neater, then I would have given the impression of modest success. As it was, I suspect I had the look of a harried editor, trying to right several off-the-rails projects at once.

We had morning meetings during which we would map out that week's issue and also do long-term planning. It was in one of those meetings, in January 2003, that we first discussed the vinegar. In China's Guangdong province, markets in some rural cities were reporting a run on all kinds of vinegar as local Chinese sought to boil the liquid in the belief that the fumes would purify the air and ward off respiratory ailments. To my discredit, the first time a staffer proposed doing a story on these rumors of vinegar boiling and of some sort of strange new disease just across the border, I turned the idea down. Anthony Spaeth, the executive editor of the magazine, said this could be like that bird flu from a few years ago. But as I sat in Hong Kong and read those awkwardly written Chinese wire-service releases, this initially struck me as one of those weird and exotic

Asian stories, that month's equivalent of Japanese schoolgirls selling their underwear or a neighborhood committee in East Java beheading a suspected witch.

Those dispatches, however, would turn out to be the first media reports about the disease that would later be known as severe acute respiratory syndrome, or SARS. The earliest press mention anywhere in the world of this new disease was in the *Heyuan Daily*, a small forty-thousand-circulation paper in a city of two hundred thousand a few hours northeast of Guangzhou, the capital of Guangdong. On its front page on January 3, 2003, this Communist Party–controlled paper published perhaps one of the least reassuring reassurances I have ever come across. It read, in its entirety: "There is no epidemic in Heyuan. There is no need for people to panic. Regarding the rumor of ongoing epidemic in the city, Health Department officials announced at 1:30 A.M. this morning, 'there is no epidemic in Heyuan.' The official pointed out that people don't need to panic and there is no need to buy preventive drugs."

Apparently, the population of Heyuan did not heed the paper's advice, as they continued to boil vinegar and flooded pharmacies, causing a run on antibiotics, available without prescription in China, and on an herbal medicine called Banlangen. Also selling briskly were black tablets remarkably similar in consistency—and flammability— to charcoal. Burning these was still another folk remedy for respiratory tract infections. Yet this proved almost as fatal as any rumored disease when several families suffocated from the thick smoke filling poorly ventilated, squalid apartments.

Citizens were panicking in China. But there was no one in Hong Kong who knew whether there was any real cause for alarm. Scanning the Chinese media gave no clues as to the extent or risk of a possible epidemic; the Chinese government, after the first week of January, banned any reporting on the outbreak.

Still, that year promised to be, for me and my staff, rewarding and not at all taxing. War in Iraq looked increasingly imminent, and most of the ambitious reporters on my staff were inquiring about covering the coming conflict by becoming embedded with American troops.

Asia, it seemed, would be a sideshow for much of the next few months. Since Afghanistan was technically on our watch—we covered everything from Japan to Iran—we had run much of that war reporting in late 2001, deploying at different times more than a dozen correspondents and photographers. This time, our war coverage would be run out of New York, and our staffers in Iraq would be reporting to that office.

The upside to this was that while our mother publication in New York would be swamped, we would be able to pick up most of our stories from the U.S. edition and so could expect to take it easy for a few months as the Iraq conflict occupied everyone's attention. It was unlikely, we all felt, that there would be any Asian news that could compete with the war for pages in the magazine.

We were mistaken.

BOOK 1
What Is It?

CHAPTER 1

- November 1, 2002
- Shenzhen, Guangdong province, China
- 7 Infected, 0 Dead

FANG LIN, TWENTY-FOUR, TOLD ME HE HAD AWAKENED TO THE USUAL cacophony: the bleat of a truck reversing; the steady, metallic thump of a jackhammer; the whining buzz of a steel saw; the driving in of nails; the slapping down of bricks; the irregular thumping—like sneakers in a dryer—of a cement mixer.

They were building—a skyscraper, a shopping mall, a factory, a new highway, an overpass, a subway, a train station—here, there, everywhere. Up and down the coast, from Shenzhen to Fujian to Shanghai to Tianjin, this was what you heard. Fang Lin had already become used to it. He had no choice, because the sound had become ubiquitous, as regular and familiar as the breath coming through his nostrils.

He had just arrived in Shenzhen, from Nanpo in Jiangxi province. The second son of a rice-farming family, he came of age during the era of reforms. The Cultural Revolution and the Great Helmsman were for him curious historical relics—Mao was the guy on the money—as relevant to Fang's life as Genghis Khan or Terra-cotta Warriors. Even the great events of his childhood were shrouded in the same obfuscating gauze of prehistory: Tiananmen represented in his mind nothing more than a square in Beijing. His parents, recalling the hardship of China's great upheavals of the fifties and sixties, were grateful to be allowed to farm their plot, raise their children, and pay their local officials for the right to slaughter their own chickens, ducks, and pigs. They even owned their ancestral plot now and could sell their harvest for cash. They'd bought a color television and were saving for a mobile phone. Having lived through decades of sacrifice and poverty, they were thrilled to be able to eat as much pork as they wanted and to watch pirated Hong Kong action pictures on their VCD players.

But if this first post-reform generation was happy to gaze at other people's better lives on their TV sets, then the second generation, Fang Lin's cohorts, were eager to inhabit those fancier images. As the second son, Fang didn't stand to inherit any of the family land. That would all go to his older brother, who was already married to a girl from the village with bad skin who seemed to go through a box of Choco Pies every day. She would become fat, Fang Lin warned his brother. But for their parents, even obesity remained a virtue. There had been no fat people before the reforms—the whole country then had subsisted on a starvation diet.

Li tu means, literally, to leave the land, to give up life as a cultivator for a nonfarming job. For Fang Lin, the decision to leave the land had been an easy one. He knew other second sons and first daughters who had gone south to make money. And it was money that mattered now, Fang Lin knew. Even Mao—or was it Deng?—had said, "To get rich is glorious." There was food in the village, but there was no money. Money was in the south, along the coasts, in the boomtowns he saw on television. China was becoming rich, but it was becoming rich around the edges while it stayed poor in the middle. For millions of Chinese trapped in the hinterlands, that meant hitting the road, hopping a bus, truck, or train to the coast and seeking employment in a factory or construction crew, restaurant or brothel. The newspapers dubbed it the Hundred Million Man March, and one boy or girl seemed to set out from the village every day to join it, especially in the early winter, after the harvest. Fang Lin borrowed five hundred *kwai* (RMB) from his brother, packed his extra shirt in a vinyl duffel bag that his sister-in-law's parents had received when they visited Nanchang with their work group five years earlier, and walked out of town to the road that ran along the river. He thumbed a ride with a truck, buying the driver a bowl of noodles at a gas station, and then caught a local bus south to Nanchang. He then paid a hundred *kwai* for an upper berth on a sleeper bus to Shenzhen. As soon as he was on board, he reclined and watched the TV embedded in the roof above the driver. But soon the bus was so thick with smoke that Fang Lin could barely make out the CCTV newscaster.

They rode for thirty-six hours, the villages gradually giving way to county seats and the rough farmland replaced by workshops and factories. By morning they were already in Guangzhou, rolling south along the Guan-Shen Highway, past multi-acre factory compounds with corrugated-roofed workshops that were bigger than Fang Lin's whole village, rice paddies and all. Entire mountains seemed to have been hollowed out for gravel and cement. There were stretches where the landscape was practically lunar, just a few stones, and hunched amid the swirling dust were a handful of shacks made of scavenged wood and cloth. Occasionally, a family farm would appear to be holding out between the encroaching factories and construction sites. Its crop— usually tropical fruit—was coated by a film of dust.

When Fang Lin looked at this, he says, he thought it was beautiful. Amazing. Progress. Soon, there would be no more farms at all. Just factories as far as the eye could see. How many people worked on a farm—one, maybe two? But in a factory, he could not begin to count.

At the central bus station in Shenzhen, Fang Lin found a red pay phone beside a cigarette stand and called the number he had for two other boys from his village, Du Chan and Huang Po, who had come south. In his thick Jiangxi accent, he asked the woman who answered if he could leave word for his friends. She told him he could leave a message, but it would be delivered only if the receiving party were willing to pay a *kwai* for the privilege.

"Why would anyone pay if they don't know who it's from?" Fang Lin recalled asking.

"Otherwise, how do we make any money from this?" she had countered. "If it's important, he'll pay for the message."

Fang Lin told her who he was and that he would be arriving shortly. He doubted anyone would ever pick up that message. After hanging up, he bought a fresh package of cigarettes and bottle of sweet lemon tea and showed an address to the cigarette seller, who studied it for a moment and then told him to follow the signs for the southern border.

Fang Lin set off on foot. He stopped every hundred meters or so and showed his address to another pedestrian. Usually, they didn't understand his accent. But they could read the paper and point him in the right direction.

He was disappointed by the buildings. He had assumed they would be taller, grander. But these were no higher than those in Nanchang, Jiangxi's provincial capital. And these roads were no wider. And the people here seemed no better dressed. The difference, he noticed, was a matter of volume. There were more tall buildings, more wide streets, and more pedestrians. There were more shops, he discovered, selling more clothes, more televisions, more VCDs, and more fake fur coats. There were more rich people. More bums. More cripples and more whores. And there were more migrants. The reason half the people he stopped to ask for directions could not help him was that many of them had just arrived themselves.

By the time Fang Lin found his way to the district of Shenzhen where his fellow villagers had supposedly bivouacked, it had already become dark and he was thirsty and hungry, having had nothing to drink since buying the bottle of tea at the station. The neighborhood was already swathed in shadow, the narrow alleys and dirt lanes obscured by smoke and steam. At one corner, he found a storefront where a woman sat behind a counter. Beside her were five booths in which five different men were shouting into red phones. He showed her the slip of paper with his contacts' names and addresses. She nodded and told him the charge would be one *kwai*. After he paid, she handed him a slip of onionskin paper on which he found the message he had left for his friends earlier that day. At least that meant he was in the right place.

After asking around, he was told by another migrant from Jiangxi that Du Chan and Huang Po had gone north, to find work in a factory that made the machines that make sewing machines.

Though he didn't know anyone in Shenzhen, Fang Lin found that numerous other men from western Jiangxi had preceded him there, and the familiar accents seemed comforting after his two-day journey. From a stall set up in a narrow alley, he ordered a plate of chicken intestines, scallions, and red peppers—a Jiangxi dish—and a bottle of beer, which he shared with two other fellows he had just met. In turn, they offered him lodging for the night—he would just have to pitch in ten *kwai* for his third of the room. And if he wanted lighting or heat, he would have to make sure to get the change from his dinner in one-*kwai* coins, for the box in the room that provided electricity by the hour.

He slipped into his sleeping pallet on the floor and turned away from his roommates in their bunks so that he could slide his red plastic wallet into his pants. When he woke up, his roommates were already gone. He noticed they had rummaged through his duffel bag and taken his cigarettes. His wallet was safe between his legs.

FANG LIN HAD ARRIVED IN SHENZHEN DURING WHAT WOULD COME TO be known as the Era of Wild Flavor. China's economic boom had been going strong for more than a decade, especially in the south, and Shenzhen, as the first of China's Special Economic Zones, or SEZs, had become the urban embodiment of that boom as well as a cautionary tale of the social costs of turbocharged economic development. The city had grown from a rice-farming village of a few thousand to a sprawling metropolis of seven million within twenty years. Each of the central government's grand plans for Shenzhen, in 1980, 1982, and 1986, had been superseded before implementation as migrants and resettlers swamped developers' ambitions. The area of Shenzhen was eventually expanded to 150 square kilometers divided into six districts straddling the border with Hong Kong. Most of Shenzhen's residents, as many as four million, had come to town illegally, not possessing the proper permits to live in the Special Economic Zone. They survived in a legal nether zone, tolerated by officials, employed by local manufacturers, and exploited by landlords, bureaucrats, and cops. The city had been designed with three million in mind, and the infrastructure was drowning in the waves of migrants who washed in every day. "The planners of the Special Economic Zone," wrote Mihai Craciun in *Great Leap Forward*, "are now devoted to improvisation and disorder." The city had no choice but to embrace chaos as a paradigm. Thousands of new buildings went up every year, 2,063 new miles of road were laid down, and 140,000 new homes were built. Adding to Shenzhen's status as a city of transients was its location as the primary entrepôt between the mainland and Hong Kong—250,000 people a day crossed this most secure internal border in the world.

Hong Kong, of course, is now a part of China, with its own mini-constitution, the arrangement known as "one country, two systems." In many ways, this has resulted in Shenzhen's becoming Hong Kong's parallel universe. It has the same Cantonese energy, the "get ahead" ethos and respect for a buck—or *kwai*. But thanks to loose laws, widespread prostitution, and dirty officials, you can get anything you want there: knockoff Chanel bags, pirated DVDs, ecstasy pills, one-night stands. Even the money, the objective of so much of this underground commerce, is suspect: local taxi drivers say that one in every twenty bills they collect is counterfeit. The city is the richest in China and also the youngest in the world, with an average age of twenty-four.

At ground level, in the shopping malls and restaurants of Shenzhen, the first impression of the boom is the sheer volume of goods and services on sale. Chinese advertising still hawks primarily based on price; there is very little aspirational marketing. Throughout Shenzhen, the plastic surgeons promise cheaper eyes, lips, and breasts, rather than better ones. And for every Dunhill or Louis Vuitton boutique, there are a thousand no-name shops operating out of retail space let by the day or week. Vast emporia the size of football fields, given over to every conceivable sort of cheap plastic toy and cut-rate cookware—plush animal zoos, narrow aluminum pan alleys—make Shenzhen sometimes seem like a ninety-nine-cent store somehow grown up into a whole city. There are enough crappy vinyl purses to outfit an entire Brezhnev-era

Soviet megalopolis, nail clippers for every toe in China. A hairbrush for every strand. Everything is on sale, all the time. Suggest a price, any price, and the vendor will meet it, beat it. Money is made on volume, and they have to sell as much as they can today because more of everything is arriving tomorrow, from factories up and down the Pearl River Delta. You see Hong Kong families returning from the Delta with all manner of household goods loaded onto wagons—new storm windows, heaters, blenders, car radios, tires. There is nothing that can't be had in the Delta for cheaper than it can be had anywhere else.

The nickname Era of Wild Flavor perfectly evoked the social, economic, and psychic dislocations brought on by this greatest mass urbanization in the history of the world, which coincided with, or perhaps catalyzed, one of the most vertiginous economic booms the planet had ever seen. A businessman friend of mine who runs factories based on the mainland recently described the Delta as a place "where more of everything is being made than has ever been made anywhere at any time." That may be hyperbole, but certainly China in general, and the Pearl River Delta in particular, has become the low-cost, cut-rate shop floor of the world. China's economy had grown through the late nineties at an average annual rate of nearly 10 percent. Entire swaths of Pearl River Delta marsh and paddy were drained to make room for factories and container ports. The country boasts 140 car companies, 90 truck companies, two dozen television firms, and 30 that make vacuum cleaners. China manufactures enough televisions to replace the entire global supply every two years. Fifty percent of the world's phones are made in China, 30 percent of the world's microwave ovens (and one would guess about 100 percent of the world's pirated CDs and knockoff Prada bags). Twenty percent of everything Wal-Mart sells is made in China, and 25 percent of Best Buy's merchandise comes from the Middle Kingdom. Look around your own home: that frying pan, blender, coffeemaker, hair dryer, sewing machine, shower curtain, doormat, flowerpot, pencil sharpener, ballpoint pen, broom, mop, and bucket. All of them made in China, and probably in the Pearl River Delta. China's demand for raw materials has driven up oil and gas prices around the world, as tankers anchored off Guangzhou, Shenzhen, and Hong Kong have to wait weeks to offload. Steel, aluminum, ore, oil—you name the raw material, and China has bid up the global price for it over the last two years.

The prevailing philosophy—that business model of "more is good and cheaper is best"—became a mantra of the Era of Wild Flavor. If you were living in the Delta, you saw that more of everything was being sold, more money was being made, more buildings were being erected. The tycoons of more—the factory owners, the landlords, the scalpel cowboys doing four dozen eye jobs a day, the real estate brokers, the pimps, the party officials—they, too, had more money than ever before. They spent it on mah-jongg and Audi automobiles and karaoke girls and bottles of Hennessy in ceramic Napoleon-on-his-rearing-steed bottles.

And they spent it on Wild Flavor, *yewei*, a key element of new China's conspicuous consumption. Southern Chinese have always noshed more widely through the animal kingdom than virtually any other peoples on earth. During the Era of Wild Flavor, the range, scope, and amount of wild animal cuisine consumed would increase to include virtually every species on land, sea, or air. Wild Flavor was supposed to give you face, to bring you luck, to make you *fan rong*, "prosperous." That expression, *fan rong*, had become the preferred phrase used in the Delta to denote anything that was cool. Wild Flavor tycoons could visit a brothel in Donguan reputed to be the world's largest, where one could choose from more than one thousand women on display behind a glass viewing wall. There were rumors that, for the right price, one could order soup made from human babies, which was believed to imbue the diner with fantastic virility. Restaurants purveyed a wider range of snakes and lizards, camels and dogs, monkeys and otters. Hunters as far away as Indonesia, Thailand, and Canada became the front end of the supply chain as Chinese traders expressed renewed interest in fresh and dried tiger penises from Sumatra and brown bear bladders from Manitoba. No one knew the number of restaurants and markets selling wild animals, but Lian Junhao, the director of Wild Animal Protection at the Guangzhou Forestry Bureau, the department mandated with regulating Wild Flavor restaurants, told me that in Guangzhou alone there were currently two thousand such restaurants. And how many were there two years ago?

"Two thousand."

When I asked whether the roundness of those numbers indicated that his agency hadn't been keeping such an accurate count on new licenses, he nodded.

"Nobody really knows how many there are."

During the one hour that I spent in his office, he issued four licenses for new Wild Flavor restaurants.

He also told me that his office was in charge of regulating which species could be sold and which had been banned. Snakes, for example, were no longer for sale in Guangzhou's Wild Flavor markets.

A half hour later, I was standing in front of a bag of writhing cobras at Xin Yuan Market. At this stall, there were twenty-five such burlap sacks of snakes, and there were at least twenty stalls at the market specializing in snakes. Cobra cost 130 *kwai* a kilogram, king snake was cheaper, at 80 *kwai*, and another type of snake—I believe it was a type of garter snake—cost about 70 *kwai* per kilo. A boa constrictor, black with red and gold stripes, was the most expensive, at 170 *kwai* per kilo. At that price, the whole snake would cost a reasonable seven hundred *kwai*, or ninety U.S. dollars.

I first visited these wild animal markets over ten years ago. At that time, a market like this one would contain about thirty stalls selling a wide variety of wild animals. During that trip, I ordered snake, turtle, and wild boar, which went from the seller to the carving board to the pot right before my eyes. We ate at a table beneath a corrugated fiberglass roof. The atmosphere of the market was genial, with kids chasing one another between the animal cages and women sitting on stools and chatting as they washed vegetables. Walking among the stalls, you could look up and see the stars.

The food had been average. The snake was bony, like eating a chicken's neck, and the turtle had been surprisingly fatty. The boar had been disappointingly flavorless. We washed the food down with some green beer.

By the time I returned to these markets a decade later, the style of dining that had been a quaint local custom had become industrialized. The wild animal trade had grown, commensurate with the rest of the economy, and those markets with a few dozen stalls had grown into vast wholesale warehouses selling millions of animals a year. In one cage at Xin Yuan, I counted fifty-two cats pushed in so tightly that their intestines were spilling out from between the wire bars. There were fifty-five such cages in this one stall. There were fifty-two stalls down this one row of vendors. And there were six rows in this one market. And there were seven markets on this one street.

A startlingly musky smell overwhelmed me as I walked between the stalls. I realized it was a combination of the feces of a thousand different animal species mingled with their panicked breaths. The range and scope of wildlife on display was a zoological chart brought to life. There were at least a dozen types of dogs, from Saint Bernard to Labrador. There had to be at least as many different breeds of house cat. There were raccoon dogs, badgers, civets, squirrels, deer, boars, rats, guinea pigs, pangolins, muskrats, ferrets, wild sheep, mountain goats, bobcats, mountain lions, three types of monkeys, horses, ponies, bats, and one camel out in the parking lot. And these were just the mammals. The avian rows sold an equally diverse range, as did the reptile rows. Predator was sometimes stacked on top of prey. Animals that had lost paws, presumably when they were trapped in the wild, were kept alive via IV drips. Because wild animals were more valuable than farm-raised creatures, it was rumored that some traders would slice off the hind paw of a civet or badger to convince potential buyers that the animal had been trapped in the wild.

I had a list of banned animals that the director of the Wild Animal Protection office had given me. I asked a vendor for the rare bird species, the monkey, the tiger.

"No problem," I was told by a smiling man with buck teeth who said he was from Guangxi Zhuang.

"What about the regulator?" I asked.

"No problem." He pointed to a fellow in a gray-and-blue uniform sitting on a white plastic chair, flicking his cigarette ashes beside a bag of banned snakes.

"Okay, how about mountain lion?"

"No problem."

"Brown bear?"

"No problem."

I decided to push my luck.

"How about panda?"

He shook his head. "You must be sick."

CHAPTER 2

■ **November 29, 2002**

■ **Penfold Park, Kowloon, Hong Kong, China**

■ **8 Infected, 1 Dead**

HONG KONG'S PENFOLD PARK COMPRISES EIGHT MANICURED HECTARES of acacia and palm trees, three small ponds, and kilometers of gently curving dirt paths that wind beneath an ornamental brick moon gate and through an elaborate hedgerow maze. On autumn and winter mornings, uniformed schoolchildren on field trips stroll hand in hand through the park, letting go of each other to point out the peacocks roosting in their aviary or the ducks and geese splashing in the ponds. Numerous migratory birds—great egrets, little egrets, and gray herons—attracted by this oasis of flora and fresh water amid Hong Kong's urban sprawl, drop in and find the regular feedings and tended gardens to their liking. They often end up staying a season or two before resuming their regular flight paths and returning to their North Korean and Siberian nesting spots.

The eleven groundskeepers who maintain the park become very familiar with the birds. During cool winter mornings, the ducks, geese, and swans of Lesser Pond like to take the sun on the lawn just south of the water. On warmer mornings, the flocks and gaggles are already in the pond when the staff arrives. The waterfowl are almost canine in their familiarity with particular gardeners and caretakers, honking and tooting as their favorite park workers—those who feed them their pellets—report for duty. Occasionally, one of these birds will fall ill. If the other animals notice before the groundskeepers can capture and take the bird to a veterinarian, the rest of the flock will turn on their stricken comrade and, as if too impatient to allow natural selection to take its course, peck the victim to death. The park's caretakers usually take little notice of these fatalities, gathering and bagging the dead bird and tossing the carcass out with the lawn trimmings and fallen leaves.

When Chiwai Mo arrived at work at six-thirty on November 30, he noticed a Canada goose listing in the gray water near the brackish algae-lined shore of Lesser Pond. The rest of the waterfowl were conspicuously sunning themselves on the lawn, spreading their wings in great, almost vulgar displays of plumage that allowed each simultaneously to reassert its place in the pecking order and dry the delicate feathers on the underside of its wings. Walking along the path closer to the pond, Mo quickly deduced that the floating bird was dead; its normally rigid and muscular neck was slack and the lightbulb-shaped head had almost dipped into the water. Scarlet rivulets of blood had run from the bird's nostrils, eyes, and ears into the dark pond.

Mo continued on to the office and equipment shed, where he slipped on a pair of thigh-high rubber wading boots. From a shelf above bags of lawn seed and fertilizer, he pulled down thick, double-ply plastic trash bags and then slid on a pair of rubber gardening gloves. Stepping from the shed back into the sunlight, he may have blinked for a moment in the glare—he never wore sunglasses, despite the long days in the sun—and headed back toward the pond. As he approached the dead bird, the rest of the geese began honking. When he took the goose by the neck, the flesh there felt surprisingly soft compared with the rigid muscularity of the neck of a living goose. The flock's protest rose to a crescendo until Mo dropped the bird into a plastic sack. As soon as the bird was out of view, the geese fell silent and went back to their preening.

Back at the shed, Mo double-bagged the carcass and secured the bag with a plastic tie before taking it to the rear of the shed and tossing it into one of the plastic Dumpsters.

One dead bird was no big deal.

Ninety-five dead birds, however, was a problem. That's how many ducks, geese, and herons would eventually die in the Penfold Park outbreak. There would be a Chinese goose fatality discovered that afternoon. By the next day, December 1, grounds staff would find four dead birds, and the day after that, they would find another four. Head groundskeeper Chiu Tsang decided to report the curious bird deaths to his supervisor, John P. Ridley, the manager of the Jockey Club's racing operations.

Penfold Park also happens to be the infield of the Hong Kong Jockey Club's eighty-five-thousand-seat Sha Tin racecourse. The Jockey Club, a 107-year-old foundation with a multibillion-dollar endowment that has built schools and museums throughout Hong Kong and maintains a legal gambling monopoly in the territory, opened the park to the community in 1979. The track that encompasses it, however, is among the most valuable and profitable racing venues in the world. Just three hundred meters as the crow—or infected gray heron—flies from Penfold Park, twenty-three stables hold approximately 1,100 thoroughbreds worth 140 million dollars on the hoof.

"What took you so long?" Ridley asked his head groundskeeper when he heard about the ten dead birds. He hung up and called Keith Watkins, the Jockey Club's chief veterinarian, and told him, "We have a problem."

Watkins had taken his post in 1992, just before an outbreak of equine influenza infected six hundred thoroughbreds and killed one, causing the cancellation of seven race meetings and an estimated 128 million dollars in lost turnover. When he received Ridley's call, he immediately went and looked out at the stables, the paddock, and the trainers fast-walking their horses around the practice post. Whatever was killing those birds, he feared, could infect the horses. He scrolled through his PDA and found the number for Trevor Ellis, fifty-five, the senior veterinary officer for Hong Kong's Department of Agriculture, Fisheries, and Conservation. Ellis's first reaction was, "Oh, shit."

CHAPTER 3

■ **November 30, 2002**
■ **Shenzhen, China**
■ **9 Infected, 1 Dead**

THAT FIRST MORNING IN SHENZHEN, FANG LIN EXPLORED HIS NEW neighborhood. Ka-Ta, or the Click, had been so nicknamed by residents after the whirring electricity meters that clicked away outside most apartments there. The Click was sixty-four eight-story buildings laid out in a perfect grid. Owing to the peculiarities of Chinese zoning laws, the Click was actually a *cun*, or "village," technically still under the jurisdiction of the farming cooperative that had been here twenty years ago when the SEZ (Special Economic Zone) was formed. City architects, planners, and politicians had virtually no sway here, as opposed to the areas zoned as part of Shenzhen itself, where such individuals ruled practically by fiat. If land is zoned as rural land, even if, as in the Click, it is now the downtown of a massive urban center, the farmers who own the land can develop their property however they like in order to maximize their revenue. This has resulted in abrupt and startling neighborhood divisions throughout the Pearl River Delta. You might be walking along a well-planned thoroughfare with wide sidewalks and potted palms every ten meters only to find that the way dead-ends into a sprawl of ill-conceived tenements.

The Click was just such an arrangement. Ten years ago, the farmers' collective that owned the land had contracted with a real estate developer to put up these white-and-pink-tiled buildings as housing for the rapidly expanding middle class. What no one had correctly reckoned was how enthusiastically and industriously the new tenants would set about renovating and customizing their new neighborhood. Because of some design quirks, the neighborhood was never likely to become a haven for middle-class families. For example, the developers had left just two meters of clearance between the buildings. As a result, the narrow, unpaved alleys saw light only at high noon every day and for the rest of the day were swathed in white fluorescence. If you were in the Click at noon on a sunny day, the sudden appearance of this yellowy natural light was almost confusing.

During the damp summer months, streams trickled between the buildings; in the winter, these areas became soggy pathways. And while the original plan had provided for commercial space in each ground-floor corner unit, enterprising tenants had rented other ground-floor flats and then simply punched their way through the brick walls facing the alleys to convert their flats into prime street-front retail space. Since urban zoning laws didn't apply to this *cun*, there was little that anyone could do to restrain such sledgehammer capitalism. Electrical lines were jury-rigged. Once a month, ChinaGas workers arrived in groups of ten to snip away illegally rigged cables that were bootlegging power. Many of the apartments had been subdivided into smaller one-room units, some without water or even windows. The interiors of the buildings had also been similarly modified, with power lines and broadband cables nailed to the walls in tangled masses that occasionally emitted bright showers of sparks. There were buckets of sand deposited at every landing in case of fire, but these were so full of cigarette butts one wondered about their efficacy at containing a blaze.

Illicit commerce thrived in these alleys. There were barber poles skirling red, white, and blue. (The barber pole, in China, very often denotes a house of ill repute.) The hookers in skintight Lycra pants and tube tops would grab your arm as you walked past. If they thought you were a foreigner, they would proffer *"amore, amore,"* Italian here, for some reason, being the language of love. There were several tiny piecework factories of three sewing machines each; the workers slept under their machines at night. There were four fellows who could repair your shoes and one fellow who converted old tires into sandals. There were a half dozen key duplicators. And no fewer than a dozen doctors in one-room practices—fifty-square-foot shop fronts usually featuring a bench covered with newspapers, a cabinet full of pills, maybe a diploma on the wall, and a stool on which the M.D. sat, smoking cigarettes. The doctors all specialized in treating venereal diseases, and a frightening few also practiced cut-rate plastic surgery. But it was easy to bypass the doctors and head straight for any of the half dozen pharmacies that did a thriving business in aphrodisiacs and antibiotics. There were the pay-by-the-call phone centers, the pay-by-the-hour hotels, and the pay-by-the-tablet ecstasy dealers. You could buy one of anything in the Click: a cigarette, a nail, a phone call, an injection, a piece of paper, an envelope, a stamp, a match, a tablet, a stick of gum, a bullet, a brick, a bath, a shave, a battery, even a feel.

Each visit to the Click would reveal new discoveries. One afternoon, I arrived to find the neighborhood in a panic. Somehow, a bag of snakes had been left open, and a dozen serpents had crawled out. Children were running up and down the narrow alleys, trying to catch the serpents before they slithered into pipes and through cracks. The hookers were standing on top of their chairs screaming, while the men smoked cigarettes and laughed. The kids would catch just six of the snakes, and for the next few months, citizens of the Click would wake up with a start as they felt the slithering muscularity of a snake passing over their feet.

On another occasion, a short man in a white T-shirt offered to sell me a gun.

IT IS NOT IN THE NATURE OF THOSE LIVING THROUGH PARADIGM-shifting economic booms to become reflective. And the Era of Wild Flavor has been a boom that is redefining how Chinese people live and even what it means to be Chinese. Life in the new cities, among these migrant communities, is anonymous, and the pace of change can seem vertiginous. The tightly knit social hierarchy that binds so many Chinese together and allows for "the harmony of heaven and earth" has been largely torn asunder. In Shenzhen, the more formal greeting *nin* has been virtually abandoned; since no one knows the social status of anyone else, every greeting is a *ni*. Who has time for ritual anyway, when there is money to be made? The mantra, as I say, is "more." Greed goes beyond good; it is everything. And the cracks and fault lines in the system are easily ignored in the midst of life-changing wealth that, even if you yourself can't quite achieve it, you know someone who has. But the warning signs are everywhere: Why are there more doctors doing breast enlargements than treating diseases? Why is a disease that was a forgotten rural plague—schistosomiasis—now appearing in downtown Shenzhen? Why is it that I have to step over a dozen cripples to reach the Armani boutique?

"Places like Shenzhen are built on sweatshops," says Zhou Litai, an attorney who represents victims of industrial accidents. "Old machinery, no training, fourteen hours a day, four hundred and fifty yuan a month, ineffective safeguards—that's the secret of China's economic miracle." Zhou is one of the very few who raised doubts during the Era of Wild Flavor. Two other Guangdong attorneys who did similar work were murdered in 2000.

Places like Shenzhen become giant cracks in a system that has become more gap than spackle. The recent arrivals try to exploit those who have just gotten off the bus. Everyone is winging it. Many residents don't even know their addresses. They pay for their electricity by the hour. Their telephone, if they have one, is a mobile. If a message is truly important, it will be text-messaged to them via their phones. And they'll be moving in a week or two or three—so what's the point of giving out an address anyway? This also makes it impossible for the police to keep tabs, especially in the *cuns*, where even the floor plans and roadways are as impermanent as sand castles. But the city, in some sense, thrives on this chaos. There are certainly risks—I was robbed in Shenzhen twice—but for the quick-witted and flexible, there are great rewards. Prices are higher here, but so are wages. The hookers are more expensive, but they are also prettier and more ethnically diverse. You may not know where you are going, but neither does anyone else. In between appointments one afternoon, I decided I would go look at a few apartments, reckoning it would perhaps be cheaper to rent a place in Shenzhen rather than commute from Hong Kong and stay in hotels for a week at a stretch. I walked into the office of a storefront real estate broker who was advertising several flats, one of them in the Click. I explained to the broker what I was looking for, and he said he had several properties that might interest me. He borrowed a set of car keys from another broker, and we set off in his rented Hyundai. The sky was its usual smudged brown color; in Shenzhen, most days it looks like it's about to rain. Yet I remember it rarely actually raining. This afternoon was no different. After a few minutes, I realized we were driving down the same avenue where I'd seen the giant VDT in which a cut-rate plastic surgeon promised a "cheaper, bigger, more womanly bust."

"Where are we going?" I asked the broker.

He explained that he wanted to show me a small apartment in Laohu.

"But why are we going in circles?"

He shrugged.

I asked my broker when he had arrived in Shenzhen.

"Two days ago," he explained.

This feeling that no one knew where he was going could add up to a sense that no one knew anything. The assumption that somewhere, somehow, there were civic forces at work—a boardroom full of wise elders subtly directing the hidden mechanisms and machinery of the city—was dispelled by a few days in Shenzhen. I've met some of those officials, blue-suited men with fleshy faces, thick lips, and unctuous smiles who dine every night on their government banquet accounts, polishing off platters of Wild Flavor. They sing karaoke with Hunanese and Shanghainese girls in slit dresses. The men who were supposed to be looking out for the rest of us were looking out only for themselves.

For first-time visitors to the Pearl River Delta, the pollution is often striking. You notice it initially as itchiness in the eyes, then a sore throat, and, finally, a sort of hacking cough that stays with you throughout your visit. Local hospitals have observed an increase in respiratory diseases that some doctors believe is related to a more polluted environment. Mao once famously stated that environmental woes affected only capitalist countries. In that regard, China has now fully qualified as a capitalist state.

Other societal ills that had previously been dismissed as ailments of the decadent West have surged. Crime is soaring in Shenzhen, with kidnapping rising 75 percent, assault 38 percent, and murder 35 percent in 2003, according to local police officials. Even more frightening, lenient rules allowing for greater internal travel have unleashed China's first waves of serial killings. Every month brings news of more killings. In Shenzhen, at the Senxin Labor Market, migrants pay ten *kwai* for admission to a vast hall packed with hundreds of brokers offering jobs and better lives. When you walk into these vast flesh markets, you get the feeling you are being swallowed, both by the volume of humanity and by the crushing weight of people's aspirations. That autumn, eleven young women who had recently emigrated from the provinces believed they had found work with a labor broker there named Ma Yong, himself a recent migrant. One of Ma's neighbors found him odd but then observed, "So many people come and go that I didn't pay much mind." Ma and a female accomplice were arrested in connection with the disappearance of the eleven women, who police believe were murdered. In Hebei in 2003, another man was arrested and suspected of killing sixty-five people in four provinces. Another man in Henan allegedly murdered seventeen boys. And all these arrests were made in one three-week period.

Throughout China's new boomtowns, and especially in the migrant-worker communities populated by transients, however, criminality rarely reaches such morbid extremes, but petty crime is becoming depressingly familiar. "Society now has blind spots that have become a haven for killers," says Ren Jiantao, head of the School of Government at Zhongshan University, in Guangzhou. In the damp dark of a migrant workers' ghetto, a killer could incubate for a long time, undisturbed. During the Era of Wild Flavor, that killer would turn out to be viral.

■ ■ ■

FANG LIN HAD CUT HIS HAIR SHORT SO THAT HIS SCALP SHOWED through. Showers cost a *kwai* each, and this was a way of saving on bathing expenses. With his new crew cut and his wide forehead, bulging brown eyes, sturdy nose, and thick lips, he appeared almost menacing, an image belied by his stutter and diffidence. At five foot ten, he was tall for a Chinese but merely average height for a migrant from Jiangxi. He had settled in a building that housed mostly Jiangxi natives, and he quickly discovered that jobs and industries were divided up based on which province the *wailai renkou*, or "nonlocal people," hailed from. Those from Henan, for example, tended to collect waste and refuse to sort through for recyclable or resellable materials. The shoe repairmen and key duplicators tended to be from Anhui. Migrants from Fujian or from other sections of Guangdong tended to sell building materials or work in construction. The Uighurs opened restaurants or set up begging networks. The men from Zhejiang tended to become garment workers. Most of the women migrants worked in sweatshops and lived on site, in dormitories, Fang Lin was told, or they worked as hookers.

For those from Jiangxi, the easiest work was found in restaurants, as dishwashers, busboys, and what they called "cut men"—fellows who spent their evenings chopping up ingredients. There were new restaurants opening daily throughout Shenzhen, so demand for service employees was steady. Within forty-eight hours, Fang Lin found a job working for a joint managed by three brothers from Sichuan.

He called his parents after a few weeks and learned that his grandparents had made a down payment on a burial site. It was a south-facing plot close to the village, he was told, with very good feng shui. His grandparents were relieved to know that they would have a most auspicious start to their afterlife. And for Fang Lin's mother, this was a tremendous source of pride: she could tell her friends that her parents would never be far from their grandchildren. Fang Lin knew what was coming next: the cost.

The site had cost six thousand *kwai*, but that did not include robes, incense, or a monk to consecrate the plot. The total would actually be eight thousand *kwai*. But Fang Lin would not have to worry that his grandparents' souls would go wandering for lack of worldly funds, his mother told him, because they could pay in install- ments. His grandparents had come up with 500 *kwai* as a down pay- ment, which meant the family had to make monthly payments of 222 *kwai* for the next four years. Even the number was good luck, his mother pointed out. As the only member of the family who earned actual cash—instead of, say, grain or rice or coupons for grain or rice—Fang Lin would bear that burden.

Fang Lin says he never really thought about what happens when you die. But he knew his parents and grandparents were adamant that one must die at home and, if possible, be buried near one's home and family.

"That's good news," he said before hanging up.

He walked along Dongmen Avenue. Already, in just the few weeks he had been in Shenzhen, the building that had stood here before—it had housed a nightclub featuring Russian girls, he remembered—had been torn down and a new one was going up in its place. Its sign promised INTERNATIONAL COMMERCE. He walked past a Häagen-Dazs and then turned down a narrow alley, on his way back to the Click, pausing for a moment before a Wild Flavor restau- rant that was expanding. Living in Shenzhen, one couldn't help but be caught up in the frenzy and philosophy of more. Fang Lin, too, dreamed of more. He aspired to smoke Panda cigarettes and slap down mah-jongg tiles while he talked on his flip phone with camera functionality. He wanted to sing karaoke with Korean girls in slit dresses and drink ten-year-old grain liquor. Like so many in the Delta, he dreamed of indulging in Wild Flavor, for that was the taste of *fan rong*, "prosperity." He read through the Heartiness and Happiness menu and wondered what camel hump tasted like. He'd had pangolin, but what about marmoset? Or badger? And what was a turtle fish? When they wrote "monkey brains," did they really mean the brains of a monkey, or was that a euphemism, like the way they wrote "phoenix" instead of "chicken"?

"Do you want a job?" a woman with gold earrings and a pearl necklace asked him.

Fang Lin, typically, didn't respond. The woman looked at him and then spoke in the flat tones of a Jiangxi accent.

"Where are you from?" she asked.

He told her.

She clapped her hands together sharply, a sudden gesture that caused Fang Lin to jump back a step.

"I have two boys from there."

Fang Lin asked, "What's the job?"

"Assistant driver," she explained.

"How much?"

"Seven hundred a month."

Fang Lin nodded.

"Come back and meet the boys."

She led Fang Lin back through the restaurant, which the busboys were in the process of setting up for dinner. He followed her through the kitchen and then through a dark, fetid menagerie of caged animals—everything on the menu was here, still on the wing, claw, hoof, or paw—out to a narrow dirt lot, where a diesel truck sat next to a pen in which two peacocks and a boar were tied to pegs driven into the earth. And sitting on concrete steps, drinking dense tea from plastic mugs, were Fang Lin's two fellow villagers Du Chan and Huang Po.

Fang Lin smiled at his good fortune. To have found his fellow Jiangxi friends and a new job on the same afternoon! Perhaps, despite his having to pay for his grandparents' burial plot, his luck was changing. The three men lit cigarettes and sat down on the steps. Du and Huang described the job to Fang. They rode with the drivers, collected the animals from local markets and farms, unloaded them at the restaurant, and then slaughtered them as the chefs called out the orders. Every night, the restaurant went through about a dozen pangolins, twenty badgers, two dozen civets during the winter, thirty-five or so snakes, and a half dozen lizards. You had to move fast with the badgers and the civets or they would take a bite out of you. In the winter, the lizards and snakes were usually too sluggish to be a problem. But just in case, the two young men wore thick rubber gloves.

They showed Fang Lin the animal pens, the peacocks, and the boar—very *fan rong*, they all agreed. Then they took him around the other side of the truck, where there was a giant bird with feathers the size of hundred-*kwai* notes.

"What is that?" Fang Lin asked.

The animal was coated in dust, and it snorted like a giant pig.

"Ostrich," Du Chan explained.

What the boys from Jiangxi had never even considered was that each of these creatures, destined for the banquet table, had a host of indigenous viruses that it shed by the millions, in feces, blood, sweat, saliva, and tears. Fang Lin and his friends would be in close proximity to microbes imported directly from the rain forests of Southeast Asia and the wild hinterlands of China. None of the men, however, knew what a virus was or what the risks were of a lethal microbe jumping the species barrier.

They knew only that they were getting a sweet deal.

CHAPTER 4

- December 10, 2002
- Tai Lung Veterinary Laboratory, New Territories, Hong Kong, China
- 20 Infected, 3 Dead

THERE ARE MANY WAYS FOR ANIMALS TO DIE IN HONG KONG. DOGS and cats are hit by cars. Pigeons and even the occasional disoriented hawk fly into the windows of skyscrapers. Beloved domestic pets pass away from old age. Chickens, pigs, and even snakes are slaughtered in preparation for a feast. The corpses of those animals not destined for the dining table are gathered every morning by doormen and caretakers, and perfunctorily trashed. Yet for every hundred dogs found mangled and marked by tire treads at the side of the road, there will be one mammalian, reptilian, or avian passing that defies explanation. Nearly every one of those suspicious animal deaths ends up in the Tai Lung Veterinary Laboratory of the Department of Agriculture, Fisheries, and Conservation. Animal corpses are scooped up from Kowloon pig farms and fancy, upscale Peak mansions, and if the fatality looks like it may be due to infectious disease, the cadaver will be bagged, frozen, and driven up to Tai Lung. There it will be slid into an airtight compartment and passed through to the dissecting lab, where Trevor Ellis's team will seek to divine the cause of death. Ellis's team performs autopsies on about 4,000 birds, 750 fish, 200 pigs, and 250 dogs and cats every year. The building itself is an airy, bright, plate-glass-and-concrete structure that looks like a military bunker designed by Frank Lloyd Wright. In addition to the necropsy lab, there are administrative offices and labs for avian virology, bacteriology, molecular biology, and fish virus, all of which open onto a central courtyard. The staff at Tai Lung are concerned, of course, about ailments that may afflict the city's population of livestock and pets, but they are perhaps even more wary of the possibility of an avian flu outbreak. Fifty percent of the lab's annual budget goes to influenza.

Avian flu is the Ebola of the poultry world, a hemorrhagic fever that induces in its victims profuse bleeding from every orifice and can turn a chicken coop into a mass of goop and feathers in just two days. Yet before Hong Kong experienced an avian flu outbreak in 1997, the disease had not been thought to be dangerous to humans. That year, however, the virus jumped species (from chickens to humans) with brutal efficiency, killing 33 percent of those it infected and eliciting an international response that (barely) succeeded in containing the disease to just eighteen human cases in Hong Kong. The 1997 outbreak required the slaughtering of more than three million chickens and ducks and the scouring of Hong Kong's wet markets and cost the local economy nearly a billion U.S. dollars. Still, Ellis and other scientists felt that it had been a close-fought battle. If the H5N1 strain of avian flu had succeeded in securing a genetic foothold among humanity, the notoriously unstable human influenza virus could swap genes with the avian flu virus and mutate into a superflu that would be as contagious as human flu but with the morbidity of avian flu. Humans would have no preexisting immunities to such a superbug. The result? Imagine a strain of influenza that kills one out of every three people infected.

Great influenza pandemics are sometimes known as slate wipers, sharp blades of disease that scrape masses of humanity from the earth. The influenza virus is believed to have first leaped from animals to humans a few thousand years ago, probably in southern China, and it has since erupted in global outbreaks every few decades or so. The last major killer epidemic, the flu pandemic of 1918, was a unique scourge, killing more than sixty million people—more than the number killed in the Black Death of the sixteenth century or during World War I. Since then, there have been smaller influenza pandemics, in 1957 and 1968, both of which originated in southern China. Influenza experts now say we are overdue for another great flu pandemic. "It's not a matter of if, but when," says Klaus Stöhr, head of the Global Influenza Programme at the World Health Organization. When another pandemic emerges, it will likely also come from southern China. Hong Kong, pressed as it is against the guts of China, will be among the first afflicted. And because Hong Kong is a global transportation hub—more than 240 international flights a day originate in Chek Lap Kok airport—the virus will reach your hometown just twenty-four hours after it reaches Hong Kong. Such a new pandemic would first manifest itself in curious signs: a few dead birds, perhaps, in a Hong Kong park.

Yoga instructors in Santa Monica and investment bankers in New York have no idea of the role that a few scientists, doctors, and public health officials in Hong Kong play in keeping them hale and hearty. With their territory flush against influenza's hottest zone, China's Guangdong province, Hong Kong's medical and scientific community must be ever vigilant against new influenza strains and possible species jumpers. With twenty-four million chickens a year coming across that border, Ag and Fish, as the Department of Agriculture, Fisheries, and Conservation is known, is the front line in the battle to keep avian flu out of the territory. Chinese officials deny that H5N1 is endemic to southern China, insisting that Hong Kong's H5N1 problem is homegrown. To avoid making disease surveillance a political issue, Ag and Fish keeps a list of two hundred poultry farms in Guangdong that are allowed to export to Hong Kong. Those farms employ veterinarians who screen the chicken and duck populations, before any containers of live birds embark for Hong Kong, bleeding birds and testing the blood for H5N1. The testing is done again at a Hong Kong government lab near the border, where the birds are reinspected and double-checked. Any birds found to have antibodies for H5N1 are turned back at the border and driven to local markets in Guangdong. China, remember, does not have an H5N1 problem. Hong Kong's surveillance system is expensive, absorbing 40 percent of Ag and Fish's entire budget, and even with the heavy monitoring,

Hong Kong has on three occasions in the last six years had to slaughter millions of live birds because of H5N1 outbreaks in local live-poultry markets.

Trevor Ellis did not mention to Keith Watkins what was most troubling about hearing of the bird deaths in Penfold Park. If the Penfold Park birds were infected with H5N1, that would mean the disease was afflicting migratory birds as well as poultry. Free-ranging geese, egrets, and herons, of course, aren't confined to pens and coops. In 1997, it was believed that some of the human cases had been acquired from bird droppings. If H5N1-positive migratory birds had taken to the skies over Asia, then death was literally falling from the sky in the form of diseased bird feces. But birds die for a lot of reasons, Ellis well knew; there were environmental factors, botulism, septosomiasis. There was no reason to jump to conclusions.

A team from Ag and Fish collected the samples from the Jockey Club on December 4, delivering the eight bagged birds to Tai Lung in the morning so that the packages would be in the airlock by the time Ellis's team arrived at 8:00 A.M. The researchers slipped into clear rubber gloves, green surgical gowns, plastic eye protectors, and paper caps and slid clear plastic protectors around their shoes before they cracked open the first sealed door to the lab and stepped, one by one, into a basin of blue disinfectant solution. The Tai Lung necropsy lab was effectively a Biosafety Level (BSL) III lab, meaning, among other safety measures, it had negative air pressure. That is, air could flow only into the lab, keeping any stray deadly microbes from escaping it. The lab's air was recirculated and expelled through a filtration system that removed any nasty microbes. To access the lab, the team had to pass through sealed double doors. Emblazoned on the outside door was the international biosafety symbol, the sinister seed at the center of swirling, barbed arms, like a melting swastika. (Ebola and smallpox are both BSL IV agents. Level IV labs require that the researchers wear pressurized suits hooked up to air hoses. As the researchers move through the lab, they have to detach and reattach those hoses. Although avian flu is generally considered to be a Level III agent, because of its pandemic potential, that categorization has become a matter for some debate in the influenza community. Some scientists and labs are already treating it as a Level IV agent.)

Ellis and veterinary officers Geraldine Luk and Lucy Bissett took their places around the stainless-steel table lit from above by a large skylight. They quickly did a visual assessment of the birds. The avian cadavers, after they were wiped down with a disinfectant solution to limit the distribution of feathers during dissection, were strikingly thin and lean; the dead ducks and geese, with their wings flopping limply and their heads dangling from flaccid necks, had completely lost their familiar, archetypal duck and goose forms. Instead, the animals, their feathers surprisingly colorless in the morning light, seemed like some sort of evolutionary dead end— some sort of winged lizard that had failed to find a niche. The scientists noted that there had been some hemorrhagic bleeding from nostrils, ears, and mouth.

Ellis placed a goose on its back with its feet toward him and grasped both legs, pulling them away from the pelvis to loosen the joints. He tightened the skin over the abdomen, cut into it with a scalpel, and removed the flap of skin covering the breast. After examining the breast muscle for decreased mass, bruising, or paleness, he cut through the ribs. The keel bone always made a satisfying crack—like a slender branch snapping—as it was pulled upward to expose the internal organs. Ellis noticed black patches of necrosis in the spleen, lungs, and, later, the brain. He also found fluid leaking from the lungs and additional discoloration and swelling in the spleen. Slicing away all the attachments close to the GI tract and intestines, Ellis removed the liver, kidney, and spleen and gently teased the lungs out of the rib cage. Biopsies were taken of each diseased organ.

That tissue was homogenized and centrifuged so that any viruses—those being the lightest particles in the test tube—would naturally stay near the top of the sample. This clear fluid was inoculated into chicken embryos, which were then stored in an incubator and candled—silhouetted against a light source—to be checked for embryonic movement every day for three days. If the embryo died, the virology team knew they had an agent fatal to chickens. That agent was then extracted from the embryo and tested for the presence of specific H5N1 hemagglutinins. If the avian flu hemagglutinins were present, the sample would leave a pinkish color—the same tint as a Cosmopolitan cocktail—around the bottom of the test tube.

Ellis's virology team soon found that whatever had killed those ducks, geese, and herons was also relentlessly effective at wiping out chicken embryos. When they extracted the embryonic fluid and tested that against the hemagglutinin reagents (antibodies that react to the presence of a microbe, essential in testing for the presence of certain diseases), they saw that lovely, almost festive, pinkish tone that indicated that they were in the presence of one of the planet's most deadly agents. All eight samples from Penfold Park would eventually test positive. Later, the team would also find H5N1 (H5 hemagglutinin N1 neuraminidase) in a gray heron that had died in the park. Ellis came down to the lab himself to look over the samples and concurred that what was going on at Penfold Park was an avian flu outbreak among migratory birds. This was the first time that H5N1 had been found in migratory birds. Ellis went back upstairs to his office.

MALIK PEIRIS WAS PULLING HIS BLACK BMW INTO ITS PARKING space at his Pok Fu Lam apartment when his mobile phone chirped. As soon as he heard what Trevor Ellis had to say, he knew that his upcoming holiday in his native Kandy, Sri Lanka, was probably going to be canceled. Peiris, fifty-three, switched off the ignition and sat in the front seat for a moment, listening as Ellis ran down the tests he had performed and the positive results he had obtained.

They hung up. Peiris went upstairs and sat on his leather living room sofa, across from his stereo and a Balinese wood carving that hung on the white wall. Compactly built and broad-shouldered, Peiris was five foot nine but gave the impression of being taller—something about his bearing, a vestige of his patrician upbringing in

Sri Lanka and his Oxford education. His brown hair was parted high on his wide forehead and had in the last few years become tinged with gray. He had a slight overbite and a pronounced jawline; as he mulled his options, that overbite began to seem even more exaggerated. When working in his lab at the University of Hong Kong's microbiology department, he wore wire-frame glasses. His careful gaze could seem almost predatory, whether he was appraising a student answering an exam question or divining which virus was growing in a cell line. His voice was gravelly and still inflected by a childhood and years of graduate and professional work in England. There was in this man, sitting on his sofa listening to his son in the next room playing on his Xbox, great intelligence, perspicacity, and cleverness. But that same intelligence had also bred ambition and a desire for professional recognition and esteem that had not yet been fulfilled. Though he had trained as a virologist, he had come to the influenza field quite late, in his forties, an age when most scientists are already established in their fields. He was still a relative newcomer and was eager for an opportunity to do research and publish substantial reputation-making work. What was happening out in Penfold Park among migratory birds, Peiris well knew, could be that chance. Disease outbreaks, Peiris sometimes thought, are what happen to lucky virologists when they are making other plans.

CHAPTER 5

- **December 20, 2002**
- **Shenzhen, China**
- **28 Infected, 7 Dead**

THE DOWNSIDE OF THE LARGEST MASS URBANIZATION IN THE HISTORY of the world soon began to appear. That many organisms the size of human beings don't migrate without creating a huge impact on their environment and on other species. In this case, the effect was felt not just by China's already beleaguered wild animals but also by the tiniest of organisms: the microbes.

In cities, we die in greater numbers and from more varied causes than we ever did as nomads or villagers. History's great urbanizations have catalyzed epidemics of emerging or reemerging diseases. The great Roman cities were ravaged by smallpox; the rich cultural flowerings of the Renaissance were also budding centers of plague; the commercial centers of England's industrial revolution were also bazaars for the swapping of cholera and tuberculosis bacilli; North America's turn-of-the-century immigrant hubs were nexuses of typhoid, tuberculosis, and, later, poliomyelitis; Africa's late-twentieth-century urbanizations may have been the catalyzing factor that launched HIV as a global epidemic. Every wave of migrants drawn by the big city has been accompanied on its journey by the Fourth Horseman. Indeed, it wasn't until the modern era that urban populations became self-sustaining; until the eighteenth century, cities relied on rural migrants to replenish populations culled by infectious disease.

The literature of the city is rife with decay and disease. Dickens, Zola, Céline, Camus, Sinclair—each has described the metropolis as a warren of corruption, filth, despair, and pestilence. Maybe there is some primordial common sense in the puritanical sermonizing against the vileness and wickedness of the city. For these places are where man has taken perhaps his greatest and most precipitous falls, not of a moral nature but a mortal one. While nomads were not free of infections, their most common diseases were chronic instead of acute. Deadly epidemics were limited and infrequent; there simply wasn't the population density required to sustain, say, a measles outbreak. Also, travel was sufficiently limited so that if a band or village was stricken by a nasty outbreak, the outbreak was likely to stay local. Most bacterial and viral infections left survivors immune, which meant that a small population would quickly develop herd immunity, preventing diseases from becoming endemic.

Once settlements pass a certain population threshold, say, several thousand, they can, and do, support most modern-day zymotics, or crowd diseases. When that happened first, probably in ancient Mesopotamia, man's history became the history of our diseases. For a microbe, a city is a target-rich environment, with slabs of human meat stacked literally one over another in apartments and houses, waiting to be consumed. If there is any conceivable way for a microbe to move from one species to another, it will find it. Of the four major modes of disease transmission—waterborne, vector borne, airborne, or direct contact—each is facilitated by urban life. Bacteria and viruses thrive in the accumulated human and animal waste. Carriers of diseases are attracted by the huge amounts of stored foods. Dirt and refuse draw germ-bearing scavengers. Puddles and cisterns harbor mosquito larvae carrying malaria, yellow fever, encephalitis, and dengue. Cats spread toxoplasmosis. Rats carry plague.

It is modern science, medical care, and the better nutrition made possible by advances in agriculture and animal husbandry that allow humans to live relatively disease-free in cities. Remove any leg of this tripod, and our happy hamlets might revert back into the dens of pestilence they were throughout most of recorded history.

■ ■ ■

CHINA'S GREAT ECONOMIC REFORMS MAY HAVE DONE JUST THAT. AT the precise moment when cities are being inundated with waves of migrants that are straining infrastructure and resources—a hundred thousand residents in Shenzhen, for example, don't have potable running water—the health care system is in a state of virtual collapse. In the past, rural Chinese could count on a few basic medical services, usually provided by the armies of so-called barefoot doctors who tromped through the countryside providing rudimentary care, setting broken bones, giving prenatal exams, and vaccinating children. This service, essentially free, helped to eradicate smallpox and sexually transmitted diseases from China and partially accounted for the near doubling of the country's life expectancy, from thirty-five to nearly seventy, between 1949 and 1990. That system, however, has been drastically scaled back as health care in China has been privatized as rapidly as the country has industrialized. Today, only 13 percent of Chinese have health insurance. For the uninsured in China, just like their counterparts in America, this system has meant the end of preventive care. Regional public health officials, also encouraged to seek profits, have similarly concluded that the return on equity from immunization drives and outbreak-response networks is minimal.

Many infectious diseases that were nearly eradicated during the sixties and seventies are rebounding in China. Tuberculosis and hepatitis B are now spreading largely unchecked, with more than 130 million hepatitis B sufferers. AIDS afflicts 1.5 million and has finally been acknowledged as a national health care crisis. Asthma, due in part to a worsening air pollution problem, is afflicting as many as one in four children in the Pearl River Delta. In the cities, doctors are seeing diseases they haven't encountered in a generation. Mao Tsetung had once declared war against infectious diseases, in particular schistosomiasis, deploying armies of pesticide-wielding workers to eradicate the host snail species and offer free checkups to those living in the Chang (Yangtze) River region. By the late eighties, partly as a result of that campaign, there were only four hundred thousand cases left in China. Today, there are more than one million, and the geographical range of the disease is again widening. Shenzhen's

CDC (no connection to the U.S. Centers for Disease Control and Prevention based in Atlanta) director, Zhuang Zhixiong, reports that he has had to contend with schistosomiasis outbreaks during the rainy season. This snail-borne blood fluke was brought into town by migrant workers and apparently found conditions welcoming enough to sustain its life cycle in some of Shenzhen's flooded alleyways.

Recently throughout the Pearl River Delta, there have been measles epidemics, meningitis outbreaks, and waves of dengue fever, malaria, and encephalitis. "If it weren't for improved nutrition," says Julie Hall, coordinator of communicable disease surveillance at the World Health Organization in Beijing, "you would be seeing even more cases."

William McNeill, in his classic book on the history of disease, *Plagues and Peoples*, estimates that measles needs a population of at least seven thousand susceptible individuals in order to maintain a chain of infection. Migrant Shenzhen neighborhoods such as Buji, Laohu, and the Click could provide many more receptive immune systems than that. Public health officials stress that it is virtually impossible to eradicate these diseases because there are so many unvaccinated migrants arriving every day, providing a perpetual supply of meat for the measles virus. Also, in migrant communities where roommates sometimes don't even know one another's names, it is very hard to trace a disease outbreak back to try to find who might have been exposed, so that they can be vaccinated or quarantined. "We basically just have to let these outbreaks burn themselves out," says Zhuang of the Shenzhen CDC.

Most migrants such as Fang Lin, of course, are even more terrified of visiting a hospital than of coming down with a disease. One night in a hospital bed could cost half a month's wages, and that's before any intravenous drips or medications or expensive tests, which many patients suspect doctors and nurses are overeager to administer and perform in order to squeeze out a few more *kwai*. Instead, when you fall ill, you drop in to one of the doctors' offices that line the alleyways of the Click, open up, say aah, and buy a box of pills. If you're really sick, you hand over a few more *kwai*, and the formerly barefoot doctor, who now has shoes but the same narrow range of treatment options, jabs a needleful of antibiotics into your ass. A report given by the

Chinese government to the WHO in October 2003 conceded the problem: "At present, health development in China lags behind economic development. The public health system is defective, the public health emergency response system is unsound and the crisis management capacity is weak."

IF THERE WAS A SOUNDTRACK TO THE ERA OF WILD FLAVOR, IT WAS bleating, high-energy techno music. The steady, rhythmic, relentlessly chirpy and upbeat dance tracks poured from every karaoke bar and fitness club and pulsated from barbershop CD players and taxicab radios. Grannies playing mah-jongg in the back of noodle shops had it cranked up to ten; old men smoking Coco Palm cigarettes in front of their knockoff-handbag stores had it blaring. During that season, the music was inescapable, an aural pink champagne bath that immersed you as soon as you crossed the border into Shenzhen. The synthetic high energy of the music was sickly sweet. If it had been food, it would have been cotton candy. If it had been a car, it would have been a cherry red Miata. Sonically tawdry yet crudely infectious, the music kept you moving, a current that propelled you forward.

Shop owners, disco touts, cabdrivers—no one seemed to know the names of any of the songs or acts. They would nod blankly when I asked which artist was playing through their Skyfox speakers. The CDs were often pirated copies of white-label releases. In other words, someone had ripped off someone who had sampled someone else. The proprietresses of the knockoff-CD stores would direct me to racks of compilations, long rows of jewel-box cases with Chinese writing proclaiming, "Feel Good Music!" "Dance Forever!" "Wild Flavor Mix!" I recognized a few of the acts: Vampire X and Shanghai Moon were two Chinese dance-music collectives that were both derivative and, in their tireless vacuousness, frighteningly appropriate. The sound was like the amyl nitrate–driven house music from some long-ago season in Ibiza or Benidorm, vaguely reminiscent of the Summer of Love, then chopped up, dipped in sugar, and injected with steroids. Here, on these narrow streets and in these muddy alleys and filthy shops, while you were stepping over this legless man or skirting that woman picking rice from a garbage can, there was something disconcerting about this much enforced feelgoodness. It felt as if the whole country were being prepared for a mass aerobics workout.

The strange thing was, when I asked Fang Lin about the music, he said he didn't really notice it. The sound had become so familiar to him, like the sound of jackhammering or drilling, that it no longer registered. In the truck, on the way to pick up the wild animals and vegetables in the markets, the driver (one of the Sichuanese brothers who ran the joint) and Fang Lin found the music a steady distraction, which they would listen to while they smoked their Honghe cigarettes. Back in the chopping room, Fang Lin would sometimes cut according to the beat, he said, but he never noticed when one song ended and another began.

The chefs, in between dishes, would stand beside their woks smoking and slugging from screw-top jars of *chuhai* grain liquor.

In the summer, the temperature would climb to 110 degrees in the kitchen and chopping room. The only good thing about the heat was that it made most of the animals sluggish.

The restaurant went through one or two hundred animals a night. High-volume creatures—wild rats, cats, and dogs—were stacked near the kitchen. Fang Lin and his colleagues would organize the animals so that the snakes—in season in late spring and summer— and other animals that might take a bite out of you were under the two fluorescent strip lights in the middle of the room. More docile creatures, which posed less of a risk as they were extracted from their cages, were kept in the corners, farther from the light. When an order was called out, Fang Lin picked through the wooden and wire-mesh cages until he found the beast he needed. There was a specific technique for retrieving each animal. For a snake, one pinned the head until the tail could be located, and that was used to swing the animal over, smashing its head against the brick floor. A lizard had to be thumped in the head with the fat part of a knife handle. Cats were easy: reach in wearing the thick rubber gloves, pull the cat out, and chop the head off; it usually took one clean blow with the cleaver. The female civets were easily enough extracted and decapitated. But the males could put up a fight and might require the pronged stick, which one chop boy would use to pin the animal to the back of the cage while another bound its legs with duct tape. Then the animal could be pulled out, killed, skinned, and made ready for the hot pot.

As the night went on and the calls came from the kitchen—duck, duck, goose, pangolin, civet—the boys pulled the animals from their cages, cut them, skinned them, bled them, separated the organs into those that were prized for their various invigorating properties and those deemed inedible or inauspicious, and then quartered the animals so that they were ready to be cooked. The chefs would further slice the animals according to the dish on order and their own recipes. Some dishes, such as civet and snake, tended to be heavily seasoned, while rat, for example, was usually skinned, basted in sweetened soy sauce, grilled, and then chopped up like an order of barbecued pork. Occasionally, an animal—a badger or, in one particularly frightening encounter, a wild boar—would get loose before the coup de grâce was administered, and the three chop boys would have to chase the animal through the chop room and kitchen and, in the case of that boar, through the dining room of the restaurant itself.

By the end of the evening, the floor of this charnel house of doomed animals would be slick with blood and entrails and feces and urine. The panicked animals released all manner of foul excretions, with their dying act most often being defecation. While Fang Lin became used to the noises—the construction, the music, and even the shrieks and screams of dying animals—he never became used to the smell. He kept a cigarette lit constantly, clenched between his lips so that the trails of smoke were always pouring into his nostrils and obscuring some of the stench.

The chefs would come out late in the evening to sit on the back steps with the chop boys. Among them were two men from Sichuan who still expressed occasional disgust at what Guangdong people were willing to eat. And there was another fellow who claimed to be from Guangxi Zhuang but who spoke with a perfect Beijing accent. Whenever anyone asked him where in Guangxi he was from, he would say, "The mountains." Whenever anyone asked which mountains, he

would make a gesture that could be interpreted as meaning either
"very far away" or "leave me alone." One evening he was gone, and the
two Sichuan chefs explained that he had stabbed another fellow in the
thigh after seeing him in the street. That man, like him, claimed to be
from Guangxi but spoke with a Beijing accent. The chop boys never
saw that chef again. It was no great loss, as the fellow hadn't been a very
good chef, and though he had claimed to be familiar with the preparation of Wild Flavor, he seemed to have made up some of his recipes as
he went along, relying on soy sauce, ginger, and garlic instead of listening to the proprietress's recommendations.

That chef was replaced by a fellow named Chou Pei, who came
from Heyuan and had been living in Shenzhen for over a year. He
was a quiet fellow, about five foot seven, with a fleshy face and spiky
black hair. Since he had come from a rural part of Guangdong, he
knew how to cook most of the dishes, and if there were any he wasn't
sure about, he could call home to double-check or consult a card in
his pocket with the proprietress's instructions.

One night, when several of the chefs and chop boys were sitting
around on the back steps, passing back and forth a jar of grain liquor,
Chou Pei looked at Fang Lin.

"You've got blood on your face," he said.

Fang Lin reached up with his hand and rubbed his cheek.

"Other side," Chou Pei pointed out.

Fang Lin tried the other cheek, and his hand came away coated
with blood. That entire side of his face, in fact, was covered with
some animal's spilled guts.

He shrugged and wiped the blood and guts off with his T-shirt.

"It's still there," said Chou Pei.

"What?" Fang Lin asked.

"The blood."

Fang Lin shrugged and wiped it again, this time smearing some of it
into his eyes. This wasn't the first time the chop boys had gotten animal
blood in their eyes or noses or down their throats. But what did it matter? Who had ever been harmed by the blood of a few wild animals?

CHAPTER 6

- **December 25, 2002**
- **Ah Chau Island, Starling Inlet, Hong Kong, China**
- **35 Infected, 8 Dead**

ON CHRISTMAS MORNING, MALIK PEIRIS AND FELLOW UNIVERSITY OF Hong Kong virologists Guan Yi and Zheng Bo Jian rode in Peiris's black BMW up Motorway 1 to Nam Chung Temple, in Hong Kong's New Territories, just a few miles south of the Chinese border. Peiris parked in the gravel lot next to the temple, and the three men stepped out of the car, unloaded their duffel bags from the trunk, and slipped on their parkas, rubber boots, and gloves. At 7:00 A.M., the air was chilly, and as they walked past the temple with its three red lintel altars and the offerings of oranges and incense laid out before a statue of Confucius, they zipped up their jackets against the damp breeze blowing from the Starling Inlet. Across the marshy pond was Ah Chau, the small island that also served as a bird sanctuary for several of Hong Kong's migratory species. The island rose in a gentle mound, with spruce, rain, and eucalyptus trees shooting leafless branches skyward above the canopy. Vultures sometimes sat on these highest branches. Below them, presumably, would be either nesting egrets, cranes, and cormorants or hundreds of dead birds.

The rumors out of China were even more alarming than usual. For example, during a November influenza conference in Beijing attended by World Health Organization officials, Chinese clinicians from Guangdong spoke darkly of an influenza outbreak that was burning through rural hospitals. "They told us they were currently having a severe influenza outbreak," Klaus Stöhr of the World

Health Organization's Global Influenza Programme had told Peiris and Guan Yi. "Guangdong officials said, 'People are dying like flies.'" Stöhr followed up, requesting samples of the influenza strain in question. What the Guangdong Health Department claimed to have isolated was a standard A-type influenza, which was unlikely to have caused the severe disease that clinicians had described.

When Stöhr inquired further, via the WHO's Beijing office, he was given a curt reply: "Everything is calm."

Though no one in China had yet linked the unexplained deaths in hospitals throughout Guangdong and postulated an epidemic, Hong Kong health care workers, through their own informal contacts, were already hearing anecdotal reports of an unexplained respiratory ailment that had their mainland counterparts worried. Joseph Sung, chief of gastroenterology and hepatology at Prince of Wales Hospital, was scheduled to attend a conference in Guangzhou the next week but was told by a physician at the Guangzhou Institute of Respiratory Diseases, "Don't come."

When he asked why, he was told, "Something bad is going on here. Just don't come."

What was it? Most of those medical professionals and public health officials who were watching were wary of an influenza outbreak, and when avian flu turned up in Penfold Park, that caused another round of dark conjecture. As anecdotal reports of unexplained respiratory ailments filtered back across the border, influenza experts around the world perked up, and local virologists like Malik Peiris and Guan Yi began to suspect that an influenza outbreak might be afoot. K. Y. Yuen, the head of the University of Hong Kong microbiology department, would discuss the matter with Margaret Chan, Hong Kong's director of health, and both believed there was a good probability that what was going on in Guangdong was the flu. If it was avian influenza, they both knew, then this would quickly become Hong Kong's problem. So far, however, there had been no instances of mass chicken infection—so far. Winter in China is notorious for bringing on a host of respiratory ailments, ranging from the common cold to asthma to influenza. When you are traveling through Chinese cities in the weeks before the lunar

New Year, it seems the whole population has a dry, hacking cough. It was entirely possible that these tales of a deadly pneumonia were fanciful, the medical equivalent of "Here There Be Dragons" scrawled at the edge of medieval maps.

Medically speaking, China was, in many ways, outside the known world. Good intelligence about what was going on in terms of infectious disease was always hard to come by, to a great extent because officials in Beijing were often themselves unaware of outbreaks in the provinces. For Hong Kong's public health team, there was nothing else to do but keep the influenza surveillance network on alert—and try to reach a Beijing Ministry of Health official who would tell them what was really going on.

For Malik Peiris, Guan Yi, and the rest of the virologists and researchers at the University of Hong Kong, the rumors of influenza in China fueled their already considerable suspicions that China was a hotbed of animal influenzas that periodically infected humans. The Chinese government steadfastly maintained a position that China did not have any avian influenza and forbade any virologists from collecting samples to be tested outside China.

Peiris and Guan Yi would have to collect their own samples, first from migratory birds—in the hopes of forestalling the later gathering of samples from sick humans. Ag and Fish had arranged for a small rowboat to be moored by the temple, and as the three men crossed the mudflats to the boat, the mudskippers and crabs feasting on algae skittered away. Offshore, where there would usually be a crane or two picking its way gingerly with backward-bending knees through the shallows, there was only still water. The team tossed their duffels into the boat before casting off and scrambling aboard. Zheng, as the junior scientist, was delegated the task of rowing as the two other scientists prepared their sample containers and wooden swabs. They were here to check out some shit, literally. If the birds were infected, they would be shedding huge amounts of virus in their feces. Later on, Peiris would wonder at the wisdom of their doing this sample collection without taking any more precautions than wearing rubber gloves.

As the boat slid ashore against the muddy bank, the three men jumped out and sank into muck up to their shins. The island was filthy with broken glass, empty cans, and abandoned Styrofoam noodle containers. Local teenagers had been trespassing on this particular bird sanctuary—and who knew if any of them were already carriers of this killer flu? The team spread out, with each man heading toward a different portion of the island, his sample case at the ready as he stooped over, looking for fresh bird droppings. Peiris slipped into the shadows beneath the tall trees, losing track of his partners. He could no longer hear the rippling of the inlet's water. The smell here was fecund, muddy, and redolent of earth and clay and mulch. He expected to see dead birds, corpses gathered in small piles beside the trunks of decaying trees.

GUAN YI HAD ALSO WANDERED AWAY FROM THE MOORING, MAKING his way between oak trees and up toward the hilly center of the island. Whatever had killed those birds at Penfold Park, and at lesser, subsequent outbreaks in Kowloon Park and in the New Territories, should also be here, on this filthy island. Guan Yi was familiar with this sort of fieldwork, having taken the lead in going into Hong Kong's wet markets during the 1997 avian flu outbreak to screen for the virus among commercial poultry. He had made himself one of the world's leading experts on the genetic mutations of various strains of animal influenza and would frequently launch into sputtering, almost hysterical warnings about how humanity was just a few viral mutations away from a global pandemic.

Guan Yi and Peiris maintained a steady, if stormy collaboration, with each respecting the other's work but also criticizing his projects with a vigilance inspired by mutual jealousy. Both were integral to the Pandemic Preparedness in Asia project, originally set up by Ken Shortridge of the University of Hong Kong and Rob Webster of St. Jude Children's Research Hospital in Memphis. Peiris was in charge of research into human flu viruses, and Guan Yi would run the animal diseases research. Obviously, since most emerging diseases start as animal diseases that jump to infect humans, the overlap between the scientists' fields was probably as great as the discrete areas themselves. And a more unlikely collaboration between two men would be hard to imagine. Peiris is thoughtful and reserved, and takes his time before coming to a hypothesis. Reporters speaking with him occasionally find frustrating his unwillingness to venture conclusions or extrapolations. He is circumspect and thorough, and that is one of the reasons why his research is above reproach.

Guan Yi, on the other hand, is famously impetuous, prone to rash utterances and fantastic claims. Spending a day with him is like hanging around with a Tasmanian devil well versed in virology. He is all bold hypotheses, boasts, and sudden conversational veers. It is remarkable that the same man who is willing to spout the wildest, most unlikely deductions is also capable of producing groundbreaking research and remarkably well reasoned scientific papers. The only personality trait the two men share, as far as I can tell, is their relentless ambition.

Lately, Guan Yi's web of contacts had been buzzing with fresh rumors about a respiratory disease of unknown provenance that was burning through Chinese hospitals. The ailment didn't even have a name, and the descriptions of its symptoms were inconsistent. Among Hong Kong physicians, scientists, and virologists with connections in China, word of a severe respiratory ailment appearing at the beginning of flu season was worrying but still inconclusive. Rumors circulated every year of killer strains, new viruses, and nasty bugs emerging from Guangdong that somehow never reached Hong Kong. Mini-epidemics were rife across the border, Guan Yi knew. Still, by speaking to his colleagues in Guangzhou, he could tell that something about this year's rumors was different. For one thing, unlike in previous years, when the early tales ranged from vivid yarns of Ebola-like hemorrhagic fevers to the more clinical descriptions of severe influenza, this year all the stories centered on this mysterious respiratory ailment. And every tale included certain peculiar details: the high fevers, the cloudy chest X-rays, and the fact that the disease seemed to be targeting medical staff.

But each time Guan Yi would call one of his colleagues on the mainland, at the Guangzhou Institute of Respiratory Diseases or

Zhongshan Number One Hospital or the Guangzhou Number One Military Hospital, the chief respiratory specialist there would tell him that he, too, had heard these rumors, but no cases had yet arrived in any of the big-city hospitals. "It's probably just the flu," one doctor at the Guangzhou Institute of Respiratory Diseases told Guan Yi. "Just this year's flu."

Guan Yi, who was raised in Jiangxi and had seen firsthand the ravages brought on by "this year's flu," knew not to be so dismissive. Born in 1962, the youngest of five children, Guan Yi grew up during the deprivations of the Cultural Revolution. As a child, he was spared the worst of Mao Tse-tung's misguided policies. Yet his family suffered as class criminals for his mother's having descended from a landowning family. His father, trained as an engineer, was sentenced to reeducation through hard labor at a factory that extracted liquor from flaxseed.

Because of the size of the Guan family, they were allowed to remain in their six-room courtyard house in Ningdu, a rural county three hundred kilometers from the provincial capital in Nanchang, although the house was subdivided to allow for more residents. Young Yi passed his days playing in a small yard behind the kitchen with the chickens and ducks, blithely unaware that his brothers and sisters, smart students who should have gone on to higher education, had been prevented by the Cultural Revolution from attending college and then, when the colleges reopened, been deemed unfit because of their family background. Instead, they would set off every day for their factory jobs and then return to teach young Yi in the evenings. He was a fast learner, and, as the youngest, he became the repository of much family aspiration and love. Even today, Guan Yi maintains the sort of confidence and self-assuredness typical of youngest siblings. He is used to being forgiven his transgressions and lavishly praised for each success as if he is still the boy getting his sums correct by the light of a lamp lit with precious rationed oil.

When the colleges finally reopened in 1978, Yi would become the first member of his family to secure a place at a university. Only now does he realize the sacrifices his family made for him: his mother removing from hiding places beneath courtyard stones treasured family heirlooms—jade and emerald jewelry and golden marriage bangles—which she would exchange for precious pork and egg ration coupons; his father's reeducation job in a workshop beating flax plants into a kind of grain substitute that party officials were distilling into a cheap alternative to rice liquor. The work resulted in vast clouds of stinging dust, which Guan Yi's father breathed in every day for four years. Guan Yi would watch his own father succumb to a respiratory disease, bronchial asthma, while he was still a medical school student in Nanchang. Later, he was certain that the reeducation work at the factory was what killed his father at age fifty-one.

Guan Yi, who would do his doctoral work in the United States at the University of Tennessee, would return to his native land that rarest of Chinese characters: the patriot who believed that only the truth, at least when it came to diseases, could save China. To that end, he built a network of contacts and sources on the mainland, which so far had provided no clear information about the rumors of a respiratory ailment cutting through hospitals in southern China. Influenza specialists around the world relied on Rob Webster, Guan Yi, and Malik Peiris to keep them apprised of any new strains or mutations emanating from southern China. Hong Kong's role as influenza sentinel is a little-remarked-upon benefit of the "one country, two systems" manifesto adopted by China upon the colony's reuniting with the mainland. The medical and scientific establishments in Hong Kong, relatively free of political meddling, could monitor, via importations of poultry and livestock, the state of animal disease on the mainland and so extrapolate possible new influenza variations. These whispers of disease in Guangdong, however, were even more alarming in that there had been no hypothesis as to what the problem could be. As far away as Geneva, World Health Organization officials were wondering if they should so easily dismiss the hearsay from southern China as mere gossip.

If it was avian flu that had achieved widespread human communicability and pathogenicity, both Peiris and Guan Yi well knew, then Hong Kong would likely become the second link in a global chain reaction that could melt down the human race. One explanation for human communicability would be the disease's becoming prevalent in migratory birds, and those birds defecating into or mingling with domestic poultry and waterfowl populations that then spread the disease to humans. Or, even less likely but more frightening, the disease had jumped directly from migratory birds to human beings, passed on via fecal droppings on win-dowsills, doorways, car door handles—anywhere—and then into the human respiratory system via inhalation or oral contact. That would explain a widespread outbreak in remote Guangdong hospitals; it would also explain the rumored fatalities.

BUT WHERE WERE THE BIRDS? MALIK PEIRIS WONDERED AS HE HIKED up the hill looking for fresh droppings. Those tended to be near nest-ing spots or under trees with large roosting populations. Finally, he saw a few dried pools of blue-gray feces. Fresh samples are preferable to stale, and as Peiris bent over, he was disappointed to discover that these seemed to have lost the goopy, chewed-gum consistency of a new deposit. Continuing up the hill, he expected to find mounds of corpses, and regretted again having no more protection than a pair of rubber gloves. If this disease had somehow become airborne and communica-ble to humans, then he might be walking into a biological killing zone.

A flurry of white and gray, a slashing yellow beak, and an eye as wide and glassy as a child's marble appeared just a few feet in front of him. Suddenly, as if they were waking up from a deep sleep, stirred egrets and herons began squawking, spreading their wings, and tak-ing flight from higher branches, with one or two swooping low to get a look at Peiris as they flew off to feed on the inlet.

Here were newer droppings, blots of cloudy-sky-colored bird shit. Peiris hunched down on the balls of his feet to swirl a cotton-tipped swab in the guano. After he had gathered a good sample, which was a little like twirling pasta onto a fork, he dropped the swab into a vial of Eagle's Minimum Essential Medium and then broke off the wooden handle of the swab and screwed a top onto the vial before dropping it back into his satchel. In all, Peiris and Guan Yi gathered about sev-enty samples. If there was virus on this island, it would be in these vials.

CHAPTER 7

- **December 26, 2002**
- **Shenzhen, China**
- **36 Infected, 8 Dead**

FANG LIN BEGAN TO FEEL FEVERISH JUST AS THE WEATHER WAS starting to turn dry. (Later, he would say it had been around the time the big Yao Ming posters advertising mobile phones had started appearing on billboards around the city.) Since arriving in Shenzhen five months earlier, he'd had several slight fevers and colds; in the Click, many residents seemed to live in a perpetual state of low-level illness of one sort or another. When the fever didn't begin to break after a day and a half, he consulted a local physician, a man with one eye and a rusty stethoscope. Fang described his symptoms—there was the fever, his muscles seemed to ache when he woke up in the morning, and he'd had nasty diarrhea for two days. The doctor nodded but didn't say anything. From behind a glass case, he removed a box of tablets.

"How much?" Fang asked.

"Fifty *jiao* each."

Fang bought four tablets, swallowed two right there, and went to work.

Riding around in the truck that day, he felt more invigorated but still feverish, and at one point, as they were tossing cages of civets, badgers, and pangolins into the truck, he felt so weak he had to sit down. The driver offered him a cigarette.

"Here," he said to Fang. "That'll make you feel better."

Fang made it through that day and night, having to pause during his work in the chop room to catch his breath, and he took frequent cigarette breaks on the back stairs. He had told his colleagues in the chop room that he didn't feel well, and though both men agreed they could pick up the slack, in actuality there was no way that fewer than three men could do the job without slowing down the kitchen. So Fang had had to keep on chopping, until the main banquet room seating was over.

The next evening, when he went out for a cigarette break and sat down on the back steps, he couldn't get up. His fever had climbed, probably to over 103 degrees, and he found that no matter how deeply he breathed, he felt perpetually winded. His body aches had reached a point where whatever position he stood or sat in, he felt as if his muscles were being pulled from his bones. Du Chan found him out on the steps and asked what was wrong.

Fang told him that he didn't know.

Soon Chou Pei came out and handed a bottle of grain liquor to Fang.

Fang took a tiny sip. Immediately, he vomited it back up, along with a stringy pool of yellow bile. He had never before experienced this combination of nausea and diarrhea.

"You need to go," Du told him.

Fang nodded but couldn't get up.

Du and Chou helped him to his feet and loaded him onto the back of a motorcycle taxi. Somehow, Fang managed to stay balanced for the seven-minute ride to his *cun*. After he was deposited at one of the alleys, he found another storefront doctor. Even in his addled state, Fang recalls, he knew he needed medical help. This doctor told him to stand still, then to stand on one leg, then the other. He asked Fang a few questions: Where did he ache? What had he eaten in the last few days? Where had he been born? Was his mother taller than his father? In what direction did his bed face? As Fang answered, the doctor laid his hands on Fang's chest and arms, squeezing gently as he ran his hands over the young man's bloodstained shirt.

The doctor pronounced Fang Lin's chi (the vital force believed by Chinese to be inherent in all things) to be unbalanced. Fang needed to reorder his life, he said, so that he would face southward more often. He also needed to drink more of a certain type of fungal tea, which this doctor sold by the can. He should also take vancomycin and propara-cetamol tablets, which the doctor sold to him for fifty *kwai*.

Fang Lin made it back to his north-facing room, but he never went back to work. The anti-febrile medication did nothing to assuage his fever, which may have spiked north of 104.5. Twice, he was unable to rouse himself from his sleeping pallet in time to reach the toilet in the hall, each time soiling his trousers. He hadn't eaten in three days, so his excrement was almost entirely liquefied. He became progressively more disoriented. Even for the fully cognizant, it was hard to differen-tiate night and day in the Click. Now, in a state of near hypoxia, he lost track of time. He was vaguely aware of when his roommates were in the room, and when he heard their steady, deep breathing, he was reminded of how shallow his own breaths had become. He found that if he moved even slightly, to roll over or sit up—to stand was now out of the question—he would be completely out of breath. He begged his roommates to open the window, a request that elicited grave silence from Du Chan and Huang Po: the room was windowless. They brought him a Styrofoam bowl of soup with pork dumplings, which Fang left on the floor. He managed to sip from cans of herbal tea.

Du Chan would bring back bottles of beer from the restaurant or plastic jugs of bright blue or orange sports drink. Fang would bend his head forward just enough to suck a little of the liquid down his throat and then would fall back exhausted and in pain. The muscle aches were so severe, he recalls, that he found staying still unbear-able, yet any movement would leave him gasping for breath. The proparacetamol would temporarily provide some relief for the mus-cle aches, but the fever never subsided. Fang found that the temper-ature in this hot little room seemed to oscillate wildly between extreme heat and terrible cold.

One of his roommates suggested he go to a hospital.

Fang squashed that idea by muttering, "No hospital. No money."

■ ■ ■

WHAT WAS HAPPENING TO HIM? FANG KNEW HE WAS ILL. BUT HE STILL assumed he was suffering from another of those respiratory infections that regularly burned through the Click. Everyone seemed to have a hacking cough of some sort; whether it was due to cigarettes, persistent asthma, or air pollution was impossible to say. During influenza season, in winter, some of those coughs turned into serious illnesses. But who ever died from a bad cough? Fang wondered. You took some antibiotics and drank plenty of fungal tea. And if you still didn't recover, you got some rest. But he had been on his back for several days now and wasn't feeling any better. Most terrifying for him, when he was conscious, was the sense that no matter how deeply he breathed, he felt that what he was inhaling was not oxygen but some other odorless, tasteless gas with similar properties but without the life-sustaining force of simple O_2. He was running out of air, yet he felt he was breathing freely.

He now had to stay perfectly still. To move was to suffocate. Stay still. And breathe. Breathe as deeply as possible. He fell into fitful bursts of sleep, angry patches of shut-eye during which he would see recurring images of the room, his roommates, the restaurant. It didn't feel like sleep but rather like a waking dream during which he would systematically re-create and recount his days since leaving Jiangxi. Bits of conversation he'd had were steadily repeated and turned over in this dark semiconsciousness. This mental exercise always left him dissatisfied and unsure upon awakening, as if he had not slept at all.

Outside, he would later recall, he could still hear the city—the clanging and banging and the steady racket of construction, the constant clomping of other residents coming and going, and the sounds of doors opening and slamming. He'd never felt so alone as he did then, hearing the working girls returning from their night's work and then, a few hours later, the day laborers and construction crews rousing themselves and setting off. Sometimes, this cacophony would blend into one dull tone, leaving him frustrated at his inability to understand what was going on.

On about the sixth day of his illness, he lost all track of his environment. From then on, there were only dark dreams and the sensation that his life was literally being squeezed from him. His muscle aches would come in steady, rolling waves and would peak as gripping cramps around his spine and in his neck and upper legs, a dreadful tightening that would coincide with a gasping inability to draw in enough oxygen. He lay still and struggled to stay awake so that he could focus on maintaining his steady, ineffectual breathing. He feared that if he fell asleep, he might forget to breathe, and that would be it. Perhaps that is what dying is, he wondered—your body forgetting how to breathe.

But he did begin to drift off, always remembering, even in his unconscious state, that he must stay still. Any movement at all, even a wiggling of toes, even blinking, used precious oxygen. That was air he didn't have. So he lay perfectly still, and in those moments between severe cramps and muscle aches, when his bowels were settled, he would drift into dark snatches of unconsciousness. But it was a cruel sleep, one that never let him forget, for even a moment, his suffering. During those naps, he would always feel very far from home and very alone. A terrifying idea began to glow in the darkness: he would die far from home, away from his family. He understood, finally, the importance of that Chinese tradition of rushing home when you were ill, even if only to pass away. And then he thought of another matter: who would pay for the cost of his funeral arrangements?

CHAPTER 8

■ January 1, 2003

■ Department of Microbiology, University of Hong Kong, Pok Fu Lam

■ 40 Infected, 8 Dead

FINDING A VIRUS IN BIRD FECES REQUIRES THAT THE SHIT HIT THE fan. The gathered waste matter must be homogenized and then centrifuged. Viruses are too small to be seen in an optical microscope. Even viewed through a transmission electron microscope—which shoots beams of electrons through a sample slice one micron thick and onto a phosphorous screen—it is almost impossible to pick out a single virus particle from the vast, almost continental landscapes that appear when even a single cell is bombarded by electrons. (Anything seen through an electron microscope is colorless because the objects are too small to reflect any light.) Trying to find one virus particle in an electron microscope slide is like trying to find Des Moines on a satellite image of the United States, when you have not been told where Des Moines is, or even *what* it is. Viruses vary in shape and size, and influenza is among the most inconstant of viruses, sometimes appearing as a long, almost wormlike tube with many spikes and at other times showing up as almost spherical. But no matter the shape, viruses are so small as to be almost impossible for human beings to comprehend. If your body were the size of, say, Des Moines, then one of your cells would be the size of a dining table. A virus would be a pea on that table.

The most efficient and cruelest killer on earth is not a lion, bear, shark, or other superpredator, but probably the smallest predator. (The difference between predator and parasite, at this point, seems to be mainly semantic.) And this killer is not even alive in the sense that we are alive, or even in the sense that a lichen or bacterium is alive. Viruses are described, simply, as active or inactive. They are little more than bits of nucleic acid (DNA or RNA) wrapped in a skein of proteins self-programmed to subvert more protein and genetic material, which they can fashion into more virus particles. A virus's mission is to replicate, and it does this by landing on a cell and attaching to specific receptors on the cell membrane, destabilizing the surface of the cell just enough to pass into the cytoplasm—the initial stages of this replication look something like the Apollo lander setting down on the surface of the moon. Once inside, the virus hijacks the cell's genetic material in order to fashion more of itself. The process has been described as taking over a factory and then refitting the assembly line to turn out a new product: in this case, more viruses. These new viruses will mature and leave the cell either by "budding," during which a few viruses at a time pop through the membrane, or by "lysis," when the cell itself ruptures and sends millions of viruses out to infect other cells. After being duped into subverting itself to become a viral factory, the host cell will deplete itself and die. When too many cells start dying, the host's immune system will react and, in the case of viruses that attack human cells, symptoms of disease and illness will appear. Just like other predators, viruses appropriate an animal's protein in order to survive. (A lion gorging on a gazelle's entrails out on the savannah is doing exactly that.) Unlike other predators, however, viruses are never sated. They don't eat; they simply reproduce. In some sense, they are the perfect life-form, only they are not technically alive. They are nothing more than manifestations of the urge to propagate. And in so doing, viruses have probably killed more of everything than anything else in the history of the world.

One of the theories to explain the biological origins of viruses is that they are degenerated or backward-evolving bacteria, the mitochondria of single-cell organisms that shed every trait save those needed to survive inside another cell. A similar view is that RNA or DNA molecules from single-cell organisms acquired the ability to evolve and replicate independently of a host cell. There are numerous other theories, including the "panspermia" thesis, which suggests that the first virus particles arrived from outer space, which is only a little more unlikely than RNA- or DNA-based life spontaneously arising in a lifeless sea billions of years ago. However they originated, viruses have remained ingenuously simple collections of proteins that can perform an astonishing number of tasks, from inserting themselves into host cells; reading, disassembling, and reassembling a host cell's RNA or DNA; and then moving toward the surface of that cell in order to float away and infect more cells. Certain viruses, such as the herpes or hepatitis B virus, can lie dormant in the nuclei of some cells for decades, while other viruses can survive in extreme heat or glacial chill. There is no living thing on this earth that isn't susceptible to some sort of virus. And while viruses are ubiquitous—there can be billions in a pinch of soil—they remain exceedingly hard to find unless they appear in huge amounts.

In order to locate viruses in a sample, Peiris and Guan would first need to separate the viral matter from the sample via the centrifuge process and then grow that virus on living matter. You don't detect viruses merely by observing them—they are too small to be easily found out—you find them when they are killing something. In order to confirm the presence of H5N1 virus, both Peiris and Guan would have to introduce their samples to host cells in which the virus might like to replicate. They both had their own methods.

As co-principal in the Pandemic Preparedness in Asia program, each scientist maintained a separate lab in the microbiology department of the University of Hong Kong. Peiris's was on the fifth floor, while Guan's was on the fourth. When they returned to the squat brown-and-gray rectangular building that backed into the green hills above the South China Sea, each took his samples to his respective lab to try to grow some virus.

Guan Yi inoculated chicken eggs, just as Trevor Ellis had done for the Department of Ag and Fish. If there was avian flu out on that island, the virus would quickly turn the chicken embryos to a mulch of blood and inchoate bone. Guan Yi had gathered thirty-six samples, which he homogenized with an antibiotic serum—he didn't want any bacteria contaminating the experiment. After separating out some clear fluid and inoculating the embryos with a ten-centimeter hypo-dermic needle, he set the eggs into an incubator.

Peiris, meanwhile, was undertaking a similar process to extract clear fluid from the bird feces, but instead of using chicken eggs, he would try to grow the virus on an MDCK cell line—dog kidney cells laid out in plastic sample trays. This was a cell line on which influenza was known to thrive. Within a few days, he should be able to see through an optical microscope if something was tearing apart these previously healthy cells. In all likelihood, that agent would prove to be H5N1, which he would isolate from those cell lines and then test against avian flu genetic primers.

Peiris and Guan Yi were reticent about revealing their findings to each other. For three days, they had been incubating, waiting and watching, checking their samples for signs of life, or death. Yet both men—Guan with his chicken embryos and Peiris with his canine kid-ney cells—had found that their host organisms had remained frus-tratingly alive. They had retested, pulling the original swabbings out of their nitrogen freezers and running through the benchwork a sec-ond time, and again the results were hearty cells and embryos.

Both men met warily in K. Y. Yuen's fourth-floor office. Neither wanted to admit that he hadn't isolated any virus. What if the other scientist had found H5N1? That would be a major breakthrough for him and an acute embarrassment for the other fellow. Finally, K. Y. Yuen, the head of microbiology at the University of Hong Kong, asked them point-blank: "Well?"

No one said a word.

Peiris eventually shook his head. "Nothing."

Guan Yi echoed, "Same results. No H5."

For a moment, both Peiris and Guan Yi were relieved—there was some small validation in knowing that their results had been dupli-cated by their rival. But Yuen looked concerned, despite the negative findings. And then Guan Yi remembered: if it wasn't avian flu, what the hell was killing those patients in China?

BOOK 2
What Does It Do?

CHAPTER 9

- **January 3, 2003**
- **Guangzhou Number Three Hospital, Guangzhou, China**
- **48 Infected, 9 Dead**

DENG ZIDE, DIRECTOR OF INFECTIOUS DISEASES AT GUANGZHOU Number Three Hospital, had just completed morning rounds with several residents when he stopped in at the third-floor Clinical Department offices to see if he had missed any calls. A nurse handed him a message from the Provincial Bureau of Health, which read, simply, that he was to appear at the bureau at 2:30 P.M. The message ended with an ominous instruction: "Bring a packed suitcase." The forty-two-year-old specialist showered in the doctors' changing room, slipped into a dark suit and black shoes, and walked back to his apartment in the complex behind the hospital. He sat down to a quick meal of pork, rice, and cabbage before throwing a change of clothes, a gown, gloves, a mask, and goggles into a small duffel bag and heading fifteen minutes by taxi across Guangzhou to the Department of Health.

He arrived at the fourth-floor meeting room a few minutes early and found several other respiratory doctors and epidemiologists all in a similar confused state as to why the meeting had been called. They took seats on wooden chairs around a conference table on which stood a dozen bottles of mineral water and paper cups. He Zhaofu, the deputy director of the Provincial Department of Health, entered the room, greeting them all and curtly thanking them for coming. As the ranking official in the room, he could dispense with anything but the most perfunctory of pleasantries; yet since this was a gathering of medical professionals and not political appointees, some courtesy was still required. The men in that room were among the most prominent physicians and scientists in the whole Pearl River Delta. Along with Deng Zide, there was Huang Wenjie, the chief of respiratory diseases at Guangzhou General Military Hospital. Next to him were three Guangdong CDC epidemiologists, led by Lou Hwei Ming; and opposite that team was Xiao Zhenglun, vice-director of the Guangzhou Institute of Respiratory Diseases.

The department had received a fax, He Zhaofu explained, from Heyuan Number One Hospital, which was four and a half hours by car northeast of Guangzhou. The director of the hospital had written under the subject line "Heyuan People's Hospital Fax to Guangdong Health Department on January 2, 2003" that two patients had been admitted on "December 15 with serious pneumonia of unknown causes. These two patients were transferred to Guangzhou Military General Hospital and Guangzhou Institute of Respiratory Diseases as their conditions deteriorated." He Zhaofu went on to say that the first patient had been admitted with fever, muscle aches, and a faint crackling sound in the lower left lung. Nothing surprising there, of course, since it was early winter, and respiratory ailments were common during the dry season. He Zhaofu continued reading aloud: "In a few days, every medical staff member who had been in close contact with these two patients developed similar symptoms. Three were in critical condition. At present, six medical staffers have been hospitalized. None of the patients have responded to conventional treatments."

While none of the doctors in this room yet knew of the mysterious deaths throughout the province that had followed this similar litany of symptoms, for Huang Wenjie of Guangzhou General Military Hospital, this sounded like an institutional cry for help disguised as a procedural document. No director of a local hospital was going to lose face by admitting to his bosses in the provincial capital that he could not handle a local health care crisis. For the director to be relaying his worries up the chain of command meant that something frightening, or at least very strange, had to be occurring in that picturesque city. But for Huang Wenjie, there was something else discomfiting about the fax. One of the patients had been transferred to his hospital—without any indication given that this was an extraordinary case. He remembered that his hospital had admitted a pointy-eared, chubby-cheeked fellow with a thick neck and a crew cut who had lost consciousness in the ambulance on the way to Guangzhou. The patient's accompanying charts indicated that his high fever and lung infection had been treated with broad-spectrum antibiotics yet had shown no sign of abating. Upon his admission at General Military Hospital and settling into bed 93, his respiratory rate reached three times the normal level even while he was lying still. Despite that labored breathing, he could not maintain 90 percent blood oxygenation. In other words, he was taking deep, frequent breaths, but no oxygen was reaching his bloodstream.

"That patient," Huang Wenjie said. "He came two weeks ago. A very strange case." What he did not reveal was the mild sense of betrayal he felt at belatedly receiving this information. The patient had been a mystery to Huang Wenjie. At first, he had assumed he was dealing with a typical pneumonia case. But the patient's rapid deterioration, combined with his unresponsiveness to any therapy, had left Huang Wenjie, a forty-two-year-old career military medical officer, struggling to come up with a diagnosis. "Pneumonia" is actually a catchall term, a roundup of symptoms rather than an explanation of the causative agent. It describes, broadly, virtually any infection of the lungs—viral, environmental, or bacterial. In the vast majority of cases, the cause is bacterial, and the patient responds to some sort of antibiotic therapy. Only in a case when the patient does not respond to standard antibiotic treatments does a clinician worry about identifying the causative agent.

In the case of the twenty-seven-year-old from Heyuan, there had been no effective course of therapy, and the patient's condition deteriorated so rapidly that within a day of his admission, he no longer recognized his own family. This sort of severe pneumonia was rare among healthy young men. But still, in China, one case was unlikely to rouse anything but passing interest. Most doctors relied instead on rotating medications until the right dose and drug finally knocked out the microbe that was causing the disease. (In the West, among most clinicians, the treatment would be much the same. If they can cure the patient without knowing what has ailed him, seldom will doctors bother to find out the actual microbial agent.)

Now Huang Wenjie discovered that the doctor and nurse who accompanied this patient from Heyuan had themselves fallen ill with this mysterious affliction and were hospitalized in critical condition.

Xiao Zhenglun from Guangzhou Institute of Respiratory Diseases then spoke up and reported that on December 22 he had accepted from Heyuan Number One a forty-one-year-old owner of a small trading company. He had no previous history of disease, but upon admission he had almost immediately gone into acute respiratory distress syndrome (ARDS), despite the antibiotic treatment he had been given. For the doctors of Guangzhou Institute of Respiratory Diseases, such a case was unusual only in its resistance to the usual treatments. The patient's condition, again, was curious but did not seem particularly noteworthy. Even when the patient was put on a ventilator and lost consciousness, the doctors at Guangzhou Institute were not alarmed. But on December 27, a chest X-ray showed an almost complete whiteout of the lungs, meaning that huge amounts of mucus had effectively blocked off large patches of the lungs, preventing them from functioning.

"How many of these cases are there?" both Xiao and Huang asked after listening to the briefing.

"We don't know," He Zhaofu told them.

He detailed their assignment: they were to head to Heyuan immediately and find out what was going on. For every man in the room, this was the first time he had ever been sent on such a mission. Deng Zide had never before heard of an epidemiological investigation of this sort, not for a respiratory ailment. It had to be a truly terrible disease if the Department of Health was rounding up a high-level team to investigate it. "And you"—He Zhaofu indicated Xiao Zhenglun with a quick glance—"are the team leader."

The thickset doctor with a shaggy comb-over stood up and gathered his suitcase. He followed the other men out the door, into the elevator, and down to the parking lot, where a white Jinbei van with a red medical cross over the front grille stood idling. They piled in, arranged themselves so that there were two men on each bench seat, and talked about what there was to eat in Heyuan.

CHAPTER 10

■ **January 4, 2003**
■ **Heyuan Number One Hospital, Heyuan, China**
■ **64 Infected, 9 Dead**

HEYUAN IS A QUIET CITY ON THE HIGHWAY BETWEEN GUANGZHOU and Jiangxi province. Tourists occasionally stay a night or two to visit a nearby picturesque reservoir renowned for its fine hiking paths. There are cable cars that ascend a few hundred meters to the top of an eponymously named mountain, from which one can hike down a lazy switchback trail that offers splendid views over the water and the steep limestone karst mountains beyond.

The city itself is a sleepy provincial center, only recently gently transformed by the massive economic changes under way throughout the Pearl River Delta. A few steel factories and glassworks have opened up along the highway, and there are many of the vast, pyramid-roofed service stations and convenience stores one sees along busy Chinese highways. Still, it is a city that reminds you, especially as you are approaching it, of how lovely China still is. As the sun sets above verdant rice paddies and old men sitting on overturned crates before their white-tiled and brown stucco houses, one recalls that much of China still lives according to an agrarian rather than industrial rhythm. Inside Heyuan itself, you still see the occasional farmer in an ill-fitting suit with a few hayseeds clinging to his unruly moptop hair gazing into store windows. And while you are driving around the city, you will unexpectedly find that the paved roads give out and you are bumping along a muddy ox path in order to get to, say, the local CDC offices.

The parking lot of Heyuan Number One is loose gravel over which the branches of tall willow trees hang down and scrape the roofs of ambulances as they roll in and out of the compound. That season, there was a red banner strung up over the entry gate that read, LET'S CLEAN UP AND BEAUTIFY OUR CITY, and many of the doctors who lived in the dormitory just inside the hospital walls had taken this message to heart and scrubbed down their tiny apartments.

A visit from a provincial medical team is a rare event for a local hospital. Most doctors and public health officials are like politicians and party members: their goal is to leave the county bureaucracy and reach the provincial government and never look back. In China, those with aptitude, ambition, or good connections are propelled toward the center, first to the county seat, then the provincial capital, and finally, for the very best, to Beijing. Those left behind rarely catch a glimpse of those who have moved on, and a return visit usually means a redressing for the locals, who figure they must have screwed up if such a high-level team is visiting.

However, in this case, the doctors at Heyuan Number One were grateful for the attention, if a little embarrassed. Around them, the city had already descended into a state of panic and hyperbolic rumor. (The simple white van in which the Guangzhou team arrived, for example, would within hours become transformed by word of mouth into three black cars full of high-ranking officials.) "People act very strangely when they think there is a deadly disease about," says Deng Zide, "especially in China. They have learned the truth is usually worse than the government will admit." And since the Heyuan officials had not admitted anything, a restless population was free to imagine a calamity. The city was experiencing its second wave of panic at the supposed epidemic emanating from the hospital. There was the compulsive vinegar buying. Parents were keeping their children out of school. Pharmacies were running out of medications as crowds gathered outside, seeking anything that could treat the flu. "All of a sudden, people were flooding in, and I didn't know why," says a clerk at the Chun Tang pharmacy. "I sold as much antibiotics in a day as I sold the whole previous year." A young female reporter from Guangzhou's *News Express*, Xiao Ping, heard from a pharmaceutical company that they were selling out of medicine in Heyuan and were sending a truck down there with fresh inventory. She hitched a ride from Guangzhou to Heyuan on the pharmaceutical company's truck and stopped by the hospital and the local CDC to ask some sensitive questions, further panicking the locals. "The city was already in chaos," says Xiao Zhenglun, the team's head. "The city leader was of the opinion that we should not report the case because it could spread more fear." Huang Wenjie of the Guangzhou General Military Hospital concurs. "People were spooked. I felt strange as soon as I got there. I had never heard of any hospital in Guangdong in which so many doctors and nurses had become infected by one disease. People feel very nervous when doctors and nurses can't protect themselves, because they wonder, who will protect the people?"

THE FIRST ORDER OF BUSINESS FOR THE TEAM FROM GUANGZHOU WAS obvious: dinner. After a short discussion, they prudently chose to avoid the shared dish of a hot pot dinner. Instead, they decided to eat a Cantonese meal and to have the server dole out the dishes in the banquet room. During the meal, they were joined by a few local doctors and given the chronology of events. The hospital, they were told, had taken precautions almost as soon as the first patient had arrived. It was impossible, however, to treat a patient with acute respiratory distress syndrome without what doctors called "zero-distance contact." Still, the Heyuan doctors were not convinced that the agent was an infectious disease. Why is it, one of them asked, that not all the health care workers who treated that first patient had fallen ill? And, although the patient's wife was sick, other members of his immediate family were asymptomatic. If this was plague or anthrax, then everyone who lived with the patient should also be sick.

There were several questions that had to be answered. Did every patient, including the two in Guangzhou, have the same disease? What were the symptoms? What was the agent? Why did physicians and nurses seem so susceptible to whatever this was? And why did they have only pneumonia symptoms and yet not respond to any standard pneumonia treatments? And, even more worrisome as several of them lingered on the edge of death, what was the mortality rate?

The possibilities raised by the doctors included legionella, influenza, chlamydia, plague, and anthrax. But if it were pneumonic plague, chlamydia, or legionella, then early antibiotics would have quickly shown results. This was ruled out. Anthrax? Also unlikely since there was no evidence of severe upper respiratory infection. What about a fungal infection? That was an interesting possibility and could allow for an environmental factor as a cause.

At the hospital, the Guangzhou team divided into two groups, one to interview patients and the other to see if there was some epidemiological link that could explain the origins of the sickness. Xiao Zhenglun and the CDC epidemiologists queried patients and staff, seeking to establish a common food or water source that could explain the infection. If the infection were fungal, or were perhaps legionella, then it could be spread through air-conditioning vents or even heating ducts. Or perhaps, Xiao hypothesized, the patients had dipped their chopsticks into a communal hot pot that had somehow been contaminated by a mysterious microbe. Each patient, however, could not recall any such banquet or shared dinner in the previous few weeks. The water supply was another possible culprit, but Xiao soon found that the hospital drew its supply from a city main that provided water to several thousand residents, and there had been no reported incidents of this disease among the surrounding households. There was no air-conditioning system, and the central heating system had not been in use yet that winter. The hospital itself was sanitary, with washrooms well attended and the nurses' station apparently well scrubbed. There were five floors and one elevator, which was used mainly to move patients around. Each ward had twenty-four rooms, and each room was shared by two patients. There were exposed bricks halfway up the white walls of each ward. The floors were of pressed stone and were mopped every evening with a mixture of one part bleach to fifty parts water. The hospital had the usual collection of rusty equipment, an X-ray machine that had probably not been working since the seventies, an early-model defibrillator, and several respirators that had been put to much use of late. The first patient had arrived in a taxi and been checked in to bed 41 on the third floor, where he shared a room. That roommate would soon become infected and sent to Guangzhou Institute of Respiratory Diseases. But what about the other patient, the one who had been sent to Guangzhou General Military Hospital?

On the intake form for that patient, Chou Pei, the admitting nurse had scribbled a few details. Chou had started feeling ill in Shenzhen, where he worked. There was nothing remarkable about that, nor about where he worked—in a Wild Flavor restaurant near one of Shenzhen's migrant-worker communities, the Click. The nurse asked him when he had started to fall ill, and he reported about five days before. He had visited Fulian Hospital in Shenzhen before boarding a bus to return to Heyuan. It was a natural impulse for a sick migrant worker to seek to return to his ancestral home; he probably believed his family would nurse him back to health. Back in Heyuan, Chou's family sent him to another hospital on the outskirts of the city, Zijin County Hospital, where doctors felt this was a standard pneumonia case and prescribed the usual broad-spectrum antibiotics. No nurse or physician bothered to ask Chou if anyone he knew had a similar ailment, and it is entirely possible that Chou, a chef at the Heartiness and Happiness Wild Flavor restaurant, may already have forgotten about Fang Lin, the chop boy from Jiangxi, who had himself fallen ill with a remarkably similar illness.

For the epidemiologists, this was disturbing information. They sat down with the Heyuan Number One Hospital director. If Chou Pei had already been to three hospitals, four including his subsequent stay at Guangzhou General Military, then who knew how many people he might have infected? The director replied that he had phoned Zijin, and that hospital had no similar cases. The Heyuan CDC had also followed up and found that the physician who had examined Chou Pei had not come down with any fever.

The respiratory specialists and infectious disease doctors from Guangzhou, meanwhile, sat down with the doctors and nurses who had fallen ill. With them, they talked through their symptoms and their speculations as to what had caused their illness. They took serum samples from the convalescing patients and gathered the remaining blood samples from the original patient. What was alarming was the range of contact these staffers had had with the infected patients. One infected nurse had worked on a different floor from the first patient and had been alone in the room with him on only one occasion. Had that first patient sneezed, coughed, or somehow expelled mucus or droplets? The nurse didn't remember. The doctor who had accompanied the first case to Guangzhou General Military Hospital was so severely ill he was unconscious and breathing through a respirator. And the first doctor who had treated Chou Pei had also become stricken by the same disease. He had somehow recovered and vividly described for the visiting physicians the rapid onset, spiking fevers, and terrifying breathlessness of the disease. "It is as if you are drowning to death on dry land," this doctor explained.

Another physician, Xie Jinkei, the second doctor to see Chou Pei, said that as soon as he saw the patient, he felt this was an unusual case. "This seemed different from any pneumonia I had ever treated," he said. "Why was this previously healthy thirty-six-year-old so sick, and not recovering? Also, there was just this strange feeling you got from treating the patient. The staff named the disease 'pneumonia without a cause.'" He told the visiting team that he felt strongly that this was a virus. Not only that, he believed it was a previously unknown agent, at least at this hospital. Xie based his statements on three observations: strong antibiotics were having no effect; the second patient had deteriorated so rapidly that this was inconsistent with known pneumonias; and, finally—and most frighteningly—the disease involved very severe blockage of the lungs.

THE TEAM WENT BACK TO THEIR HOTEL WITH FULL NOTEBOOKS AND a sense that whatever had happened here was a medical curiosity unlike any they had ever encountered. The good news, they all agreed, was after that first cluster of cases that had inspired the hospital director to transmit that panicked fax to the Guangzhou Department of Health, there hadn't been any additional cases. And so far, despite the nervous populace, there was no evidence that the disease was spreading in the community outside the hospital.

The team's assignment required writing up a report of this incident, which meant they would have to work through the night to complete the document before returning to Guangzhou. For Xiao, the sixty-year-old deputy hospital chief who had been appointed team leader, this also meant forging a consensus among opinionated medical professionals. Yet his greatest attribute was his avuncular manner. He had a large, fleshy face from which his dark skin sagged asymmetrically, giving him the appearance of a tired old bear. But there was an implied threat to his easygoing manner. If one crossed Xiao at the wrong time, he could suddenly lash out like an old bear, to win an argument or to silence a garrulous dissenter. The Zhongshan Medical University graduate had been an exchange scholar in pulmonary and critical care at the University of Washington in the mid-nineties, and while in the States, he had been impressed by the frequent and sometimes quarrelsome exchanges between junior and senior medical staff. He knew that such discussions could lead to more accurate diagnoses. But he was also aware that he had just a few hours to come up with a document that would be read by some of the most senior medical personnel in his province, and perhaps even Ministry of Health officials in Beijing.

Xiao suggested that the group settle down in a conference room on the second floor of the hotel, order some tea, and begin working through some conclusions. He appointed Deng Zide of Guangzhou Number Three Hospital to do the actual preparation of the official document, which they would all later sign.

But from the start, there was some disagreement. Huang Wenjie of Guangzhou General Military Hospital believed that the agent was *Chlamydia*, a bacterial infection that often induces respiratory disease. Xiao Zhenglun still wouldn't rule out Legionnaire's disease, and since he was the team leader, his opinion was hard to ignore. The CDC team said they ought to just skirt the issue of the agent and focus on the crucial issue of why there appeared to be no disease outbreak outside the hospital. Deng Zide, who was sitting opposite Xiao, believed they were dealing with an influenza outbreak. Huang scoffed at Deng's notion of influenza, waving dismissively and pointing out that influenza would have caused upper respiratory tract symptoms. Deng, an infectious-disease specialist, wondered out loud if Huang understood that there were a much wider range of influenzas than perhaps Huang had treated in his military hospital. Tempers rose, and Xiao saw that they were not getting close to consensus. After all, they were here on a bureaucratic as well as a medical mission, and the provincial government would be assessing the team on its conclusion. In one evening, Xiao explained, they could not be expected to take samples and conduct virological or bacterial tests. But they could, at least, give some sort of answer to what had transpired in Guangzhou and Heyuan.

The group agreed that the illness was a kind of pneumonia but could not decide on the agent that had caused it, except that it was non-bacterial. Finally, Deng Zide suggested calling it atypical pneumonia.

The team then launched into a discussion of symptoms, with some debate over whether or not to include diarrhea, which only some of the patients had exhibited. In the end, they settled on four symptoms that each of the Heyuan patients had manifested: fever, respiratory problems, cloudy chest X-rays, and decreased white blood cell counts. They then drafted a short list of infection-control suggestions and recommended better ventilation for those patients convalescing at Heyuan Hospital.

The document they prepared was titled "The Preliminary Report on the Investigation and Consultation at Heyuan People's Hospital." It begins, "A team of six people arrived at Heyuan People's Hospital on the afternoon of January 2. The group interviewed staff from Heyuan Health Bureau, Heyuan People's Hospital and Heyuan CDC, reviewed the personal data of each patient and conducted a thorough physical examination and interview of each patient. The group also inspected the physical environment of the infectious diseases ward of the hospital. The preliminary conclusion is an outbreak of Atypical Pneumonia, probably caused by an atypical pathogen."

The report goes on to give the four agreed-upon symptoms as diagnostic criteria, and notes that the provincial CDC had gathered serum for antibody testing for *Chlamydia* and influenza. This document, short enough to fit on one sheet of A4 notebook paper, is the first official medical report on the disease that would soon become known as SARS.

Where this document went and who saw it would later become a source of dispute among Chinese officials. What is certain, however, is that the team that investigated this outbreak did not feel that what had happened in Heyuan was of global medical significance or that that hospital had been the first way station for the first new disease of the twenty-first century. A few doctors had fallen sick after treating a patient, and a hospital director had panicked. This had been a small-scale hospital outbreak, nothing more—a medical curiosity, not a crisis.

"By the time we left, it had been five days since the last new infection," says Deng Zide. "It seemed that the worst was over." The team concluded that there was no reason to panic. They were eager to get back on the road to Guangzhou.

IN THE VAN ON THE WAY BACK, HOWEVER, DR. HUANG WENJIE BEGAN coughing and running a high fever. The rest of the team teased him, saying he had come down with this atypical pneumonia. Huang dismissed their jibing but vowed to do a blood workup upon his return to Guangzhou, to measure his white blood cell count.

As Deng Zide thought over the events of the past twenty-four hours, he found himself still bothered by something about the outbreak in Heyuan. As an infectious-disease specialist, he saw thousands of cases a year, but he recalled encountering only once before a situation similar to what had happened in Heyuan. As the van wound along the Heyuan River, past peasants walking back from their fields for lunch, he told his fellow doctors about an expedition of the month before to investigate an outbreak, this one a solo mission.

He had been summoned by a doctor whose sister was a nurse at the Number One Hospital in Foshan, a booming industrial town an hour from Guangzhou. He arrived there at ten-thirty on the morning of the twelfth and immediately visited the ICU on the fourth floor, where, through a window, he could see two patients on respirators. One was Mr. Zheng, he was told, and the lady was his aunt. Both patients had this strange lung ailment. Mr. Zheng had been admitted in November with a slight fever as the only real symptom. He had gone to a pharmacy for some tablets and then to a local clinic to see a village doctor. Still, his fever had persisted. Since he was a relatively wealthy man, his wife had insisted that he go to the hospital to get treated. Doctors at Shiwan Hospital, happy for another paying customer, admitted him and performed a chest X-ray, which did not reveal anything. The doctors elicited a tentative diagnosis of rickettsial pneumonia. Then, after two days, Mr. Zheng was joined at the hospital by his middle-aged aunt, who was running a high fever and in whose lungs doctors could hear scratchiness. After taking a chest X-ray that revealed some shadows, doctors diagnosed her as a tuberculosis patient. They transferred both patients to Number One Hospital. Within just days of arriving there, however, Mr. Zheng went into pulmonary and respiratory failure. He was transferred to the ICU on the twenty-eighth and was then intubated—that is, a tube was inserted down his breathing passage that would pass oxygen from a ventilator and directly into the lungs. The day before, his auntie came down with a similarly high fever. Within five days, Zheng's uncle was also admitted with a high fever, and then Zheng's wife, and finally his daughter, all apparently suffering, doctors now believed, from the same illness.

The consensus among local physicians was that this was bacterial pneumonia. But why, then, were various broad-spectrum antibiotics employed to no effect? Deng Zide, as a well-known infectious-disease specialist, was called in because of his colleague's family contact at Number One Hospital. "When I saw the patients, they were all sedated and intubated," Deng told his fellow team members. "My first thought was that this was very strange. This disease was obviously highly contagious—there was a whole family in the ICU—and the patients were not responding to antibiotics. Also, those chest X-rays were so cloudy that this did not appear to be a bacterial lung infection."

At the time, Deng had no way of determining if this was a virus. Most hospitals do not have virology departments capable of isolating respiratory viruses. "But I had the impression that this was a viral infection." He advised the local staff to increase the dose of corticosteroids and also administer antiviral medications. As a clinician, Deng made his main objective to cure the patient; he would worry about identifying the etiological agent only if standard treatments weren't effective.

Deng Zide had returned to Guangzhou that afternoon, and a few weeks later, he heard that Mr. Zheng had spent a week on a respirator before miraculously recovering. The rest of his family were also convalescing. There were so many respiratory patients throughout southern China that winter that Deng hadn't given any more thought to that peculiar afternoon in Foshan until today, driving back from another, even more curious mission.

"Do you think it was the same disease?" asked Xiao Zhenglun.

"I don't know," Deng said, "but there were so many similarities."

The implications of another cluster of cases were terrifying. Who knew how many hospitals might be dealing with similar outbreaks? And those were the hospitals where the insured and better-heeled patients ended up. Among the general population—the migrant workers, the factory hands, the rural peasants—there could be many more cases that would never reach the public health system at all. If this was a new disease—highly unlikely, but possible—then there would be no metric in place to track it. You would have doctors, like the team in Foshan, improvising treatments as they went along and never informing anyone at a higher bureaucratic level of what was going on.

China had official protocols for the reporting of infectious disease. Certain ailments, such as plague, dengue, cholera, and polio, were mandated as state secrets, and local medical staff were under orders to report them only up the chain of command. But there was no such system in place for diseases that did not fit into existing categories. So a new disease could wander through hospital after hospital, and the only way the provincial government might even become aware of it would be if a fax was sent by a nervous hospital director, as had happened in Heyuan, or if a local nurse with *guanxi,* "connections," called in a big-city hotshot to help out a local hospital, as had happened in Foshan. Most likely, such cases would be quietly dealt with at the local level, even if that meant a few patients dying. Better to handle it yourself, a local hospital director might reckon, than lose face with your superior. Even more worrisome, one of the CDC staff reminded everyone, was the fact that at least two of the patients under question had been exported to Guangzhou. Who knew how many others might have been transferred by smaller clinics seeking to offload their medical problems onto the big city hospitals? "There could be other cases already in the system," said Huang Wenjie. The team rode in silence for a while, through the long shadows cast by the winter sun, the air becoming progressively fouler the closer they drove to polluted Guangzhou and the Pearl River Delta.

THE NEXT DAY, THE *HEYUAN DAILY* **PRINTED THIS STORY BY A** reporter named Li Jianhua on its front page:

EPIDEMIC IS ONLY A RUMOR

The Director of Heyuan People's Hospital announced there is no epidemic. Since yesterday, due to this terrifying rumor of a serious infectious disease, there has been a rush to pharmacies to purchase certain antiviral tonics. This irrational purchasing has driven prices of these drugs to ridiculous levels; a tonic that usually costs ¥10 now costs ¥450. Antibiotics have also become more expensive, the price rising to ¥30. Yet no matter the price, as of 9 P.M., these medicines were sold out at most pharmacies. Until yesterday morning there were long lines waiting to buy these drugs with customers purchasing up to 10 boxes each.

This rumor also has parents keeping their children home from kindergarten. Authorities from the Old City Kindergarten told this reporter that not only were many youngsters not in school, but some parents had during the school day taken home those children who were there. In the Central Kindergarten, two classes contained a total of 20 kids, less than half the usual attendance. Kindergarten officials said they had also heard the rumors of a disease but didn't believe it. But just in case, they added, they had prepared a cold elixir tea to ward off any sickness. Primary schools were not affected. This reporter went to four primary schools and was told that only 10 students stayed home.

People's Hospital of Heyuan received two patients from Zijin Hospital on the fifteenth of last month. The patients were transferred to Shenzhen and Guangzhou. Specialists from hospitals in Guangzhou were sent to Heyuan to help in the treatment. The Hospital Director said that after the meeting of provincial experts it was proven that the disease is a very common disease: atypical pneumonia. This disease is not infectious and is caused by changing weather. The symptoms are high fever, coughing and spots in the lungs. This disease is not similar to any communicable disease identified by the government so there has been no reason to report it to provincial authorities.

A doctor on duty at another hospital, the People's Hospital of Yuanchen, said staff there didn't receive any notice from higher officials that there was a communicable disease outbreak. He dismissed such reports as rumors.

This reporter spoke with the family of one of the patients and was told the symptoms were high fever and that the patient was conscious and in stable condition. It is worth remembering that it is now influenza season. There have been several medical staff from Heyuan People's Hospital who got infected by this disease but the situation is not serious. The Hospital Director has warned that there is no reason to panic.

One senses in the story the skepticism of the concerned reporter bumping up against an editor's desire to make this an innocuous service story. The early paragraphs on drug prices and school attendance are meant, somehow, to make the story the Chinese local daily equivalent of news you can use—much breaking Chinese news is dispensed in this indirect manner. And as with many Chinese newspaper stories, one has to read through to the second half of the story to see what is actually going on and that the "rumor" described in the headline is not a rumor at all but in fact a disease outbreak, of "atypical pneumonia" no less, at a local hospital. Even more remarkable, the article speaks of a team of provincial officials being sent to "help" in treating this "rumor." The reporter's own reassurance that the disease is not "similar to any infectious disease identified by the government" rings hollow, as does the kicker's echoing of the lead, a stern reminder that there is no reason to panic. Astute readers, however, would immediately glean that there probably was a real disease to be concerned about. Otherwise, why would there be actual symptoms given and, amazing for a Chinese news story, real information about an official medical visit that was not of the ribbon-cutting or medal-giving variety. The story was both an attempt to quell rumors and a game effort by the reporter to slip the inconsistency of the official version of events past party officials. Most Chinese reading this article might have accepted that "changing weather" was a factor in causing the disease, but they would certainly have stopped short at mention of spots on the lungs, which in any society or culture sounds sinister.

Other reporters, such as Xiao Ping of the *News Express*, would return to Guangzhou and write stories that downplayed the significance of the disease outbreak, quoting experts who said that atypical pneumonia was a very mild ailment and that rumors of an infectious-disease outbreak were false. At any rate, by the end of the first week of January, provincial party leaders, seeing no reason to incite panic with the New Year approaching, and being as unsure about the nature of the disease as the reporters writing these vaguely worded accounts, barred the media from publishing any more about what had happened in Heyuan and Foshan. "I was told by my editor," says Xiao Ping, "that we couldn't publish anything on this atypical pneumonia for a while. We decided to gather what information we could and wait for an official announcement [about the disease] so we could run it." In the meantime, her paper ran a story on January 10 reminding readers that during "this time of changing seasons, one should take care not to catch a cold." That was virtually all the press this mysterious new disease would get for the rest of January.

CHAPTER 11

- **January 15, 2003**
- **Guangzhou, China**
- **131 Infected, 14 Dead**

THE FIRST TIME I VISITED GUANGZHOU, OVER A DECADE AGO, IT seemed that the entire American industrial revolution—that whole period of America's retooling, urbanizing, and mass-producing that commenced after Appomattox and continued through World War I—was happening in just a few moments right before my eyes. The bustle and energy of the Pearl River Delta was already astonishing and, anyone could see, was merely a starting point for China's own industrial revolution. There was every conceivable new business launching, and already there were equity markets in Shenzhen and Shanghai, along with the established bourse in Hong Kong, that were helping mainland firms raise ever more capital to invest in ever more factories, machines, and equipment. I remember a businessman I met at the time—a fellow who was then scheming to open a stock exchange devoted exclusively to private-sector companies in Guangzhou—telling me that China would change from within. "When businessmen have all the power," he said, "then we can fix the system, even the party."

What he could never have counted on was how the party would so efficiently coopt businessmen like him. Former Communist Party general secretary Jiang Zemin's Three Represents, a vague blandishment of a political ideology that completes the transformation of Chinese Communism into klepto-capitalism, has opened the party to businessmen and capitalists, provided they are successful enough. One has to give credit to the plutocrats in Beijing—men and women who lived through the Great Leap Forward, the Cultural Revolution, and the Tiananmen Square massacre—for being so pragmatic as to see that the party could no longer bludgeon or banish all opposition. A far more effective method of ruling was to make Chinese Communism so encompassing that even the most venal owner of the means and modes of production would see no reason to take issue with the regime. In exchange, the government has implicitly promised that China will be an orderly place to do business. As long as the party remains in power, a captain of industry is free to pursue pretty much whatever bountiful ventures he wishes. The party, provided that a capitalist pays off the right local officials and pays his taxes, will even turn a blind eye as that same capitalist despoils the environment, gouges the competition, and, most apostate in a nominally Communist state, exploits the proletariat.

The result has been that the country looks transformed on the outside—as long as a decade ago, the neon wattage of new karaoke bars in cities like Shanghai and Shenzhen already put one more in mind of Times Square than Tiananmen. But where it counts, deep in the state's synaptic network, where data is traded and exchanged, the party still rules, not with an iron fist but with a discreet proctor. While the party's ideology has withered, its authority has not. And the way it maintains control is still through the selective dissemination of information. The press remains muzzled; the Internet is strictly partitioned between sanctioned sites and those that are off-limits to mainland computers. (Try accessing, for example, TIME.com the next time you're in China.) CCTV, the government-run television conglomerate, which enjoys a massive viewership—perhaps the largest TV audience in the world—is strictly bowdlerized. Even SMS messages are now monitored as they pass through the phone networks, lest anyone disseminate antigovernment messages from phone to phone.

The result is a propaganda system whose first impulse is to avoid mention of any uncomfortable issues or topics and to squelch any real news—such as a disease outbreak—that might not fit in with the government's promises of order, peace, and prosperity. Bad news, be it a coal-mining disaster, a bad harvest, or a toxic-waste scandal, is very rarely reported as it's happening. The state news agencies and propaganda ministries may take several weeks to figure out how to play it so as not to upset the perceived order.

For most of my terms as deputy editor and editor, *Time Asia* was also banned from newsstand distribution in China. The inciting article for this had been an early 2001 story about Falun Gong, the outlawed religious movement that was then withering in the face of a brutal crackdown. Our piece had included a photo of Hong Kong protesters carrying a sign that read, DEATH TO JIANG ZEMIN. The general secretary himself had supposedly taken offense and ordered our proscription. A small number of copies of the magazine were still permitted into China for our controlled, vetted subscriber list—foreigners living in China or politically connected Chinese who the government felt could withstand what mainland officials believed was *Time's* biased coverage. It is a small tribute to the integrity of "one country, two systems" that we were able to publish in Hong Kong even after we were effectively blackballed on the mainland.

My introduction to negotiating the byzantine corridors of Communist Party power came during periodic trips to Beijing, when I would meet with officials at the Ministry of Propaganda or the Ministry of Post, Telecommunications, and Press to find out how we could again appear on Chinese newsstands. The *New York Times*, *Newsweek*, the *Wall Street Journal*, and CNN, I would point out, were now free to distribute in China. Why should *Time* be singled out for censure? I would spend most of these trips trying to ascertain whether or not the official I was meeting was even in a position to influence our situation. Usually, I would leave with an order to deliver two or three years' worth of back issues. The deputy assistant vice-minister whom I had spoken with would promise to review our China coverage and give us an opinion on how to get back into circulation. But since our stay in the penalty box had been ordained, reportedly, by the party secretary himself, it seemed unlikely that any lower-ranking official would see any political mileage in recommending that *Time* be permitted back onto the mainland.

■ ■ ■

FOR THESE OFFICIALS WHO SIT IN THE CONTROL ROOMS OF THIS information-making machinery, the data they receive is usually accurate and reliable; the officials just very seldom bother to tell the people—anything. The Communist Party runs two communication systems, each with a very different mission. One collects data and sends it up the bureaucratic hierarchy. This information is supposed to be accurate, objective, and reliable. Its quality and quantity increase according to how high a position an official occupies. But no matter who is receiving this information, it can be passed on only one way: upward. Otherwise, it must be kept secret. For example, during the nationwide student demonstrations in 1989, state news agency reporters throughout the country were writing thorough accounts of local student activities that were sent to the top leaders in Beijing. As was later revealed in *The Tiananmen Papers*, based on a cache of government documents smuggled to the West and published in 2001, this information was remarkably accurate. The sense of crisis that gripped the Chinese leadership compound of Zhongnanhai during the student protests at Tiananmen was based to a great extent on the scope of the unrest nationally. For many international observers, on the other hand, how widespread the crisis was would become apparent only as dissidents fled to the West in the subsequent months and years.

In the other, parallel system, information flows from the top down. This, according to Perry Link, professor of East Asian Studies at Princeton and among the leading scholars on contemporary China, "is the open, public information that party leaders have decided that people below them may—and in some cases should—know about. It might or might not be solid, but it should never harm the interests of the leaders." The task of a local official is, therefore, to pass information upward and to pass down only what has been approved at the next-higher level. Seeking to dispense sensitive information is not only uncommon, it's simply not an option. For example, it took several minutes for me to explain what I meant to a public health official in Heyuan when asking how he would secure the release of sensitive data that had not been cleared. "There is no reason for me to ask someone at [the provincial level] if I can release sensitive information," he finally told me. From his difficulty in comprehending my question, I deduced that he had simply never thought about the matter.

Such great care regarding the husbanding of information has reinforced a previous generation's inveterate secrecy. In China, very often, the default official position on almost every issue is to maintain secrecy. The reporting of seemingly innocuous data, such as steel production figures or the inflation rate, has ended officials' careers, landed visiting Taiwanese and Hong Kong scholars in jail, and shut down newspapers. An unexplained new disease was precisely the type of information—potentially destabilizing and possibly indicative of official incompetence—that the system would automatically squelch.

There was, however, an even more compelling reason that officials were unwilling to speak out about this curious new disease and editors were not eager to publish stories: they didn't want to be executed. In China, information about infectious diseases is still deemed to be a state secret. Divulging data regarding infectious-disease outbreaks, even statistics on the number of patients, could make one a defendant in a treason case. Though seldom enforced, this law is a vestige of Mao's paranoid fantasies of what American and Russian propaganda could make of a disease outbreak in China. (Ironically, Mao actually made great progress in the fight against infectious disease, even while he almost starved his population through preposterous economic programs such as ill-conceived mass industrialization. His work in curtailing diseases like smallpox, polio, and schistosomiasis falls into the "70 percent good" scorecard that Chinese historians finally gave to Mao.) Thus, for any journalist or public health official, there was simply no reason to go public regarding this curious incident in Heyuan or the rumors of a new disease that were sweeping through parts of Guangdong. Those few stories that were published were dressed up as reporting on a rumor, not a disease.

THE DOCTORS AND EPIDEMIOLOGISTS WHO WERE AWARE OF THE Heyuan outbreak and of the earlier incident in Foshan were living in a privileged bubble of information. This small group of doctors, nurses, and patients knew, at this point, more than anyone else in the world about SARS.

For the team that returned from Heyuan, the first matter was to determine what the agent was. But virological testing of the samples brought back from Heyuan hadn't turned up any likely infectious agent. And the CDC labs in Guangzhou had come up with negatives in their antibody tests of several types of influenza.

Yet as a week went by, and as what happened in Heyuan and Foshan increasingly began to seem like isolated events, the various doctors went back to their routines. "We have thousands of unexplained respiratory cases a year," said Xiao Zhenglun, the team leader. "This is China. You don't worry too much about just a few." And it is human nature, even among highly professional physicians, to ignore a problematic question—what is the agent?—unless forced to answer it.

But on January 20, the Heyuan team was reconvened at the Department of Health. For the past few days, rumors of outbreaks in Shenzhen, Jiangmen, and Shunde had been bombarding SMS screens and pagers throughout Guangdong and had incited another, if milder, round of vinegar buying and pharmacy clearing. The deputy director of health confirmed for the team that there had been another outbreak of this mysterious disease, in Zhongshan, just two and a half hours south of Guangzhou. The director of the Zhongshan CDC was reporting twenty-six cases in three hospitals that matched the symptoms given in the report filed by the team from Heyuan. The disease was highly contagious and had already infected thirteen medical staff.

By now, the team's earlier Heyuan report had been passed up the chain of command to the national CDC and the Ministry of Health in Beijing, where top officials, including the deputy minister of health, Ma Xiaowei, received a brief synopsis of the findings but did not review the actual document. The situation was deemed to be of enough interest that, upon word of the incident in Zhongshan, a team was dispatched from the national CDC to join the local team in the investigation. (No information was passed to any international agencies. Both the World Health Organization and doctors in Hong Kong were still dealing strictly with rumors.)

The outbreak had also piqued the interest of Zhong Nanshan, the chief of the Guangzhou Institute of Respiratory Diseases. Zhong had treated his hospital's first case and had found it both curious and ominous. His interest in the matter automatically lent the investigation greater urgency. There is a certain arrogance endemic to top Chinese doctors. These, after all, are the brightest boys who managed to win places in their provinces' most prestigious schools and then even more coveted slots in medical colleges. In some counties, one boy a year might go to the provincial capital to enroll in medical college. While the education is rigorous, it is in the tradition of Chinese education to cosset those who have made it through the meritocracy to the elite levels of academia. (During the Cultural Revolution, of course, this was turned upside down as those mandarins—future doctors and scientists and engineers—were sent to perform reeducation through labor in communal farms far from their universities.) From day one of their five-year term at a medical university, future doctors are trained in physiology, anatomy, biology, and the basic sciences. Chinese doctors insist that their training is more rigorous for being more focused. And certainly those educated at the top medical schools, after completing their three-year residencies, are every bit as accomplished as their counterparts in the West.

Zhong Nanshan, sixty-eight, a native of Jiangsu province, had ridden that rigorous system all the way to Beijing, where he graduated from Beijing Medical University. He had made his name as a respiratory-diseases specialist and rose to fame for having treated many of China's top leaders, including the pack-a-day smoker Deng Xiaoping. Already among the best-known doctors in Guangdong, within the year, because of his clinical work fighting SARS, he would arrive among the most famous in all of China and be touted as a possible future minister of health.

With his wire-frame glasses, strong jawline, chiseled features, and wavy pompadour, Zhong Nanshan looks more like a Hollywood action hero than a physician. Unlike so many Chinese, he will stare you down while he speaks to you. I remember one afternoon, while I was sitting in his office, he used the acronym "ARDS." At the time, I didn't know what it stood for and asked him to explain.

"Oh." He stopped and glared at me over his glasses. "So I have to be very simple with you."

To get some idea of how exceptional Zhong Nanshan is, consider his office. Most Chinese physicians' offices tend to have barren walls. If the doctor has been lucky enough to have had his picture taken with a high official, there will be a blown-up photo of that occasion. Or, perhaps, there will be an anatomical chart of the human body or respiratory system, with pennant captions indicating the names of organs or lung tissues. In Zhong Nanshan's office at the Guangzhou Institute of Respiratory Diseases, across the street from the Pearl River, is a huge color photo, taken just a few years ago, of a shirtless Zhong himself, his buff and ripped chest glistening in the sunlight as he wades into a river. He is certainly the most intimidating physician in China—one World Health Organization official described him as being "like a god." Among his many attributes, he was once a world-class athlete, and he still holds the Chinese collegiate record for the four-hundred-meter hurdle. He remains in such impressive condition that he looks as if he could equal that time today.

Zhong had taken an interest in the case, and as he studied the fax describing new cases in Zhonghsan, he overheard one of the CDC team members saying that this was probably influenza; he slipped off his glasses and said, "I've seen plenty of influenza cases. If this is the same disease as the patients [have] in my hospital, then this is no influenza."

NONE OF THE HEYUAN TEAM HAD EVER HEARD OF BEIJING'S SENDING a squad out to the field to investigate a disease outbreak. "This meant that the leaders were taking this very seriously," said one public health official in Guangdong. "There was understanding at a high level that something very strange was going on here."

Indeed, the situation in Zhongshan was more ominous than what the first team had seen in Heyuan. The cases here had spread throughout three hospitals, and the disease was now at large in the community; in Heyuan, the epidemic had been confined to one hospital.

Su Guoqiang and his younger brother Su Qingshan ran a Wild Flavor restaurant in Beijiao. The restaurant sold the usual range of exotic fare, including cats and masked palm civets, in a vast low-ceilinged space that shared a parking lot with several other, similar establishments. Both brothers used to ride in the restaurant's truck to wholesale markets, including a wild animal seller in Zhongshan. They would load into the back of the truck about ten thousand *kwai* worth of Wild Flavor; the restaurant sold six palm civets every day. Older brother Guoqiang hadn't been sick for as long as his younger brother could remember. Yet he would come down with this mystery ailment first, his fever spiking to over 102 for three days in a row before he checked in to a local clinic on January 20. Doctors there had no means of treating this previously healthy man who appeared to be suffocating right before their eyes. His lungs, according to his chest X-ray, were a pool of mucus. The doctors hurriedly transferred Guoqiang to Zhongshan Number Two Hospital. A few days later, doctors there, having never heard specifics of this mysterious new disease, were unable to stabilize the new patient. He was intubated upon admission and pre-scribed antifebrile medications, which had no effect. Guoqiang went into severe hypoxia and become so disoriented that he kept repeating that he and his brother needed to clean the restaurant for the Chinese New Year. "I felt this was a new disease," said Dr. Li Jianguo of Zhongshan Number Two Hospital, "but I didn't know what."

"It was terrible watching him in that state," says younger brother Qingshan. "He would open his eyes once in a while. He never looked frightened, just angry."

Younger brother himself soon came down with the same disease and became disoriented and hostile because of oxygen deprivation. His fever stayed at 103 for more than seven straight days, and his blood oxygen saturation level—the amount of oxygen his red blood cells were carrying from his lungs to his vital organs—was already at just 80 percent of capacity. Levels below 90 percent are considered fatal. Unable to stabilize him, doctors admitted him to the ICU alongside his brother. Qingshan would spend twenty-three days in the hospital—nearly a week of that amid the green-tiled walls of the intensive care ward in the respiratory department on Number Two Hospital's ninth floor. He would miraculously recover. When his lungs had stabilized and he could breathe unassisted, he was told that on January 28, 2003, his older brother had become one of this mysterious new disease's earliest fatalities. Older brother Guoqiang left behind one nine-year-old daughter and a teenage son.

By then, doctors at Zhongshan Number Two Hospital and several other hospitals in Guangzhou had admitted dozens of similar cases— yet they still had not received any official explanation as to what was going on, or medical advice as to how to treat this deadly disease.

THE TEAM THAT ZHONG NANSHAN LED DOWN TO ZHONGSHAN WOULD prepare another report, this one also written by Deng Zide. The team had inspected local hospitals, pored over patients' charts, and interviewed doctors, nurses, and patients. This time, Deng and the rest of the team knew, the situation called for far more serious measures and analyses than were performed during the previous trip to Heyuan. Though the manifestations of the disease had not changed, it was now clear that this was a wide-ranging outbreak. There were cases popping up all over Guangdong, and the locals had even come up with a nickname for the disease: *ling ren zhi xi de*, or "breath taker."

From 9:00 P.M. until 3:00 A.M., the team met in a conference room at the 139 Hotel in Zhongshan. Deng Zide had asked the hotel to send up a typist. The woman who arrived, he soon ascertained, was not up to the task, as she took about an hour to type half of one mistake-riddled page. Deng took over himself and laid out the framework for the document, using the previous Heyuan report as an outline. This time, everyone in the room agreed, a more substantial set of recommendations was necessary. What was going on in Zhongshan represented an acceleration of the outbreak. Foshan, Deng explained, had been the first community outbreak. Heyuan was the first hospital outbreak.

"Zhongshan," Deng says, "was both."

The document the team produced gave symptoms, incubation periods, and possible treatments, including a recommendation for the use of steroids to reduce lung inflammation. "This disease," the report read, "is highly infectious and is possibly airborne." It went on to recommend isolation of patients and stringent infection-control measures, including the sterilization with chlorine and ultraviolet light of infected areas. Medical staff who had to be in contact with patients, the report warned, "should wear masks, goggles and gloves and be aware that the disease can be passed on by flying particles. Pay close attention to sterilizing hands after contact." Dated January 21 and signed by the "Guangdong Provincial Health Bureau Zhongshan City Investigation Experts Group," the report was precisely the sort of document that frontline hospitals throughout Guangdong needed as waves of infected patients were now being admitted to hospitals throughout the province. This was a report that could save lives. The great pity—and scandal—is that it was labeled *neibu,* or "top secret." Which meant that the physicians who were at the greatest risk of being exposed to this disease were prevented from seeing a copy of the report. They did not know how to treat their patients effectively; they also could not adequately protect themselves from falling ill.

Instead, the report was distributed only to top officials at the provincial level, party leaders, and to a few well-connected hospital directors and department chiefs. The document was hand-delivered in a plain manila envelope with the TOP SECRET stamp prominently displayed above the addressee's name. Only the addressee could dare open such an envelope; no secretary could risk the dressing-down or possible espionage charge that would accompany his reading the envelope's contents. And even many of those who were allowed to read the document, and who could have distilled and disseminated the information so that it would be useful in the medical community, were on vacation for the Chinese New Year. In some cases, the document sat for days unopened on hospital office desks while medical staff just a few floors below were becoming infected with this deadly new disease.

CHAPTER 12

■ February 1, 2003
■ Zhongshan Number Three Hospital, Zhongshan, China
■ 268 Infected, 28 Dead

JIAN YAN HAD WHAT CHINESE CALLED A LOOK OF GOOD FORTUNE: A fleshy face, wide, round eyes, and black hair with silver streaks. But on February 1, when his wife checked him in to Zhongshan Number Three Hospital, physicians there immediately detected in his labored breathing a distinctly unfortunate sound: a crackle as he inhaled, which indicated severe damage to his right lung. He was already on the verge of delirium, one of the physicians, Cao Hong, the chief respiratory specialist, would note on Jian's intake form.

IN GUANGZHOU, HOSPITALS WERE BY NOW BECOMING FAMILIAR WITH these cases: patients in their thirties or even younger suddenly stricken with a breathlessness so debilitating that they were reduced to gasping hulks within twenty-four hours of first feeling feverish. Local hospitals lacking respirators or infectious-disease specialists were transferring these patients by the dozen up the medical establishment to larger hospitals. It was exactly the scenario that the team returning from Heyuan had feared weeks earlier: a stealth outbreak was under way, with hundreds of patients already clogging a system that was not aware it was crashing.

Had the doctors at Zhongshan Number Three Hospital, for example, known what the investigative team had uncovered—or been privy to the reports that had been prepared on those early cases in Heyuan and Zhongshan—then they might have taken greater precautions. As it was, they quickly deduced that this patient was already in acute respiratory distress. "I gave him twenty-four hours to live," recalls Dr. Cao Hong. The decision was made to intubate the patient.

Orotracheal intubation, as the process is known, required that Dr. Cao insert into Jian's mouth a flashlightlike device called a laryngoscope, which also serves to suppress the tongue. Dr. Cao bent over the patient and strained to see down the oral cavity to the glottis, the flap of skin covering the larynx, as a nurse massaged Jian's fleshy neck to suppress the involuntary vomiting that often accompanies the gag reflex. After Dr. Cao established that he could see the glottis, he inserted a long, flexible, clear plastic endotracheal tube and slid it down the throat until it reached the base of the tongue. There, the glottis, which normally serves as a protective mechanism to prevent exactly this type of intrusion, had to be finessed and then forced aside so that the eye of the tube and the accompanying inflatable cuff could slide through the windpipe, or trachea, and into the lungs just above the carnia, the point at which the main stems branch off into the right and left lungs. Though there is a blunt, almost pleasing logic to the simplicity of this procedure—bring air to the lungs—in practice it is fraught and often frustrating for physicians, who usually have to make several attempts before safely inserting the tube. It is also one of those medical procedures, like breaking an improperly healed bone to reset it, that can't be refined into something more delicate; the process is violent, brutally simple, and practically the definition of an invasive procedure—foreign matter is shoved deeply inside the chest cavity. The rudiments of the treatment were first described in the eleventh century, in *The Canon of Medicine,* by the Arab medical chronicler Avicenna. There was "no harm from inserting a long reed or anything similar around which a piece of cotton-wool is attached. . . . Perhaps also insert a tube made of gold or silver into the pharynx to assist breathing," he suggested. In 1869, a German physician, Franz Kuhn, looking to administer anesthesia more safely and efficiently, would push the process further, so to speak, by inserting the tube all the way into the lungs themselves. Between 1880 and

1887, Cleveland-born physician Joseph O'Dwyer would devise a series of tubes that could be pressed into the throat of a child suffering from diphtheria—the disease causes a membrane to form that can completely occlude the larynx, suffocating the patient. The tubes were of great value to patients suffering from what was then a common childhood killer and to some syphilis patients, who suffered from strictures in the breathing passages. However, with the advent of antibiotics, this symptomatic relief became less frequently necessary. It wouldn't be until after World War II that O'Dwyer's tubes, combined with the external pumps developed by Kuhn, would become standard procedure in emergency rooms and trauma wards around the world for patients in acute respiratory distress. To this day, orotracheal intubation remains perhaps the most dangerous procedure that could be described as common, and it is considered among the most complicated that any emergency medical technician will ever have to perform. No one keeps accurate numbers on intubation fatalities, but in the United States, about two hundred people a year die from this basic procedure going wrong. In China, according to one physician, the number is certainly in the thousands.

Numerous complications can ensue during an intubation. Even anesthetized or delirious patients possess enough involuntary muscle response to gag, cough, spit, and convulse at this intrusion. Often, by the time a patient has reached the point where he requires intubation, his throat may already be bruised, scarred, or swollen from coughing, infection, or trauma, making locating the glottis and identifying the vocal cords, as opposed to the esophagus, where the food goes down, especially difficult. If a doctor repeatedly tries and fails to pry open the epiglottis, then this might add to the swelling and complicate subsequent attempts. Patients have died or suffered from brain damage because doctors could not find their breathing tubes in time or mistakenly intubated their digestive systems instead of their lungs. Inevitably, successful or not, the process causes massive gagging and coughing, producing phlegm and mucus in abundance.

Because Jian had been previously intubated while at a local hospital, his throat and glottis were already scarred and slightly swollen, making identifying the proper opening that much more difficult. On his first try, Cao couldn't get the tube under the glottis and had to start the process over again, first suctioning out a considerable amount of blood and mucus and then ventilating the patient with a heavy dose of high-concentration oxygen to keep up blood oxygen levels. He told the nurse to remove the patient's oxygen mask after a dozen pumps of the ventilator bag, and then he suctioned out the throat again and slid the flashlight back into the patient's mouth. This time, he managed to hitch the breathing tube under the glottis, mainly by prodding until he felt some give, then suctioning blood out again and peering down the oral cavity for visual confirmation.

Jian Yan started bucking and retching, his barrel chest rising up from the bed so that when he landed on the mattress, it shook the IV tubes and monitors around the operating theater. Was he convulsing? Dr. Cao opened both the patient's eyes to see if they were rolling back. No. Jian was simply reacting to the foreign matter in his throat. Still, Dr. Cao could not see if the tube was in or not. Two nurses held Jian down as the doctor pulled hard on Jian's jaw and peered down his throat.

At that instant, Jian coughed up a massive amount of bloody mucus that, according to Cao, "was like a fountain of virus." Both nurses felt mucus land on their cheeks. Dr. Cao's shirt was covered with the goo. He wiped the mucus from his forehead and eyes with a gauze pad that was at hand and bent over the patient's mouth again.

"It's in!" he shouted. He inflated the cuff with five milliliters of air and then attached the respirator hose and nipple to the adapter end of the tube as air with a high concentration of oxygen was sent coursing into Jian's lungs. Upon completing the procedure, the whole team scrubbed down and quickly moved on to the next patient, another sufferer from this new ailment.

Dr. Cao eventually stabilized Jian Yan, pumping in antifebrile medication and starting a dose of corticosteroids to keep the autoimmune system from overresponding and exacerbating the symptoms. Besides the explosion of mucus he had fired out during the intubation, Jian was otherwise, as far as Dr. Cao could tell, simply another victim of this dreadful new disease that was causing panic throughout the province. Breath Taker, or Breath Stalker, as some were calling it, had reached the status of provincial legend. Several specialists had ventured down from Guangzhou to help with treatment or to observe these new cases. They ranged from the usual respiratory specialists and infectious-disease experts to acupuncturists, herbalists, and even a Qi Gong practitioner— all of them drawn by the medical curiosity and eager to offer their own treatments. Among those who visited was a nephrologist—kidney specialist—named Liu Jianlun, a taciturn man with wire-frame glasses, combed-over gray hair, and a cherubic smile. He conferred with Dr. Cao on how to increase the lymphocyte count of the patients, and read the charts of several patients, including Jian Yan.

Even more worrying to Liu and other medical professionals was the increasing occurrence of this disease among medical staff. Three nurses in the hospital had already fallen sick, and at least one physician had taken sick leave with something that seemed very similar to what was afflicting these patients. Already, Dr. Cao had ordered all members of his department to sleep in their offices or in local hotels, and to avoid contact with their families.

Newspaper reporters from *City Weekend, Southern Express Daily,* and other Guangzhou newspapers had called repeatedly asking for information about this new disease and what could be done to prevent it. "At that point, we weren't told not to talk about it," says a doctor at Guangzhou Number Three Hospital. "But we really didn't have that much to say about it either. Because we didn't know what it was. There was a sense that the less said about this the better, at least until we heard something from the provincial authorities." Newspaper editors, for their part, still didn't dare publish anything about the disease. What is certain is that Zhang Dejiang, Politburo member and general secretary of the Guangdong Communist Party—effectively the real ruler of Guangdong—knew about the disease by late December, when the special investigation was launched in Heyuan. No Western journalist has ever interviewed Zhang, and the few Xinhua reporters who speak with him do so only in platitudes. His own CDC officials say that they briefed him about the outbreak. But even they were still unsure about what exactly was going on or how widespread the outbreak really was. A few high-ranking health officials, including the mediagenic Zhong Nanshan, were already convinced that they were not dealing with an influenza outbreak, and they were scoffing at the notion that this was a bacterial disease. Yet until the party line was established, the provincial government was unwilling to make any sort of public statement confessing its own ignorance.

For Dr. Cao, this meant a gradual increase in his ward's population, until he had two dozen patients and no more empty beds. Finally, he received word that all the patients with mysterious respiratory ailments were to be sent to four hospitals in Guangzhou. At last, the central government was responding.

Cao filled out the transfer document for Jian Yan and gave it to a nurse to slide into the transparent plastic sheets that accompany patients throughout the Chinese hospital system. The patient they were to wheel out would eventually become famous as the Poison King, the first known superspreader of this new disease. He was now on his way to Guangzhou.

CHAPTER 13

- **February 2, 2003**
- **Guangzhou, China**
- **300 Infected, 20 Dead**

LATE JANUARY IN GUANGDONG PROVINCE TENDS TO BE DAMP, FOGGY, and chilly. If you are used to Hong Kong's smoggy swelter, which in January lets up just a few degrees so that you need a light jacket in the evening, then the upper Pearl River Delta's frigid winter can feel practically Siberian in comparison. That winter was particularly cold in Guangdong and lent the Chinese New Year celebrations that much more familial glow as relatives gathered to welcome the impending Year of the Goat. This was a year, many agreed, to play it safe. After the bounty of the outgoing Horse, and the Snake before that—both years as prosperous as the Chinese zodiac had promised—the Year of the Goat was generally viewed as a time to consolidate and settle accounts. There would be war in the Middle East—you didn't need to consult one of the oracle readers by the train stations to predict that. The United States was preparing to invade Iraq. While most Cantonese didn't see what this war had to do with them, it contributed to the atmosphere of caution, of "wait and see." Shopgirls and assembly-line workers dreamed of getting married, but under no circumstances did they wish to sire a child this year. Children born in the Year of the Goat, especially girls, were destined for lives of hardship. In the days before the New Year, women desperate to deliver during the auspicious Year of the Horse were flooding into hospitals for C-sections rather than risking having their child born an unlucky Goat. Still, the opinion of those who paid attention to such matters was that this would be a sluggish year. But a pause in the breakneck prosperity might be healthy—some yin for the expansionist yang. Hadn't the last few years—Dragon, Snake, and Horse—given the region previously unimaginable riches and prosperity? People were all eating better. Dressing better. A lucky few were driving new cars.

The physicians, in particular, had enjoyed the previous few years. Their income had steadily increased as they were allowed to bill patients for a wider range of treatments and medications. Hospitals had added all sorts of high-tech testing and diagnostic equipment, which spouted additional revenue streams. And more and more Chinese, especially here in the booming south, had slipped into what passed for the middle class in China, earning about two thousand dollars per year, and were now prosperous enough to afford basic medical care, even if it was no longer subsidized by the state.

Deng Zide, the doctor who had first investigated the Foshan case and had been on the team sent out to Heyuan, was among those physicians who had found in this new era, the Era of Wild Flavor, greater personal satisfaction and professional success than would have been possible just a decade before. He was a ranking physician at one of Guangdong's elite hospitals, and, as such, he had enjoyed a lavish New Year's feast, provided in part by the government, through his hospital, which had given him pounds of smoked pork, duck, tofu, several kinds of melons, two cases of beer, six bottles of grain wine, and a box of sweet bean cakes from the best shop in Guangzhou. Deng had shared the feast with his parents, who lived on the outskirts of the city, and then smoked Panda cigarettes, also provided by the state. Doctors such as Deng Zide were treated almost as high officials, and he had come to enjoy these perks. As he watched his father fly a kite with his son at a local park, where during the week migrant workers waited to be picked for odd jobs, he was as optimistic as a man could be at the start of a new year. Later, after he and his family had lit incense and burned replica televisions, cell phones, gold coins, and automobiles at a local shrine for their ancestors, they caught a ride back home in a minibus, all three of them falling asleep after the transfer so that they had to scurry off the bus when Deng woke up and discovered they were already back at the hospital.

All across Guangdong, doctors were returning from Chinese New Year to find their wards crammed with patients suffering from respiratory ailments, usually misdiagnosed as pneumonia. "We were now hearing regularly about these new cases," says Xu Ruiheng, deputy director of the Guangdong CDC. "Doctors were calling each other and asking what this was, what might work, what was going on."

On the last hour of the last day of January, Deng Zide's department received an emergency fax from the Provincial Health Department announcing that Deng's hospital, Guangzhou Number Three, was one of four designated to receive atypical pneumonia cases. The other hospitals were Zhongshan Medical University, Guangzhou Number Eight Hospital, and Guangzhou Institute of Respiratory Diseases. The document stated that public health officials had determined that centralizing the cases would lower the risk of an outbreak—not that any government official was willing to concede that there was any chance of that at all. This was, the fax implied, merely a precaution.

Almost as soon as Deng put down the fax, a nurse in her mid-twenties rushed into the Clinical Department offices and told him that the hospital had just received two patients in acute respiratory distress. Deng looked up at the duty chart on the wall and noticed that he was one of only two doctors in his department already back from New Year's. As he rushed down the hallway to the intensive care ward, he recalled the facial expressions of the first cases he had seen in Foshan. They had been ciphers, complete blanks. The doctors in Foshan had explained to him that the patients had been irritable a few days after the onset of the disease, with the decreased oxygen flow making them short-tempered and slightly delusional, like cranky Alzheimer's sufferers. Then, after that period of hypoxia, their faces would lose any trace of expression as the facial muscles sagged. Those in the later stages of this new disease lacked even the energy to infuse their facial features with the muscularity that made them identifiable. All the patients, men and women, their hair covered by caps and their blue lips wrapped around plastic breathing tubes, looked like limpid pools of flesh. They were blanks.

As Deng Zide slipped on his mask and bent over these new patients, he noticed that same sag and elasticity. The pulse was weak although still measurable, but looking at these new admittees, he could tell they were already gone. Even more alarming was the age of one of these patients: eleven years old.

CHAPTER 14

"WE USED FIVE HUNDRED MASKS THAT FIRST NIGHT, THREE HUNDRED gowns," says Deng. "We ran out of supplies. But we had to continue working. We had to try to stabilize these patients. The party leadership knew we had these patients."

At 10:00 A.M. the next day, after Deng caught a few hours of fitful sleep on his office floor, he received a call from Zhongshan Number Three Hospital. They wanted to transfer a patient to Guangzhou Number Three. "No, no," Deng told them, "you don't need to do that. We already have a dozen of these patients."

"You're the big brother hospital," the doctor on the line told him. "We are the little brother, so he's yours."

Deng tried to think of an excuse. He needed a few hours' more sleep. In fact, he says, he had never felt so tired. He was almost dozing off on the phone.

"He's already on the way," he heard the physician say.

When Deng put down the phone, he could see the elevator doors opening, and out came the Poison King. The patient's look of good fortune had by now largely dissipated. His face had taken on the familiar featurelessness that Deng had seen in the earlier cases, but his skin had turned bright red, as if he had been scalded. And his eyes, despite their sag, looked angry. The man was stout, and, even supine in a hospital bed beneath sheets and pricked by IV lines, he exuded tremendous physical strength. He had been a seafood wholesaler, his forms said, running a shop at the Thousand Flavors Seafood Market in Guangdong. Deng says he had once visited the place, a sprawl of aquariums, fish tanks, and marine carcasses on ice. You could buy any form of aquatic life there, and Deng had purchased a half dozen in-season hairy crabs one year for a fair price, only to find out later that they were knockoffs of those succulent freshwater crustaceans so popular in winter months. The genuine article, raised in a special lake near Shanghai, were sweeter and less briny than those farmed elsewhere. These had been disappointingly salty.

Deng focused on the patient's charts. Under "Diagnosis," it just said AP, atypical pneumonia.

A handwritten note accompanied the new admittee. In it, the doctor in Zhongshan reported that he had six medical staff stricken with the same disease as this patient. Whatever the patient had, the doctor wrote, it was highly contagious. But for now, the Poison King was stable.

Deng went back to the ward, where the eleven-year-old was lying still and terrified under a mask. Deng rolled his head on his neck, trying to unknot his aches and pains. It had to be stress that was making his neck hurt. He was also shivering slightly. Probably just a cold. Anyway, he did not have time for his own touch of la grippe while he worked to stabilize Little Xu, as he called the eleven-year-old who lay on his back, mouth gaping open around his endotracheal tube. The boy had arrived with a ten-day history of fever. In addition to antiviral medication, he had been placed on a respirator and given intensive steroid therapy. He had awakened earlier that morning, the nurse told Deng, but it wasn't clear that he understood where he was. But how could an eleven-year-old comprehend the enormity of this anyway? Patients in severe states of hypoxia and with blood oxygenation levels as low as Little Xu's tend to be aware of only one thing: terror. Little Xu had good reason to be frightened, even in his unconscious state, with tubes rammed down his throat and into his arms and his brain now running on about 10 percent of the oxygen it required to function.

Deng continued his rounds as usual, checking in on his other patients. He found himself losing patience with those complaining of minor respiratory maladies or petty asthma cases. You live in the most polluted country on earth, he wanted to tell them, how could you expect not to wheeze or cough once in a while? You want to see a real respiratory problem, check out Little Xu. He even shouted at his ward's business manager, a glorified clerk, really, whose job it was to quartermaster the hospital, collect money from delinquent patients, and screen any prospective new patients to assess their ability to pay for treatment—or, in some cases, determine just how much treatment they could afford. Because of the disposable gowns and masks the hospital staff had been going through treating just these patients, the business manager hadn't been able to keep them resupplied with N-95 particle masks. "How can I do my job?" Deng recalls shouting at him.

The manager looked away from his department head and apologized, saying he would do better.

Deng went back to check on Little Xu.

He was dying, lapsing into multi-organ failure. The boy was turning yellow, a symptom of jaundice. His systemic inflammatory response syndrome, an immune reaction to the presence of whatever microbe was ravaging his respiratory system, was impeding his lungs' ability to "load" his red blood cells with oxygen, thus robbing the liver, heart, and brain of oxygen. Deprived of precious O_2 for extended periods, the organs one by one start selectively shutting down certain functions, like an overmatched chess player sacrificing rooks to save his queen. Eventually, the liver stops cleansing toxins from the blood and metabolizing red blood cells, and the brain eliminates conscious thought to ensure that involuntary synaptic functions—the beating of the heart, the processing of waste by the intestines—can continue. At a molecular level, the body's cells stop functioning and replicating because they are lacking one of the basic building blocks required for any activity. For a doctor reading these signs, it's like watching a windup toy sputter to a halt. To reverse this, to wind the crank even a little bit, Deng was joined by the nephrologist, and they commenced a series of four intermittent liver- and kidney-support treatments, employing dialysis machines and charcoal-and-anion exchange filters that could nourish the blood and at least keep the patient from developing sepsis as the liver stopped cleansing the blood of toxins and bacteria.

These desperate measures proved only a temporary salve, as respiratory function decreased to the point where convulsions began. Finally, late the next afternoon, Little Xu went into cardiac arrest. Deng had never seen a child succumb so quickly to pneumonia. As he and a nurse worked to resuscitate the boy, pounding on his chest to restart his heart, he cautioned the nurse not to break his ribs.

"You want a dead boy with good ribs," he would always remember her saying, "or a live boy with broken ribs?"

As it was, at around 7:00 P.M., they had a dead boy with broken ribs.

AFTERWARD, DENG CLIMBED THE STAIRS TO HIS OFFICE, BECOMING short of breath by the time he reached his floor. He was forty-one years old, in fine shape, a regular badminton player and, in the summer, a swimmer. But now his whole body ached, and he'd had a headache for about two days. He felt irritable, frustrated, confused. He popped a digital thermometer in his ear and was surprised to see that he had a fever of 101 degrees. He swallowed a few paracetamols and phoned his wife, suggesting she take their son to her parents' house. "I may have come down with something," he told her. "I'm afraid it might be contagious." When he returned home that evening, he fell into bed but could not sleep, instead lapsing into feverish visions of Little Xu and the Poison King and remembering how just before he left the hospital he had heard that one of his nurses was running a 103-degree fever. There had been the journey to Heyuan, and it was like he was back in that van, driving along the winding country highways, past the green mountains and the slow-moving placid streams. His body ached as he imagined the van driving and driving, each bump and pothole jarring his body with another bruising ache, never getting any closer to finding the source of the disease.

At dawn, he took his temperature again: 103. His phone rang. It was another doctor telling him he couldn't make his rounds. He had a fever of 102. "I've got one hundred and three," Deng said. He told the doctor to call around and see how the rest of the staff was doing.

"Four doctors are sick," the other doctor told Deng when he called back, "including you. And four nurses."

By then, Deng was so fatigued from sitting up in bed holding the phone to his ear that he lay back again. Eight staff down in only three days of treating this disease. And this was just one hospital. Before drifting from consciousness, he wondered about the other hospitals, the rural clinics, the villages. Who knew how many cases were out there?

CHAPTER 15

■ February 4, 2003
■ Guangzhou Institute of Respiratory Diseases,
  Guangzhou, China
■ 345 Infected, 33 Dead

BY EARLY FEBRUARY, THE HEAD OF THE GUANGDONG CDC, CHOU
Yenfou, and the head of the Guangdong Provincial Health Department, Huang Qingdao, admitted to each other and to other health officials, including those higher up the bureaucracy in Beijing, that this
was an unusual occurrence. Still, they quietly reminded their colleagues, it was more a curiosity than an epidemic. Respiratory diseases
were common, pneumonia was endemic, and, at any rate, outbreaks
burned through Chinese provinces. Remember, Guangdong had
roughly the landmass of Germany. How alarming would a few hundred
cases of pneumonia be if they were spread out through all of Middle
Europe? However, these officials were nervous about one possibility:
though there were at least 250 cases—that they knew of—they still
could not confirm if this was some mutated, and therefore potentially
catastrophic, new type of influenza with pandemic potential. Among
the frontline clinicians such as Deng Zide, there was already emerging
a dispirited consensus that they were dealing with a virus of some sort,
perhaps a mutated influenza. Hospital virology labs had not had any
luck in identifying the microbe—there had been very few samples
taken, and there were no reagents available in Guangdong for certain
influenzas. The director of the Guangzhou CDC, Du Lin, received
vials of nasal-pharyngeal samples from several Guangzhou hospitals in
early February. "We thought it might be the flu," she said. "It was the

right season for it." Her labs centrifuged the material and introduced it to several cell lines but were unable to identify any cytopathic effect. Du's chief virologist, Ma Peng, working with an electron microscopy specialist, took samples from several cell lines but failed to find any abnormal cell pathology. Even when they employed human fetal cell lines, a practice that is not allowed in the United States, they were unable to observe any destruction of the cells. Du Lin still suspected that this was a bacterial infection of some sort.

Zhong Nanshan, the mediagenic chief of the Guangzhou Institute of Respiratory Diseases, told the head of the Guangdong CDC and the Health Department that he doubted this was a bacterial infection because it did not react to any broad-spectrum antibiotics, though he could not rule out the possibility of some mutated, antibiotic-resistant strain. Identifying the agent was certainly a worthy goal, Zhong reminded his colleagues, but what was necessary now was that the treatment protocol be established and especially that health care workers be told to practice barrier infection control and administer light doses of steroids and heavy antivirals. This disease, whatever it was, was burning through hospitals and taking down doctors and nurses at a faster rate than anything he had ever seen. Already, more than fifty health care workers around Guangdong had been stricken. Even in the height of flu season, he had never heard of this many being infected this quickly. (During influenza season in Guangdong, health care workers take Tamiflu, an anti-influenza drug, prophylactically.)

Zhong Nanshan urged public health officials to act quickly. The public was already panicking; the run on antibiotic and antiviral medication that began in Heyuan and spread to Zhongshan had now swept the whole province. "We are dealing with something completely new," he told the head of the Guangdong CDC. "We have never seen patients deteriorate this quickly."

OUTSIDE ZHONG NANSHAN'S GUANGZHOU INSTITUTE OF RESPIRATORY Diseases, a six-story, gray-and-brown ferro-cement structure with a bank of overcrowded elevators that stop at every floor, is a huge bronze statue of a lunging, cartoonishly muscled worker swinging a sledgehammer. The statue is a striking example of socialist realist statuary, pushing the genre to almost Botero-like extremes of physique and sexuality. Sometimes, when I visited the institute, hoping to catch Zhong Nanshan in between appointments—he seemed to spend most of his time lecturing government officials about how to better deal with respiratory diseases—I would imagine that this was a statue of Zhong Nanshan, wielding that sledgehammer to smash infectious diseases.

During the winter of 2003, that plaza by the river with its gaudy statue would become crowded with peasants and workers seeking some sort of protection from this dreaded disease. Perhaps no rumor travels faster than word of an impending plague. It is a primal, human fear; almost as soon as the possibility is implanted in the mind, some survival instinct transforms the idea into a motivating terror. A few came because they had relatives or friends in the institute, others because they suspected there was some information available at the hospital run by the most renowned doctor in Guangdong. Newspapers in Hong Kong—the *Sun* and the *Hong Kong Commercial Daily*—had carried reports about the curious disease outbreak across the border, mentioning the run on vinegar and the selling-out of certain medications. These newspapers, while not available in Guangdong, were disseminated in samizdat versions through Internet chat rooms, and they confirmed what the locals had known for a while: hundreds were falling ill—and those were just the patients who had made it to hospitals. Who knew how many were languishing in urban slums or rural hovels, unable to afford or reach first-rate medical care? "There is a fatal flu in Guangdong," read a text message passed along more than 125 million times, according to Guangzhou's *Southern Weekend* newspaper. Where or what this thing was simply was not clear, and this fed the worst sort of rumors.

One that was circulating was of a People's Army biological warfare experiment gone wrong. Another was of a chemical weapons trial by the United States, seeking to test its arsenal before invading Iraq. Text message rumors passed along these and other equally implausible explanations, and fed an atmosphere of anxiety and tension. If there was a deadly outbreak, citizens wondered, could they count on the government to tell them about it? This was a government, after all, that had never really admitted to the massacre at Tiananmen Square or to the disastrous collectivization schemes that resulted in the starvation of millions.

The impulse, then, was to channel your panic into an area you imagined to be productive: socking away vinegar, buying Banlangen, eating plenty of winter greens, keeping the windows tightly sealed. Or, for those who were returning home for Chinese New Year, simply staying away from the cities. "People were overcome with Delusions," wrote Daniel Defoe in *Journal of the Plague Year*, "as they had a Notion of the Approach of a Visitation, all their Predictions run upon a most dreadful Plague, which should lay the whole City, and even the Kingdom waste; and should destroy almost all the Nation, both Man and Beast."

Word was passed to the National Health Bureau in Beijing in late January. By early February, Ma Xiaowei, the national deputy minister for health, was planning to visit Guangdong personally to monitor the situation. "We knew that something extraordinary was going on," says the deputy minister. "But we didn't know what exactly it was. That's why I felt I needed to visit."

While the deputy minister may have felt he needed to visit Guangdong personally—a remarkable and rare occurrence in itself—what is even more surprising is that no one bothered to notify Hong Kong, literally next door to Guangdong, about what was really going on in the province. "We just didn't think about Hong Kong," says Xu Ruiheng of the Guangdong CDC. "That was our mistake."

CHAPTER 16

WINTER IN HONG KONG BRINGS RELIEF FROM THE SWELTER OF SUMMER and autumn. The humidity level drops, and on the occasional clear day it reminds me of Christmas in Los Angeles. The evenings are chilly and require a light jacket; the days are sweater weather. Because our apartment on The Peak—the mountain that overlooks Hong Kong's central district—lacked central heating, we had to roll the space heaters out of storage in December. My older daughter, Esmee, almost four, had outgrown her nursery school, and we began the process of securing her admission to a proper kindergarten in the fall. My younger daughter, Lola, was just one and taking her first, tentative steps.

One morning, I was walking with Esmee on the Lugard Path around The Peak. It was a Monday morning—I worked Tuesday through Saturday at the magazine—and the trail was not crowded. As I held Esmee's hand and slowed my pace so that we could walk in step, she asked me about breathing. She had just become aware that this was something she did. And she understood that she was sucking something called air or, as my wife told her, O into her mouth or nose when she did this. She wasn't sure where this O went, but since she had just begun swimming and had put her head underwater, she now had an intuitive feel for the consequences of being deprived of it. This had prompted a new appreciation for the importance of breathing.

An even larger idea that was now creeping into her consciousness was death. She wasn't sure what it meant, and I believe she still equated it with sleeping. But now she had a notion that was troubling her.

"Daddy," she asked. "Do we keep breathing when we die?"

"No, we don't." I looked down at her: brown hair tied in a ponytail, olive eyes, her tiny nose pointed like the tip of a strawberry, her mouth turned into a frown.

"We don't even breathe?"

"Nope," I told her. "When we die, we sort of just stop. It's like we're asleep but we never wake up." I regretted relying on this typical parental explanation of death.

"But you told me we breathe when we sleep," Esmee pointed out.

"That's the difference," I told her. "Dying is like sleeping without breathing—or dreaming."

"Daddy." Esmee tugged on my hand. "Why is she wearing a mask?"

A woman in a tracksuit was speed-walking along the path toward us. We were walking on the section of path that runs along the central and then the western district before veering behind another mountain and emerging alongside the South China Sea. The woman wore a surgical mask with the straps tied around the back of her head. Her eyes above the mask appeared focused on keeping her pace. But they were also wary and suspicious. Living in Japan, I had frequently seen people wearing such masks, especially during the winter, when respiratory ailments and common colds were in season. I had worn them myself when I lived in Kugenuma, outside Tokyo. Japanese feel that the masks serve two purposes: to keep the respiratory tract moist and to prevent the spread of infection to others.

Yet this was the first time I had seen one in Hong Kong. This woman did not have the feel or look of a visiting Japanese—her tracksuit was slightly frowsy in a manner that I didn't associate with a Japanese tourist—so I guessed she was a local.

"I don't know," I told Esmee. "Maybe she has a cold."

As is so often the case, I felt I had not given my daughter a satisfactory explanation.

■ ■ ■

HONG KONG, IN THE WINTER OF THAT YEAR—AS ACROSS THE BORDER the Era of Wild Flavor was unfolding with gaudy excess—was living through a contrasting season of self-doubt and diffidence. While China boomed, Hong Kong withered. Property prices were depressed. Retail sales were down. Many of the best and brightest were leaving the city for Vancouver, Sydney, or Beijing. The city was stratified and hierarchical, and lacked the Wild West exuberance of the mainland. The reunion with China in 1997 had not ended Hong Kong's history as a colony but had instead transferred fealty to a new colonial overlord. Hong Kongers had traded one distant capital, London, for another, Beijing; they had traded one imperial language, English, for another, Mandarin. Most of the Special Administrative Region's citizens speak Cantonese, a language indigenous to southern China that is as distinct from Mandarin as Spanish is from French.

For Hong Kong natives, it seemed that the boundaries and horizons had somehow contracted. Even their dreams had been downsized. The postwar, pre-handover generation had been inspired by the notion that anyone in Hong Kong could become rich; just look at the many refugees who had escaped China to become billionaire Taipans: Li Ka Shing, Lee Shau Kee, and Run Run Shaw among them. That dream, however, was slowly dying. Today, many of those educated in Hong Kong's public schools were linguistically shut out of the mainland economy and were forced to take whatever work they could find in Hong Kong's service sector, where wages had been depressed since the late nineties. While the city still looked like an urban bauble perched between Victoria Harbor and The Peak, a closer inspection revealed a city that was choking on its own fumes and strangled in red tape. The city needed dynamic leadership.

Instead, it had Tung Chee-hwa, the scion of a shipping magnate and, in many ways, a perfect personification of what ailed Hong Kong: Tung was a rich kid reduced to pleasing a bunch of Communist cadres. Neither charismatic nor intelligent, he enjoyed only one attribute, as far as the public could see: he had been born wealthy. Tung was that unique political figure: a toady who managed to disappoint even in his toadying. He strove to please his real master, Beijing, yet would be so overzealous in his misinterpretation of what his overlords wished that he would end up simultaneously failing his master and doing a disservice to Hong Kong. China, of course, is not a democracy, and Hong Kong never has been. Tung Chee-hwa was selected by an executive committee of eight hundred prominent, mostly pro-Beijing officials, businessmen, and professionals, and he never projected confidence so much as a spectacular ignorance. When public opinion was against him, he would stand by an unpopular decision to, say, allow more landfill in the harbor or delay the democratization process, not because he felt it was the right thing to do but because he simply wasn't aware that most Hong Kongers were opposed.

He had presided over Hong Kong's decline, and it seemed he had no idea that that decline was even happening. I used to see him taking his constitutional along Bowen Road. In his white T-shirt with a tracksuit top and jogging pants, he appeared more barrel-chested than he did in the suits he wore during his rare public appearances. His silver crew-cut hair and round eyes made him look like a cross between an oversize badger and Pete Rose. His mouth seemed perpetually frozen in a slack O, as if he had just had a trivial but pleasant surprise. A bodyguard always trailed behind him, taking one long stride to keep pace with three of the chief executive's little steps.

But in some sense, I empathized with Tung. For, just as he directed Hong Kong at the sufferance of his overlords in Beijing, so too did I run TIME Asia only as long as my bosses in New York would grant me the privilege. We were both appointees who perhaps had grand ideas upon taking our respective jobs but who had discovered that, in the process of running our remote little corners of great empires—in his case China and in mine Time Warner—we did not have the free hand that would have made the task rewarding. If the chief executive was unable to project anything but obliviousness to the travails and struggles of his fellow Hong Kongers, I believe I may have appeared just as unsuited to my own job. After all, I had been a writer for my entire career and had never managed anything more than a YMCA basketball team, yet here I was, charged with guiding a staff of seventy through one of the most difficult and arduous Asian media environments of the last twenty years. As the boss, I had to receive a steady procession of staffers seeking more money, days off, and explanations of why their stories had been spiked, cut, rewritten, or ignored. There were complaints about sexist and racist comments in the office. Photo editors would lament that their computers were obsolete; designers wanted new software; production staff demanded better scanners. Advertising people wondered if I could speak at a client's lunch. The marketing department asked if I was available for a reception in Tokyo or Delhi. The easiest response to almost every request, I quickly concluded, was to say yes and then do nothing. If the issue was truly important, then someone would remind me of it. Otherwise, it would just go away by itself.

But Tung Chee-hwa's Beijing was my New York. If he could be second-guessed by more powerful men in a distant city, then I could be reduced to blathering repeated apologies if my bosses at TIME Inc. were displeased with one of my decisions. At least Tung Chee-hwa was rich, a scion of great wealth and a historical figure of sorts. If he had ever given an interview, there would have been any number of media outlets, including my own, that would have been grateful for the privilege of airing the man's thoughts.

In my own small way, I tried to nudge the magazine into covering the relentless energy of urban Asia. My editors and I commissioned a host of renowned talents: William T. Vollman, Paul Theroux, David Mitchell, William Gibson, Pramoedya Toer, Ma Jian, Jan Morris, Chang Rae Lee, Pankaj Mishra, and V. S. Naipaul, among others. Frequently, the stories were lovely extensions of *Time*'s coverage of the region; occasionally, as in Vollman's review of a Nick Tosches pamphlet on opium smoking, they seemed ostentatiously risqué for the sake of demonstrating this new boldness. (Vollman would also write wonderful stories for us, such as his account of crossing China by train, which captured the movement and dynamism of China's massive internal migrations, and another of a Philippine numbers runner making his nightly speed-fueled rounds.) Our own reporters branched out into a range of topics that *Time* had previously dismissed as too fringe. One of our correspondents raised the ire of human rights organizations when he bought the freedom of two underage Thai sex slaves while writing a story about slavery. Evan Wright, who would later win a National Magazine Award for his *Rolling Stone* series on the war in Iraq (subsequently collected into the bestseller *Generation Kill*), wrote a chilling story about an English bar hostess who was murdered in Tokyo. I wrote a piece about methamphetamine abuse in Asia, in which I discussed my own extensive history with the drug. (That story was anthologized in the first *Best American Nonrequired Reading* collection.) The magazine we were producing was becoming ever more eclectic, but at the expense, some of my staffers and critics complained, of our basic mission as a regional newsmagazine. I would disagree but then had to admit that it was the great feature stories that excited me more than the news gathering.

But in some sense, hadn't we become a sideshow? The only story in the world that mattered since 9/11, at least for international media outlets such as *Time*, was terrorism. We had broken plenty of scoops regarding terrorism in Asia, notably printing the confessions of an Al Qaeda terrorist obtained by the CIA but leaked to our Jakarta stringer by local law enforcement authorities. But now that the Bush administration had so successfully refocused everyone's attention on Iraq, our role had been diminished. China, which just two years ago had seemed like the emerging Big Story, especially in the wake of the Hainan Island spy plane incident, was no longer of great interest to my bosses in New York. The impending leadership change in Beijing, in which Jiang Zemin was stepping down as party general secretary, to be replaced by Hu Jintao, could not compete for pages with the oncoming war. So if we were a sideshow, I reasoned, let's have a little fun and run the sorts of stories no one would expect to see in *Time* magazine.

But already, in China, the beginnings of the only story in the world that would rival the Iraq war in 2003 could have been discerned—if I had been a more acute journalist and editor—by rumors of plague and disease that already had thousands of Chinese on the run.

MY OWN FAMILY WAS FULLY ACCLIMATED TO HONG KONG, AND THE pollution and decline and demoralization of the city hardly seemed to affect us. Our lives, like those of many expatriates sent by multinational companies, were insulated from the hardships of many of the native Hong Kongers by housing allowances, health insurance, home leaves, and domestic helpers. But the pathos of Hong Kong would manifest itself in curious ways. A Chinese woman who worked for the magazine sent me a note saying she wished we could be friends but then added that this was impossible because I was a foreigner, and foreigners traveled all over the world while Hong Kong people were stuck here. I thought about that for a moment: they really were stuck. They didn't even have Chinese passports; instead, the departing English had issued Hong Kong passports that looked like British passports but gave their holders none of the rights or entitlements allowed British citizens. In their tiny apartments, with diminishing prospects, in a city whose role on the world stage was receding—we read every day about how Shanghai had usurped Hong Kong as China's international city—to what did this city's citizens have to look forward? It is the problem of city-states, I suppose, to be bereft of the sense of optimism that a hinterland provides. In New York, Paris, Moscow, even Tokyo, there is at least the illusion that if you fail, you can strike out for parts unknown and make a fresh start. In Hong Kong, there was Shenzhen, where the only thought on most people's minds was how to get to Hong Kong.

Yet, after the handover, Hong Kong was supposed to have a hinterland, a great expanse where one could start a new life. So China had usurped the English, put the kibosh on democracy, and saddled the city with Tung Chee-hwa. But it was booming, and the mainland's opportunities should have become Hong Kong's as well. "At some point," a Hong Kong lawyer told me, "there has to be some upside to reuniting with the motherland."

The downsides were becoming ever more apparent. The way the one-country, two-system structure worked, there were so few official ties between Hong Kong and the mainland that any hope of securing real information about what was going on across the border lay with Tung Chee-hwa. And he was spectacularly unsuited to the task of delving into an issue such as an unexplained disease outbreak next door. Remember, infectious diseases were state secrets in China, and Tung Chee-hwa was far too timid to risk a reprimand from the Politburo for asking about such a sensitive matter.

Instead, it fell to his director of health, Margaret Chan, and secretary for health, welfare, and food, E. K. Yeoh, to find out what was going on in China. The structure of the relationship with the mainland, however, required that any official public-health-related issue be raised through the Ministry of Health in Beijing. Any conversations with doctors or public health officials across the border in Guangdong could only be unofficial. Considering that four hundred thousand people a day crossed the border between Hong Kong and Guangdong, it is remarkable that the highest health officials in Hong Kong had virtually no relationship with their counterparts across that border. When they tried to get in touch, they had to scramble even to come up with phone numbers for the relevant officials in Guangzhou. "We had no information from official sources that we could use," says Yeoh. "We tried to contact Guangdong officials, but there was no response. We had to try to understand what was going on in China. We tried the standard lines but got nothing."

The assumption among Hong Kong's public health officials was that if there was a disease outbreak raging across the border, then they would be alerted by the Ministry of Health. "We are a part of China now," Margaret Chan would say. "They can't treat us as a foreign city."

But they did.

CHAPTER 17

- **February 5, 2003**
- **Pok Fu Lam Garden, Hong Kong, China**
- **357 Infected, 35 Dead**

GUAN YI, ONE OF THE VIROLOGISTS WHO LED THE EXPEDITION THAT investigated migratory bird deaths on Ah Chau Island, smoked constantly, sometimes lighting one cigarette from the butt of his last. He frequently switched brands, going from Mild Sevens to Winstons to Pandas and then back again to Japanese cigarettes. Sometimes, when I visited, he would give me a pack of whatever he was smoking. "Take them, try these," he would say, offering up a Chinese brand I had never heard of. "They're from Yunnan, but not so bad." And he would light our cigarettes, and we would smoke and talk about respiratory diseases.

One afternoon, as he was sitting watching his large-screen TV—the news networks at that point seemed to be covering nothing else but the countdown to war—he received a call from one of his colleagues across the border, a doctor at a respiratory disease clinic in Guangdong. There was something very bad going on, this doctor was telling him. They were getting flooded with patients who were presenting symptoms of severe respiratory tract infection, high fever, myalgia, headache.

"We have no idea what this is," the doctor told him, "but we are frightened. My colleagues have been falling ill. And these patients are dying."

Guan Yi, who views the whole world as a potential petri dish for influenza, was sure that this had to be some sort of mutant influenza.

He ran through the checklist. Winter. Southern China. High threshold of human-animal interaction. "Can you send me samples?" he asked.

Of course not, the doctor replied. That could get him arrested, or worse, for exporting state secrets in the form of respiratory swabs. He could never get his department heads to go along with that.

Guan Yi, who had returned home to eat lunch, finished his last sip of tea, snuffed out his cigarette, and slipped on a beige vinyl windbreaker. He lived on the eleventh floor of an apartment block on Primrose Hill, near the University of Hong Kong microbiogy department at Queen Mary Hospital, just a few miles from Malik Peiris. Unlike Peiris's complex, which was a more typical expat's flat—spacious rooms, generous balconies, ample parking—this was a far less capacious accommodation for Guan Yi, his wife, and their son. Their block of flats was typical of the housing developments in which Chinese families tended to live—four apartments to a floor, four rooms in each apartment, each room barely big enough for a single bed. There was a Wellcome supermarket on the ground floor and a 7-Eleven next door.

As soon as he hit the street, he hailed a taxi. Sliding in, he buckled his seat belt, as he always did. (Not a driver himself, he was dubious about the safety of internal-combustion transport.) Arriving back on campus, he walked down to the basement of Queen Mary Hospital's micobiology unit, where K. Y. Yuen, the head of the University of Hong Kong microbiogy department, was sitting in his makeshift office. (Yuen's formal office was in the university's microbiology department, but he found it more convenient to keep a cluttered office here, near the hospital labs and research assistants.)

His slouching posture in a white lab coat over a cardigan sweater belied his role as one of Hong Kong's most knowledgeable infectious- and respiratory-disease experts. He spoke in a soft voice, almost a mumble, that forced you to lean in to hear him. It seemed he had been having the same informal conversations with colleagues across the border as Guan Yi. "When you have young, immuno-healthy adults dying of pneumonia, you have to worry about what is going on," he would later tell me. "You have two differential diagnoses: a mutated H5N1 or a mutated H3N2 or H3N1 or an adenovirus." (He was referring to several types of zoonotic, or species-jumping, influenzas and a common respiratory virus.)

The son of a dental assistant, Yuen had grown up in Hong Kong's impoverished western district when that area of Hong Kong was even more ragged and squalid than it is today. When he was a boy, he shared a two-room apartment with his parents and three brothers, and they had been among the luckier families. Some families shared their flats with four other families—twenty or thirty people in three or four tiny rooms. Many of these flats lacked running water or electricity, and much of a family's life—cooking, laundry, bathing—was communal. Diseases had burned through Hong Kong's poor neighborhoods when Yuen was a boy—the cholera epidemic of 1963, the Chinese flu of 1968—and many of his neighbors had been stricken, and a few had perished, in those outbreaks. Yuen had seen firsthand how a densely populated area could amplify a disease. When he graduated from the University of Hong Kong's medical school and interned in the United States, at the Fred Hutchinson Cancer Research Center, he decided to dedicate his career to keeping Hong Kong safe from infectious diseases. His fierce commitment to Hong Kong was similar to that of a soldier defending his own village—only, in Yuen's case, the enemies were microbial rather than martial.

When Yuen heard about the unexplained pneumonia cases across the border, he thought first about the epidemics he had witnessed as a boy. Then he recalled studying the 1894 bubonic plague epidemic, which had reduced Hong Kong's population by a third. At the time, the *Yersinia pestis* bacteria, the causative microbe, was unknown, making appearances of bubonic and pneumonic plague seem like mysterious cosmic calamities. "Bubonic plague," wrote Edward Marriott in *The Plague Race*, his account of the 1894 outbreak, "exploding out of China, had torn through the island colony and now the houses lay shuttered and boarded-up, the only movement through the narrow alleyways the slow shuffle of soldiers, doctors, scientists." That global meltdown is today viewed as the start of the Third Plague Pandemic, which would infect much of the old and new worlds, causing a notable outbreak in California in 1900. We are now living through the Fourth Pandemic, which most recently, in 1994, killed hundreds in Surat, India. With the advent of modern antibiotics, if treated early enough, the disease is usually not fatal. But in Hong Kong in 1894, infection by plague was still tantamount to a death sentence, and those who suffered disproportionately were the Chinese underclass. The main hospitals, the European Government Civil Hospital and the Chinese community's Tung Wah, were quickly filled to capacity. The colonial government pressed into service a battered hulk, the *Hygeia*, to serve as a floating hospital moored three hundred yards off the Admiralty waterfront. This "hospital," vowed James Lawson, superintendent of the Government Civil Hospital, would be designated specifically for infected Chinese. The *Hygeia*, a mastless, dark, slit-windowed vessel drifting ominously offshore, quickly became the object of intense speculation as hundreds of supposedly plague-stricken Chinese were sent to her, never to return. (In one case, a healthy pregnant woman was mistakenly sent to the death ship.) The ship, the Chinese said, was really a laboratory where the most awful experiments were conducted. One popular rumor swirling through the cantons and narrow alleyways was that the English were perfecting a cure for the plague by removing children's livers and using the restorative powers of these internal organs as a remedy for sick colonials. Chinese parents began withdrawing their children from school before they could be kidnapped by English soldiers for this purpose. Many English could and did flee, as did the Japanese, French, and others of the colonial class. Chinese fled back to China, carrying the disease into the hinterlands. Other Chinese stayed and died by the tens of thousands.

Several of the world's foremost scientists, notably Japanese bacteriologist Shibasaburo Kitasato and Pasteur Institute–trained French-Swiss scientist Alexandre Yersin, converged on Hong Kong. They engaged in a race to isolate the agent, with Yersin emerging as the more successful microbe hunter after Kitasato hastily published a paper in which he claimed to have found the causative bacteria and that it was not the same as that responsible for bubonic plague. Yersin would not only correctly isolate the bacterium—hence its

Latin name of *Yersina pestis*—but would also be the first to cure a plague victim with serum taken from a convalescent patient.

That had been the first great race for a microbe undertaken in Hong Kong. One hundred years later, K. Y. Yuen worried that perhaps they were on the verge of yet another microbe hunt. Any new agent, he realized, especially a mutated influenza virus, could do similar damage in the crowded, fetid alleyways of Kowloon or in his native western district.

"So what's going on across the border?" Yuen asked Guan Yi.

CHAPTER 18

- **February 10, 2003**
- **World Health Organization Headquarters, Dongzhimenwai, Beijing, China**
- **393 Infected, 40 Dead**

"WE TRY TO TAKE SERIOUSLY EVERY REPORT OF AN OUTBREAK," DR. Henk Bekedam, the head of the World Health Organization mission in Beijing, would tell me later. "But this is China. Do you know how many rumors you hear of influenza or plague or even anthrax? In a country this big, there is no way to follow up on each individual piece of gossip."

Perhaps the main task of the World Health Organization's Global Outbreak Alert and Response Network (GOARN), based in Geneva, was to sort through the dozen or so credible rumors received every day. Each morning at 9:00 A.M., WHO epidemiologists Mike Ryan and Tim Gruen would go through and discard most of them, forwarding a few to the appropriate officials. On February 5, they forwarded a report on the outbreak of possible avian flu in Guangdong to Klaus Stöhr, the head of the WHO Global Influenza Surveillance Network. The report came in the form of an e-mail from the son of a former WHO staffer who was currently teaching English in Guangdong. The young man reported that there was some sort of new disease in Guangdong and that folks down there seemed "pretty spooked." What concerned WHO officials was that this report seemed to concur with speculation that some sort of influenza outbreak was afoot. Still, the officials could confer only through back channels with their Chinese counterparts.

By Monday, February 10, a report had hit ProMed, the Harvard-based infectious disease surveillance website dedicated to tracking potential outbreaks and read by health care specialists and public health officials around the world. "Have you heard of an epidemic in Guangdong?" asked an e-mail sent in by Dr. Stephen O. Cunnion, a retired captain in the U.S. Navy Medical Corps. He forwarded a note passed along to him by a colleague. It read, "An acquaintance of mine . . . reports that the hospitals [in Guangdong] have been closed and people are dying."

ProMed tends to deal in frightening and very real disease outbreaks. Swine fever, classical (Brazil), hantavirus pulmonary syndrome (USA), anthrax, human and livestock (China)—these are typical headings in the daily ProMed digest. A posting there meant that doctors, scientists, nurses, and journalists from Harvard to Hong Kong were turning their attention to Guangdong. For the WHO's Hitoshi Oshitani, based in the group's western Pacific regional office in Manila, the posting meant a flood of inquiries. As the head of Communicable Disease Surveillance and Response throughout Asia, he knew that if this was coming from Guangdong, then the greatest likelihood was influenza.

"What's going on?" he asked Alan Schnur, the country coordinator for the Communicable Disease Surveillance and Response Department and effectively the second in command of the WHO's China mission. "We should get a team down there."

Schnur, always cautious by nature, advised that they wait for the Chinese government to respond and provide more information. Later that day, WHO officials in Beijing would have even more reason to be alarmed. A posting to the Chinese website Boxon.com read, "non-official sources have confirmed the panic in Guangdong province. In the last two weeks, hundreds of people in Guangzhou and Zhongshan have been infected by strange decease [sic]." A doctor from Renai Hospital, in Guangzhou, reported on the website that according to a confidential Public Security Bureau fax, "280 people have died . . . the disease has been confirmed to be very infectious, many doctors are affected."

That same day, Hitoshi Oshitani in Manila received an e-mail from a parasitologist named Carlo Urbani, based in Hanoi, Vietnam.

He said he had heard from the French embassy in Hanoi that they had had reports from their consulate in Guangzhou that there was an outbreak of a severe respiratory ailment and that many had already died. "Everyone suspected influenza," Oshitani would tell me later. "I thought immediately that this was a new influenza." If this was influenza, then the fact that there were already hundreds of cases being reported and hospitals closing down was alarming. A standard A-type influenza would not ravage Chinese hospitals, certainly not in Guangdong, one of the most sophisticated provinces in China. Oshitani had visited those hospitals, and he knew the doctors. They were very good, perhaps among the most experienced in the world at dealing with respiratory ailments. If this really was influenza and it was burning through hospital wards, then it had to be avian flu. He knew about the H5N1 outbreaks in Penfold Park, in Hong Kong, and about Malik Peiris's and Guan Yi's investigations into that outbreak. What if this were a shape-shifted avian influenza subtype that had achieved efficient human-to-human transmission? That would mean that he was now receiving reports of the front end of a global pandemic that could . . .

The first journalist to call Hitoshi Oshitani was Mary Anne Benitez of the *South China Morning Post*. "What is it?" she asked.

"Whatever it is," Oshitani could only reply, "it's bad news."

CHAPTER 19

- **February 10, 2003**
- **Guangzhou Provincial Health Bureau, Guangdong, China**
- **394 Infected, 40 Dead**

THE THREE OFFICIALS TAKING THEIR SEATS BEHIND A LONG, NARROW desk covered in a white tablecloth in an ornate conference room at the Guangzhou Provincial Health Bureau headquarters appeared ruddy and in fine fettle until the cadmium television lights came on and seemed to wash away their fleshy good health, leaving them pale and slightly confused as they began their press conference. Xu Ruiheng, the deputy director of the Guangdong CDC, and Zhong Nanshan, the mediagenic director of the Guangzhou Institute of Respiratory Diseases, were joined by Provincial Health Bureau chief Huang Qingdao in what would be the first official public pronouncement on the epidemic sweeping Guangdong. Of the three, only Zhong, with his high cheekbones and stern gaze, appeared comfortable in front of the cameras. He leaned back slightly in his chair, waiting for his colleagues to start. A spokesman for the Health Bureau stood up and read a brief introduction, then quickly turned the proceedings over to Huang Qingdao.

There had been considerable internal debate in the provincial government as to the wisdom of disseminating any public information regarding this atypical pneumonia. Some officials, notably Deputy Governor Lei Yulan and Party Secretary Zhang Dejiang, felt that giving out any information was counterproductive, certainly before doctors understood the cause of the disease. Also, in a China where business now trumped almost every other consideration, no one had wanted to jeopardize the Chinese New Year holiday, when hundreds of millions of Chinese spend vast sums on food, shopping, and travel. "The most important vacation in the life of the Chinese people, the Spring Festival, was coming. We didn't want to spoil everyone's happy time," Feng Shaomin, director of foreign affairs for the Guangdong Health Department, told John Pomfret of the *Washington Post*. "You can imagine how people would have reacted if we had told them about the disease. They wouldn't eat out, nor would they go shopping or get together with family members and friends. If we had done it earlier, it would definitely have caused chaos."

There was simply no incentive for releasing the information earlier. "The Chinese government prefers stability at all costs," says George Lim, former editor of the Guangdong newspaper *Twenty-First Century World Herald*. "Bad news like earthquakes, riots, and contagious diseases is not freely reported by the media because officials see spreading this sort of news as being bad for stability. If the public finds out that a new infectious disease is spreading rapidly, there is likely to be chaos."

Consider, then, the decision by party secretary Zhang Dejiang to allow a press conference as a partial admission of defeat. Now that health care workers and public health officials around the world had picked up chatter on ProMed about a possible epidemic, Guangdong party officials had no choice but to speak out and seek to avoid further panic. Even if Guangdong newspapers weren't covering the epidemic, the fact that newspapers outside China were now starting to write speculatively about it was causing anxiety among locals. "We heard so many crazy things about this sickness," said Guangzhou taxi driver Luo Tam. "Everybody was buying vinegar or medicine. I didn't even know why I should buy vinegar, but I went to buy some, and the stores were sold out. And that's when I became really nervous."

Seeking to capitalize on the panic, a spokesman for Roche Pharmaceuticals' China subsidiary even called a press conference, announcing that the company's Tamiflu medication was effective at curing avian flu. This sent shockwaves through the local population and added to the alarming gossip reaching Hong Kong and the rest of the world.

THE GOAL OF THE THREE OFFICIALS NOW SWEATING UNDER THE lights was to calm nerves. But the agendas of each would turn out to be very different. Huang Qingdao, as the ranking official present, would handle most of the conference and give a situation report. Xu Ruiheng, as the epidemiologist, would talk about the course of the outbreak, and Zhong Nanshan, as a clinician, would talk about what was known of the etiology of the disease itself. From the start, the conference sought to soothe rather than inform. Zhong Nanshan and Xu Ruiheng were startled to hear Huang Qingdao say that the outbreak was now completely under control. Both men knew that this was not the case, and that health care workers in Guangzhou, even as Huang Qingdao spoke, were becoming infected with a disease that had an alarming mortality rate—as high as 20 percent by some estimates. Instead, Huang was saying that most of the patients from Zhongshan and Heyuan had recovered—technically, this was true, but a thoughtful listener may not have been too reassured to hear that "most" patients hadn't actually died. Huang went on to explain that of the 305 cases reported so far, there had been only five fatalities. This last number was not even technically true—there had been at least a dozen fatalities in January alone. Furthermore, Huang pointed out, the government's decision to delay the release of any information to the public had made sense because "atypical pneumonia is not a listed infectious disease that you have to report by law. That's why we don't feel it's necessary to make it public. But now that everyone is talking about it, we have decided to talk."

He emphasized, repeatedly, that there was no need to panic. "As to the alert system, the national infectious-disease law states clearly what to do. And we will act in accordance with this law. It won't affect social stability."

It is widely understood by those who cover China, including the Chinese media itself, that official press conferences are almost never notable for the news disseminated. Instead, they are useful as indicators for what topics may now be written about or discussed in Chinese newspapers and magazines or aired on television. Almost immediately after the press conference, a flood of articles about the new disease began appearing in Chinese newspapers, each of them starting with Huang Qingdao's reassuring words but then launching into reports on the situation in Guangzhou hospitals, how doctors were faring, and speculation as to the real cause of the disease. The *Xin Kuai Daily*, for example, ran an interview with a local professor, Cai Lihui, who noted that after the 2001 attacks on the World Trade Center, and as anthrax letters passed through the U.S. postal system, the U.S. government had quickly released what information it had. Cai pointed out the contrast between what had happened in the United States and what had happened in Guangdong: "the way the government has dealt with atypical pneumonia shows its inability to deal with the crisis. "

Party Secretary Zhang Dejiang quickly shut down coverage of the disease, telling one of his deputies that "if the newspapers are not going to responsibly cover the matter, then why should we let them write about it?" It would be over a month before Chinese newspapers could again begin reporting on the disease outbreak. By then, more than three hundred Chinese would have died of SARS.

CHAPTER 20

- **February 11, 2003**
- **University of Hong Kong Department of Microbiology, Pok Fu Lam, Hong Kong, China**
- **400 Infected, 41 Dead**

FOR VIROLOGISTS, EPIDEMIOLOGISTS, AND CLINICIANS AROUND THE world, the February 10 press conference was more notable for what it didn't say. As they scanned reports in newspapers and medical websites, they didn't find any mention of whether or not the illness was influenza. In Hong Kong, six daily newspapers carried reports on what was then being called an atypical pneumonia outbreak in Guangdong. Reading those stories, Malik Peiris, Guan Yi, and K. Y. Yuen were stunned: a new, fatal respiratory disease, most likely a virus, was emerging in southern China, for all intents and purposes influenza's home field, and not one of the public health officials present could say reliably if it was influenza, or possibly avian influenza? Peiris, Guan, and Yuen knew that the labs in Guangdong were not as thorough or meticulous as the best Hong Kong labs. Guan had spoken to Zhong Nanshan, who had told him that he didn't believe this was influenza, but "Yi was convinced from the start that this was bird flu," Zhong told me later. As for the explanation that this was atypical pneumonia, that was like describing a species of animal as an atypical mammal. It had no real meaning, save a description of symptoms.

If this was avian influenza, as Guan Yi and others feared, then that meant that the disease had not only successfully jumped the species barrier, it had also achieved efficient human-to-human transmission.

A global pandemic could already be well under way, the three men agreed, and all they now knew about it was what had been revealed in that superficial press conference.

"We need samples [from the mainland]," Guan Yi said again.

On the phone, K. Y. Yuen and Klaus Stöhr had already discussed the sensitive matter of securing samples from mainland hospitals. Both men knew that it was highly illegal and possibly unethical to gather samples and take them across the border to Hong Kong. Yet all the epidemiologists and virologists who had heard about this disease would have at least considered compromising ethical and legal standards to get their hands on some respiratory swabs from infected Guangdong hospitals. There was a larger issue at stake than local laws or international agreements concerning the transport of medical samples. In that clear fluid medium surrounding those swabs could be the proof that the world was heading toward a biological Armageddon—or perhaps the best chance modern medicine had to save humankind.

The only scientist with the connections—and gumption—not only to enter into China but to convince a Chinese hospital to hand over precious infectious-disease samples was Guan Yi. "Yi is impetuous," says Rob Webster, the head of the St. Jude Children's Research Hospital Animal Influenza Program and Guan Yi's mentor. "But that's also what makes him so talented. I was calling them [Guan Yi and Malik Peiris] and talking about what was going on in China. We knew it was flu. We needed someone who could go up there and get it."

Guan Yi and his assistant, Zheng Bo Jian, spoke with Zhong Nanshan at the Guangzhou Institute of Respiratory Diseases, correctly assuming that Zhong Nanshan was one of the few Chinese doctors in Guangdong who wouldn't worry about violating state security laws if it meant possibly saving lives. Perhaps because of his high-level connections, Zhong could take chances that other physicians wouldn't. And Guan Yi knew how to appeal to Zhong's vanity by couching his request to come up and gather samples as an offer to "do some lab work for the respected Zhong Nanshan," as Guan Yi put it.

As long as Zhong Nanshan agreed, then Guan Yi felt that he should be the one to go into China and collect the samples. K. Y. Yuen would let Margaret Chan, Hong Kong's secretary of health, know that his department was planning to gather some samples, but he would leave deliberately vague his methodology. If Guan Yi were arrested or captured, the Hong Kong government would have plausible deniability. "I knew," Margaret Chan would tell me later. "After all, I am a scientist, so I have an interest in what my fellow scientists are doing. I was the secretary of health. Of course I had to know. But I want to remind you that it was very important to me that we do everything we can to protect Hong Kong."

Guan Yi set out from Hong Hom Station on the morning of February 11, wearing a gray suit and carrying a black canvas duffel bag. With his long bangs and wire-rim glasses, and speaking to his son about college applications in rapid-fire Mandarin on his cell phone, he looked and sounded like a typical Chinese businessman shuttling back and forth between, say, his factories in Guangdong and his bankers in Hong Kong. His mission, however, was far more sensitive—and illegal. If apprehended, he would probably be charged with espionage, which carried a possible death sentence. The Chinese government has never been forthcoming with infectious-disease samples, or in terms of collective science generally. If any infectious-disease microbes were going to be legally exported from China, then they were going to be piggybacking a ride in the respiratory tract of an already infected human being.

The ride to Guangzhou from Hong Kong takes about two hours by the fast Kowloon-Canton Railway train, on which Guan Yi was riding. The countryside is lush and green as the train glides through Hong Kong's new territories, and then quickly turns brown and desiccated once it hits Shenzhen in mainland China. A traveler must pass through customs and immigration in Hong Kong, and then again upon arrival in Guangzhou. The inspection tends to be far more rigorous for Chinese than for foreigners—for most Chinese, even securing a passport still requires some low-level political or bureaucratic string pulling. But Guan Yi, appearing as he did to be a prosperous mainlander, elicited little suspicion when he crossed the border, and on this trip, he says, he didn't feel any more nervous than usual. After all, he wasn't doing anything illegal on the inbound journey.

He hailed a taxi at the Guangzhou East Railway Station and rode it west, to the Pearl River. There, the driver turned north to cruise along the bank for a while before depositing Guan at the huge statue of the valiant socialist-realist virus smasher in front of the Guangzhou Institute of Respiratory Diseases. In the hospital, he scrubbed down, slipped on a gown, gloves, a mask, a cap, and goggles, and then entered the fourth-floor atypical pneumonia ward. Only two of the nurses were willing to join him, the rest having been frightened away by the contagiousness of the disease. Guan Yi's first thought was that these patients were even sicker than he had imagined.

Zhong Nanshan didn't believe in endotracheal intubation as a treatment, relying instead on what he called "mask intubation"—the placement of an oxygen mask over a patient's nose and mouth, attached to a line of concentrated oxygen. Shunning the messy, invasive process of endotracheal intubation kept to a minimum the risk of health care workers' coming into contact with potentially infectious droplets and particles.

And though he had commenced treatment with mild doses of corticosteroids, Zhong Nanshan had still seen two-thirds of his ninety patients deteriorate to critical status within four days of their admission. His ICU was overwhelmed by patients, forcing him to set up makeshift ICUs on the sixth floor. Correctly analyzing that among the characteristics that made this pneumonia atypical was heightened contagiousness, Zhong had moved quickly to isolate patients, setting up wards specifically devoted to atypical pneumonia. One unforeseen and unfortunate consequence of this, however, was that a few patients admitted with more conventional respiratory ailments but who had the high fever and cloudy chest X-rays consistent with this new atypical pneumonia had been placed in these new wards. If they weren't infected before they were admitted, then they were certainly infected a day or two later. It is unknown how many patients became infected in this way, but it is quite normal in the earliest days of an outbreak for new cases to be clustered among patients and medical staff.

For Guan Yi, this was his first exposure to the symptoms of this latest outbreak. He had seen avian influenza patients before, in 1997, but then only eighteen people had been infected. Here was a ward with eight rooms and four patients to each room, all of them in critical condition. Most striking was the depleted look in their eyes. They didn't move their eyes to follow him as he passed through their rooms, the way most patients did. Nor were they asleep. Instead, they gazed upward, the act of breathing requiring all their strength and concentration. One SARS patient would later tell me that he had to be careful not to blink too much, because the act of closing and opening his eyes used to tire him out.

"Listless" was too vibrant a word to describe these cases, Guan Yi thought. They seemed to be the living dead. A few of the nurses were already unwilling to work with them, requiring the hospital to institute a lottery system to schedule shifts. Throughout the province, there had already been more than eighty health care workers infected; in this hospital, there were fourteen.

Guan Yi gathered his sample swabs and his vials filled with fluid medium and accompanied a nurse to the first of the patients. She elevated the back of the man's bed, removed his oxygen mask, and ordered him to open his mouth. Tentatively reaching a wooden stick with a Q-tip-like bulb at the end into the oral cavity, the nurse dabbed at the patient's tongue twice, removed the swab, and handed it to Guan Yi. He broke the top of the stick and dropped the swab into a vial of medium. He didn't say anything, but he could already see that this wasn't going to work. In order for it to be an effective screening, he would need mucus and phlegm from farther down the patient's throat, as well as some nasal aspiration. The nurse was too frightened to gather anything but the faintest of saliva samples.

At the next station, he told the nurse to step aside as he held a female patient up, ordered her to open her mouth, and then began slapping her back gently to encourage her to expectorate mucus. This time the swab was pressed so far down the patient's throat that when Guan removed it, it was coated with a satisfactory blob of mucus that glistened under the white lights. He managed to take twelve swabs, each of which he sealed in a vial of suspension medium so that they appeared almost like miniature moth cocoons preserved in formaldehyde.

Back up on the sixth floor, after he had taken off his mask and gown and scrubbed again, he met with Zhong Nanshan, once more stating his belief that this was probably influenza. Zhong Nanshan believed it was possible, but if it was influenza, it was curious that this was mainly a lower respiratory tract infection. Most influenzas had more pronounced upper respiratory tract presentation.

"We'll find it," Guan Yi promised. "Whatever it is."

He sealed the samples with clear tape and then carefully rinsed the outside of the vials with an alcohol solution. He wanted to make sure that the cases appeared as pristine as possible. After laying the samples on a bed of dry ice inside a small Styrofoam container, he placed the container in his canvas duffel and set a towel and newspaper over the biohazardous materials. As he rode in a taxi back through Guangzhou to the train station and then climbed the broken escalator and walked through immigration and then customs, he found himself craving a cigarette. His samples under his arm, he slipped into a side stairwell and lit a Mild Seven and inhaled deeply, grateful, after what he had seen at the hospital, for his healthy lungs.

Finally, a female announcer called the 6:30 P.M. KTV back to Hong Kong, and Guan Yi mounted the stairs and got in line. Up ahead, a police officer and a customs official were conducting a second passport inspection and random baggage checks. Guan Yi feigned indifference as the officer asked a woman in a cheap-looking dress to step out of line. The customs official walked with her a few meters from the train and asked her to open her suitcase. Guan Yi can occasionally affect a look of extreme impatience—many Chinese of a certain rank and stature can suddenly take on this air of stern annoyance. In Hong Kong, Guan Yi seldom deployed this combination of squinty gaze and smirking lips wrapped around a cigarette. Here, he gave the cop the full heat of his bored annoyance as he tossed his cigarette aside and reached into his jacket pocket to show his passport one more time.

The customs officer was back now, having found nothing of interest in the woman's suitcase. It was a beauty treatment, Guan decided to say if the vials were discovered. But now that didn't seem believable. Contact lens cases? He thought about the swabs and the clear fluid; the blobs of mucus on the swabs would make that explanation unlikely. Ah, how about saying this was a fertility treatment? That might work.

Guan Yi indifferently looked over the customs official. It was an indolent glance, one intended to reveal that while his eyes may have fallen on the customs officer in his gray uniform, he was not bothering to think about what he was looking at. The weight of the black duffel on his shoulder could be, Guan Yi knew, literally the weight of the world. The virus in those sample cases could alert the planet of a biological disaster, or it could get Guan thrown into a labor camp for the rest of his life.

The policeman returned Guan Yi's passport, and he boarded the train and found his upstairs seat. He was about to place the satchel in the overhead baggage rack but then thought better. He might be carrying enough virus to infect the whole planet. He kept it on his lap as the train clanged south into the night.

CHAPTER 21

- **February 11, 2003**
- **World Health Organization Headquarters, Geneva, Switzerland**
- **400 Infected, 41 Dead**

THE UNITED NATIONS WORLD HEALTH ORGANIZATION HEADQUARTERS in Switzerland comprises a three building compound at the end of a horseshoe drive on a grassy rise overlooking Geneva. The glass-and-steel architecture of the sleek, low-slung buildings still conveys the postwar optimism that birthed such international organizations as this one. In its somber, earnest modernity, the WHO campus also comes across as a factory of sorts, an assembly line of well-intentioned ideas and sound public service. This is a building whose exterior and interior are intentionally sparse, as if to imply that every nook and cranny is being efficiently utilized so as to make the world a better place. The organization was established in 1948 with the stated objective of "attainment by all peoples of the highest possible level of health"—even if in the intervening years such noble goals have been increasingly compromised by politics, economics, and the reality of a planet where many governments and heads of state choose preservation of power over protection of public health. During that winter of 2003, the United Nations as a whole had been deemed virtually irrelevant by an American president who had decided to forgo UN collaboration as he prepared to invade Iraq. Republican politicians had been vilifying the UN for decades and had now finally succeeded in portraying it as an unwanted obstacle to Pax Americana. The United States regularly withheld its UN funding; a large segment of the American people were increasingly unsure of why the UN was even necessary in a world where American power seemed a better vehicle for keeping order. Yet with the WHO, the UN had a sterling example of the sort of advancements that only a multilateral organization could achieve: start with the eradication of smallpox and the containment of several Ebola outbreaks and continue through the organization's ongoing campaigns against polio, influenza, and AIDS, and you are only beginning to understand its significance and global scope. Ironically, as American troops were preparing to storm across Iraq's borders in an act that disregarded the United Nations General Assembly and made that agency seem as irrelevant as the League of Nations, the UN World Health Organization was about to prove itself indispensable in the fight to save the world from a global epidemic.

Comprising just fifteen hundred full-time employees, the organization borrows its muscle, for the most part, from top-flight hospitals, laboratories, and universities around the world. To be seconded to the WHO is a great honor for many scientists and public health officials. "Not too many people ever turn us down," David Heymann, executive director, Communicable Diseases, once told me. Spending a few days at the World Health Organization is a bit like how it would feel for a baseball fan to spend a week in spring training with a major league baseball team: you are constantly reminded of your own inadequacies, not because everyone else is showing off but because being very smart in a systematic and scientific way is what these people do. These are all the brightest boys and girls drawn from 192 nations and kept here because of their brains and abilities.

That means that not only must an executive director such as David Heymann be able to keep up with the smartest kids in the class, but he must also direct individualistic intellects on specific tasks and problems, and deal adroitly and sensitively with governments that are occasionally reluctant to collaborate. Born in Pennsylvania, Heymann holds a bachelor of arts from Pennsylvania State University, a master's from Wake Forest, and a diploma in tropical medicine from the London School of Hygiene and Tropical Medicine. He spent thirteen years on assignment from the U.S. Centers for Disease Control and Prevention in sub-Saharan Africa as a medical epidemiologist, working, in particular, on the Ebola outbreaks in the Congo in 1976 and 1977, and in 1995 he directed the international response to the Ebola outbreak in Kikwit, Zaire.

It was Heymann, along with Guénaël Rodier, WHO director of Communicable Disease Surveillance and Response, who in the late nineties conceived and constructed the framework for the WHO's Global Outbreak Alert and Response Network (GOARN). The network formalized a procedural response to reports of disease outbreaks: "systemic collection of reports or rumors of new outbreaks, outbreak verification, communication of confirmed facts to selected partners and the world at large, and containment, including coordination of international assistance when required," wrote Heymann. The WHO, for a few decades, had been moving away from eradicating infectious diseases as its primary mission. After all, during the fifties, sixties, and seventies, some of the great scourges of humanity were beaten back. Mass campaigns using penicillin had been launched against sexually transmitted diseases. Water drainage and insecticides had eliminated malaria from some areas in which that disease had been endemic. As living standards improved and better drugs became more widely available, even tuberculosis was banished from many countries. As the WHO celebrated its twentieth anniversary, in 1968, it seemed that many of the world's deadliest diseases were either in retreat or, even more amazing, might have been eradicated—as smallpox would be just a few years later. As a result, the WHO began to turn its focus to other areas, especially to societal ailments such as malnourishment and behavioral diseases such as heart disease and lung cancer.

By the late nineties, however, it was clear that the optimism of the early antibiotic era had been premature, as a host of diseases were emerging and, in some cases, reemerging. "Infectious diseases have resurged to an extent that again merits the highest level of international concern," wrote Heymann in a paper for the medical journal *The Lancet*. Population growth, rural-urban migration, and the inadequacy of sanitation and other infrastructural shortcomings contributed to a multitude of disease outbreaks. Plague, cholera, tuberculosis, typhoid, and yellow fever were again causing epidemics in countries where such diseases hadn't been reported since the 1970s. Even more pressing, as far as the WHO team was concerned, human encroachment into new ecological zones had increased the frequency with which new diseases, previously confined to animals, were jumping the species barrier to infect people—Lassa fever in Africa, hantavirus in North America, and Chagas disease in Latin America are all examples of recent zoonotic infections. Between 1996 and 2001, there were no fewer than forty outbreaks of emerging or reemerging diseases. Perhaps the most ominous, Heymann and his colleagues pointed out, was the possibility of interspecies transmission of influenza virus. Hong Kong in 1997 had been a close call. The virus had jumped and killed but had somehow not achieved efficient human-to-human transmission. The world may have been lucky on that one, the WHO team knew, but would they get that lucky again?

The reality was that GOARN consisted of three full-time medical professionals working out of two offices on the first floor of the WHO's annex building. While first-rate epidemiologists and public health officials were periodically seconded to the team, its vaunted intelligence "network" was the really careful monitoring of 110 existing networks, such as ProMed. To a great extent, GOARN relied on media reports and informal contacts with public health officials and doctors around the world to stay abreast of disease outbreaks. In most cases, the WHO would find out about an outbreak only after it had already run its course. Again, in those cases, the world had been lucky.

That winter and spring of 2003, the WHO's network was facing its first real-time test against an emerging disease in one of the most crowded and politically complex countries on earth. For David Heymann, Klaus Stöhr, and Mike Ryan, what was going on in China was about as frightening as epidemiology gets. Heymann and Ryan had watched Ebola slash through jungle villages and crash rural hospitals. Those had been remote areas, where access to roads, much less international airports, had been limited. For that microbe to piggyback a ride out of the jungle required several days' walking and then a long ride into a city, and even from there, African cities were not yet as well connected to global transport routes as was, say, southern China. And with Ebola's rapid onset of symptoms, the likelihood that a patient could get that far was diminished. He or she would probably bleed out before leaving the jungle. In those instances where Ebola or Ebola-like hemorrhagic fevers escaped sub-Saharan Africa, as with the Marburg virus in Germany, it had hopped along with infected monkeys. To this day, the host species for Ebola virus remains a mystery.

But Heymann had learned enough in working to curtail those outbreaks to have an inkling of what to expect the next time a virus jumped the species barrier and achieved human-to-human transmission. The last time a fatal virus emerged from an animal reservoir and achieved widespread human-to-human transmission, HIV had caused a global AIDS epidemic. Since then, no virus had emerged that achieved efficient human transmission. But it had been at least twenty-five years since AIDS, about thirty since Ebola. Humanity was enjoying quite a run of luck. "I know firsthand what an emerging virus outbreak looks like," Heymann has said. "It's not a pleasant experience. But we've learned that if you do everything you can—throw everything you can at it; get cooperation from the media, the government, from international agencies; you educate the people quickly; get infection control and screening in place and on the ground—then you have a chance to stop one of these things before it becomes a pandemic. But you have to act fast."

But hemorrhagic fevers such as Ebola, as terrifying and rapid as they are, are actually easier to interrupt than respiratory viruses. While new African diseases were frightening in their ferocity and high mortality rate, a novel virus emanating from southern China, such as a mutated influenza, would be more lethal because of its potential for becoming airborne. The average influenza patient infects four to eight other human beings. If a newly mutated influenza has the same mortality rate as the avian influenza of 1997, then that would mean that two or three of those eight would die. China, the WHO Communicable Diseases team knew, was potentially a hotter zone than Africa.

Heymann listened to Stöhr and read the ProMed reports and reflected that this sounded alarmingly like Ebola. A disease with a high mortality rate burning through hospitals and infecting health care professionals—only, this one seemed to be moving faster than Ebola. Also, even more troubling, the Chinese government did not seem especially eager to let in a WHO team. The government's press conference, intended to allay international concerns, actually reinforced for Heymann the necessity of getting a team to Guangdong as soon as possible. "This really looked like this was H5N1 that was becoming different, that was behaving differently, than it had behaved in Hong Kong [in 1997]," says Heymann. "We had to be prepared for the worst. If you look at the Asian flu outbreak of 1957; that spread within a very short period of time." In 1957, an influenza outbreak that started in southern China in March reached every corner of the globe by June.

That evening, Henk Bckedam sent a fax from the Beijing WHO offices to Director-General Liu Peilong of the Ministry of Health, in which he stated that "since 10 February this office has been receiving reports of an outbreak of respiratory illness in Guangzhou." The fax requested more data than the February 11 press conference had provided, specifically the number of new cases and deaths per week since the disease had appeared, the location of cases, when exactly the outbreak had started, some comparison with the usual rate of respiratory illnesses in previous years, clinical symptoms, age of each death and range of cases broken down by age group, ongoing surveillance activities, and information on ongoing efforts "to isolate the causative pathogen. One of the main issues of concern to the international community is the lack of identification of the cause of the cases. It is important to urgently identify the pathogen causing this illness. WHO is ready to provide required support for China on this outbreak, including providing rapid laboratory diagnostic support through WHO collaborating centers to urgently identify the cause of the disease. . . . WHO is ready to work with the MOH on this."

Heymann had seen a copy of this fax and had been teleconferencing with the Beijing WHO office and the Manila regional headquarters daily. Though there was some skepticism regarding the Guangdong press conference, there was still a faint hope that the Chinese government was telling the truth. Perhaps whatever it was had come and gone, said one WHO official hopefully. That, Heymann added, was not very satisfying from a scientific point of view, but "we didn't have data to refute what the Chinese government was saying." Epidemics burned through remote Chinese provinces every year, the WHO Beijing team knew, and it was impossible to get the Chinese government to allow a WHO team to check up on each of them.

China, in not being eager to invite a WHO team in to investigate, was acting just as any of a number of sovereign states would have when confronting what it viewed as a domestic health problem. How likely would the United States be to invite a WHO team, led by Chinese scientists, to investigate a plague outbreak in California? "You don't get anywhere pushing the Chinese government," said Heymann. "I mean, even though they were very slow, you still get more information from them than you do from some other countries."

The WHO team in Geneva, in particular Klaus Stöhr and his Global Influenza Surveillance Network, had already increased its vigilance for a novel influenza virus after the earliest reports from Guangdong. "With the exception of AIDS," Stöhr would write later, "most new diseases that emerged within the past two decades or that established endemicity in new geographical areas have features that limit their capacity to pose a major threat to international public health." What he was now hearing from China, on the other hand, sounded like a major threat. Klaus and David Heymann decided to issue a pandemic alert, based on the possibility that what was going on in southern China was influenza. The primary purpose of the alert was to heighten the surveillance for a possible new strain of influenza. The four collaborating WHO influenza centers, in Tokyo, Atlanta, Sydney, and London, then began preparing vaccines and reagents. The mobilization of the influenza lab network was largely a symbolic gesture, and Heymann knew it. "There were only two hundred and fifty thousand doses of influenza vaccine in the world," Heymann said. "And none of those would probably be very effective against an H5."

CHAPTER 22

■ **February 12, 2003**

■ **University of Hong Kong Department of Microbiology, Pok Fu Lam, Hong Kong, China**

■ **405 Infected, 42 Dead**

HONG KONG'S HEALTH DEPARTMENT, ON FEBRUARY 11, HAD ENHANCED its own surveillance for all pneumonia cases, in particular screening those patients who had recently traveled to Guangdong. Dr. Wilina Lim, chief virologist for Hong Kong's Public Health Laboratory Center, was receiving about fifteen hundred samples a month as part of the regular influenza surveillance network. Yet so far that winter, nothing extraordinary had appeared.

Meanwhile, across town, Malik Peiris and Guan Yi had centrifuged and tested the samples smuggled back from Guangdong. Guan was doing cultures in chicken eggs, taking the material from the swabs, spinning that down, and then inoculating the eggs. Peiris had introduced the material to canine cells in the hope of observing some kind of cytopathic effect that would be consistent with a virus replicating and destroying host cells. The respiratory secretions had come from infected patients, both Peiris and Guan Yi knew, at the epicenter of whatever this outbreak was, yet the cell lines—cultured cells—they were introducing remained defiantly alive. Peiris then began raising the possibility that this might not be flu. Surely, if it was, they would be picking up some significant viral replication. Yet there was nothing.

Guan Yi, also unable to culture any influenza from the samples he had been testing, still insisted that this was most likely influenza.

Back at St. Jude Hospital in Tennessee, Rob Webster, the mentor to both men, was dismissive when told by each of them that they had been unable to identify an agent. "You mean to tell me you can't find what it is?" he responded over the phone. "Get back in there and find the damn culprit."

Peiris and Guan could agree on only one thing: there was no culprit, not in these samples. They were about to give up on the screening when Peiris received a call from Wilina Lim. "We have a positive for A-type H5N1 in a father and son who were admitted to Princess Margaret Hospital on February eleventh." In other words: bird flu.

The family, Lim explained, had a travel history in Fujian, close to Guangdong. They had supposedly contracted the disease there, come back to Hong Kong, and only then sought medical attention. Lim was in the process of doing the cell cultures herself. But it appeared very likely that she had confirmed human H5N1 cases coming out of China.

"The Guangdong outbreak has to be H5[N1] as well," Guan Yi told K. Y. Yuen and Malik Peiris. Peiris agreed that it was likely but not definite. He was much less willing to jump to conclusions, preferring to keep his conjectures private. Until he saw a virus growing in cell lines and had positive PCR (polymerase chain reaction, a test that detects the presence of viral antibodies) results, he would not state what he thought was the agent. He telephoned Geneva and Manila, to alert WHO officials that Wilina Lim had found H5N1. The information had the immediate effect of stoking everyone's worst fears: China was in the throes of an avian influenza outbreak. Guan Yi packed a bag and headed back on another microbial espionage run into China, determined to collect more samples.

CHAPTER 23

- **February 17, 2003**
- **World Health Organization Headquarters, Dongzheminwai, Beijing, China**
- **461 Infected, 47 Dead**

MINISTRY OF HEALTH DIRECTOR LIU PEILONG'S RESPONSE TO WHO China chief Henk Bekedam's fax of February 12 had been highly unsatisfactory. The WHO, in its next communication, on February 17, was particularly concerned that the Ministry of Health's response hadn't included "some of the basic laboratory and epidemiological information requested." While the MOH had ruled out anthrax, pulmonary plague, leptospirosis, and hemorrhagic fever, it had significantly not mentioned influenza. Now, Bekedam replied, "International experts are concerned about the possibility of a new strain of influenza, which could first be detected as atypical pneumonia." He went on to write, "there is still no information on where we are in the course of the outbreak. If the peak of the epidemic curve in several cities in Guangdong was in early February, it is extremely unusual that there would be no new cases so soon after control measures were started." The WHO was skeptical about the information they were receiving. Trust between the WHO and the China MOH would become an early victim of this epidemic.

Bekedam faxed Liu Peilong two days later and asked for permission to send a team to Guangdong and Fujian. The ministry did not respond, instead insisting that the epidemic had passed. Liu Peilong made a counterproposal, suggesting that the MOH bring a team from Guangdong to Beijing to brief the WHO on the status of the outbreak. "When you are dealing with China, you have to think about progress," says WHO official David Heymann, "and we saw this as progress. We were going to get information, which is all you can ask for." The WHO's Hitoshi Oshitani started packing for Beijing. (He would be joined there by Keiji Fukuda, the chief of the epidemiology section for the influenza branch of the Centers for Disease Control and Prevention in Atlanta. Fukuda had worked closely with the Chinese CDC in the past and already knew several of the Chinese scientists and influenza experts.) Before departing his three-bedroom Manila house, Oshitani took a swim in his rectangular swimming pool. He often liked to swim a few laps to clear his head. He and his wife, a teacher at an international school in Manila, were both athletic, having met while rock climbing in Sendai. They were a surprisingly adventurous couple, having lived in Lusaka, Zambia, where Oshitani had worked for the Japan International Cooperation Agency (similar to USAID), setting up the country's first virology lab. There, Oshitani did his first investigations into cholera and influenza outbreaks and realized that he preferred field-work to lab work. In his posting at the WHO's Manila office, he found that he spent almost half his time traveling, investigating outbreaks throughout Southeast Asia and the Pacific region. Now, as his wife was preparing to move back to Tokyo with the children—they wanted their sons to go to high school in Japan—Oshitani was packing to leave again, this time to Beijing on a mission that, he told his wife, felt a little bit different.

As he swam his stately backstroke, he considered what he knew: European consulates in Guangdong were shutting down and had stopped issuing visas over a week ago. There was avian influenza in Hong Kong with origins in southern China, with 305 cases admitted by the Chinese government and health care workers among the patients. Klaus Stöhr and David Heymann of the WHO in Geneva had already been discussing the possibility of a Global Alert—but based on what scientific evidence? "You can't make a decision without scientific evidence," Heymann had insisted. The Global Outbreak Alert and Response Network constructed by Heymann and the WHO was now being put to its first real test. Scientists, virologists, laboratory workers, and doctors all over the world were turning their attention to southern China. Ironically, while the rest of the world focused on the Middle East—the television in Oshitani's den broadcast nothing but news on the impending Iraq war—the story that was now occupying the bulk of the mindshare for the best and brightest medical minds in the world was whatever this new bug was in southern China.

Oshitani had hardly given a thought to the coming war. As he stepped out of the pool and was reaching for a towel, he was thinking that as long as whatever this thing is stays contained in China, then we have a chance.

BOOK 3
Where Does It Come From?

CHAPTER 24

- **February 21, 2003**
- **Hotel Metropole, Kowloon, Hong Kong, China**
- **479 Infected, 49 Dead**

SIXTY-FOUR-YEAR-OLD LIU JIANLUN—THE GUANGZHOU NEPHROLOGIST who had helped treat the Poison King and numerous other atypical pneumonia patients in Zhongshan Number Three Hospital and Guangzhou Number Three Hospital among others, was boarding a bus near Guangzhou Central Railway Station. After putting in repeated double shifts working to stabilize several patients on the ICUs of several Guangzhou hospitals, he was taking a few days off from the feverish pace. He had a kindly, round face that was surprisingly congruent with the circular frames of his thick glasses. His slight mustache gave the impression of a somewhat lax approach to shaving—even his colleagues weren't sure if his mustache was intentional or not. (Mustaches remain quite rare among Chinese.) He wore his silver-streaked hair in a comb-over, and sometimes, when he spoke, he had the habit of ducking his head slightly, as if to bend through some too-small doorway. He had never reached the pinnacle of his profession, a position as department head of a major hospital, yet he had managed to sustain a prosperous career through one of the most tumultuous periods in Chinese history; the fact that colleagues twenty years his junior still called him to consult on particularly acute cases was tribute to his earnestness and diligence in keeping up with modern medical practices. In his gray trousers, white button-down shirt, and brown cardigan, with his red suitcase mounted on wheels, he seemed a typical Chinese tourist heading off to visit Hong Kong. The ostensible purpose of the visit was a friend's daughter's wedding, but he also saw the trip as a chance for a little shopping, mah-jongg, and karaoke.

He had been feverish for a few days, with some mild body aches and a dry cough, causing him to regret not having taken Tamiflu prophylactically. Still, if he had the flu, it was a mild case. He did not consider the possibility that he had the same disease as those he had been treating; he had taken the precaution of wearing N-95 particle masks with the most severe cases. At any rate, the vague worries had quickly vanished when he woke up the morning of his scheduled departure for Hong Kong feeling slightly better. A few days' R & R in Hong Kong might be more rejuvenative than bed rest, he reflected. He took a seat on the bus for the border crossing at Lo Wu. At the border, he changed the SIM card in his phone from his Chinese service to a Hong Kong account and called his sister, who lived in Hong Kong. He was again feeling a bit feverish but swallowed two Panadol and tried to nap on the bus. After telling his sister that he had decided to make the journey despite feeling under the weather, he hung up. Then his dry, hacking cough recommenced. The cough was surprisingly unproductive, the doctor noticed, and he was a little distressed at feeling severely out of breath after each fit of coughing. Still, bus riders hacking and coughing in China are exceedingly common. The other Chinese on that bus would later recall nothing exceptional about this passenger. Mainland tourists frequently book package deals that include transportation, accommodation, and a few meals. Until recently, mainlanders hoping to visit Hong Kong were required to travel on such packages, which were a boon to middle-tier hotels like Kowloon's Metropole.

The fourteen-story, ferro-cement hotel is just a few kilometers from the luxurious Peninsula and Regent, which overlook Victoria Harbor, yet in décor and service it is light-years apart. The lobby is a barren space surrounding a narrow fountain next to a stairwell. Four elevators, each with a capacity of thirteen persons and measuring approximately forty square feet, serve each floor. There is a karaoke bar downstairs that usually fills up around midnight with mainland tourists. Each room features a queen-size bed with a green-padded headboard, beige smudge-patterned wallpaper, and a television with eight channels of pornography. With its somewhat out-of-the-way location, in Hong Kong's Jordan district, and discreet street entrance, the hotel serves not just mainland tourists but also Hong Kong Islanders looking for a quiet spot to rendezvous with their mainland mistresses.

Room 911, where Dr. Liu set his bag down on a gray armchair, looked out over a Caltex station. Opposite the window was a reproduction of a watercolor of a rural river with a thatched-roofed village in the distance. Liu still felt fatigued and considered taking a nap. Instead, he slipped on his jacket and went to the elevator bank, where several other guests were already waiting. In the elevator, he was seized by a coughing fit so severe that the other passengers would later recall being alarmed. Still, the doctor was determined to head out for a shopping trip. Amazingly, despite again feeling feverish and, now, nauseated, he made his way to Nathan Road and eventually bought a new windbreaker from one of the discount shops there. Upon his return to the hotel, he felt so weak he had to sit down in the lobby and then lean against the wall in the elevator for support. On the ninth floor, he would vomit in the hallway before reaching his room. There, he would pass out in his bed, unable to call his sister to tell her he was canceling dinner.

Disgusted guests noticed the odor first, then the viscous, yellow stain on the brown carpet. Hotel staff moved quickly to clean up the mess, but by then, at least sixteen other guests had made their way past the pool of vomit on the floor. Three of them were stewardesses from Singapore sharing rooms 938 and 915. Another was a seventy-eight-year-old woman from Toronto staying in room 904. Another was a forty-nine-year-old garment manufacturer who would check out of room 910, across the hall from Dr. Liu, and then fly to Hanoi on February 23. One other guest was an American visiting from the United States, who would entertain a local friend, twenty-six-year-old K. K. Li, in room 906.

The doctor would wake up early the next day feeling extremely feverish, with severe body aches and shortness of breath. He was still in his street clothes from the day before when he rode down in the elevator, again sharing it with several other passengers, and ventured out onto the street. Though it is just two hundred meters from the Metropole Hotel to Kwong Wah Hospital, walking that distance in Dr. Liu's condition indicates his tremendous strength but also what must have been acute terror. He had seen firsthand what this disease did to patients, especially the elderly. And by now he knew that what he was suffering from was no common flu. The shortness of breath, the aches, the fever, the agitation and impatience. Though self-diagnosis is often risky, and frequently mistaken, in this case Dr. Liu was quite sure of his assessment. When he arrived at Kwong Wah Hospital's Accident and Emergency Department, the examining physician, Dr. Wu Chun-wah, based on the patient's dangerously low oxygen saturation levels, ordered him admitted to the intensive care unit. When Dr. Gordon Watt, chief of pathology, heard that this patient was a physician visiting from the mainland, he instructed the nursing staff to prepare an isolation room. Dr. Watt had heard about the outbreak on the mainland; he ordered all medical staff to wear N-95 masks, gloves, and gowns when treating Dr. Liu.

Dr. Liu's symptoms were fever and shortness of breath. The diagnosis, by Dr. Wu, was severe pneumonia. Upon his admission to the ICU, just before he was intubated, the patient told the treating physicians and nurses that he was a physician from Guangzhou and that he had treated numerous patients who had this so-called atypical pneumonia. "It's terrible," he gasped. "It's very contagious." Dr. Wu had seen the Department of Health advisory regarding increased surveillance for atypical pneumonia. He reiterated the order that all medical staff take protective measures. The hospital faxed a report on the case to the Hospital Authority of Hong Kong. The Hospital Authority would notify the Department of Health of the case two days later.

Dr. Watt called his colleague and former medical school classmate K. Y. Yuen of the University of Hong Kong, the doctor who had grown up in the poor western district of Hong Kong, to consult on this curious case. "By the time I saw [Dr. Liu], he was in terrible shape," says Yuen. "He had been intubated for five days. His lungs were already whited out. He had been on antibiotics, antivirals—nothing was working. We couldn't save him."

Yuen also had a strange feeling about this patient. At the time, he didn't mention it to anybody, but he now says, "It was a strange case, one that gave me a strange feeling. I see dozens of patients a week. Why did this one seem strange?"

Perhaps it was the difference between theory and practice. They had heard of the disease across the border and had read on ProMed and in the newspapers of the panic and fear in Guangzhou and of the health care workers stricken, but this was the first time they were this close to the disease. As he conferred with Dr. Watt and offered suggestions regarding treatment and infection control, Yuen saw the look in Dr. Watt's eyes. It was frustration. There was nothing they could do to save this man. His lungs were failing. His blood oxygen saturation level was dropping below 90 percent. He was going to die, K. Y. Yuen knew, and for a physician in a modern hospital used to being able to stop most life-threatening diseases, especially among healthy adults such as this one, he had an unfamiliar feeling of helplessness. Even more confusing, Yuen learned, a team of doctors from the mainland had visited Dr. Liu in the hospital just the day before, ostensibly to wish him well. They had met with the hospital's physicians and conferred with them on Dr. Liu's condition and had asked for and received a copy of the test results of specimens drawn from him. Upon receiving the results, they quickly prepared to leave for the mainland. When asked for information about the atypical pneumonia situation in Guangdong, the three visiting doctors demurred, saying that the situation was under control. The visit of a team of mainland physicians had spawned a host of rumors that mainland hospitals were seeking information from Kwong Wah Hospital about the still-unknown agent. Yuen admitted to Watt that he too would like to know the agent, but, so far, he had had no luck in isolating the microbe. He collected nasal-pharyngeal samples—snot on a swab, basically—along with lung tissue biopsies and hurried back to the University of Hong Kong.

CHAPTER 25

GUAN YI WAS PACKING ANOTHER ROUND OF THIRTY-ONE SAMPLES from China, these even more carefully gathered and prepared than the previous group.

As he said good-bye to Zhong Nanshan, he was given some stunning news: another Chinese scientist had discovered the agent. A senior microbiologist at the National CDC's Institute of Virology, seventy-one-year-old Hong Tao, had announced that the causative agent was *Chlamydia*. *Chlamydia* is most notorious for a particular strain that causes the common sexually transmitted disease. However, two variations of the bacterium can also cause opportunistic respiratory infections, particularly among patients with compromised immune systems. In this instance, Hong Tao announced that he had detected *Chlamydia* through his electron microscope in the lung tissue of two deceased patients. Hong would later write in the *National Medical Journal of China* about finding "*Chlamydia*-like particles" in seven patients. However, he admitted he had failed to isolate the microbe, thereby making genetic characterization impossible. More disconcerting, antibodies to known *Chlamydia* species did not react with these samples, causing Hong to hypothesize that he had discovered a new strain of *Chlamydia*. (Scientists at the U.S. Centers for Disease Control and Prevention would notice later, in electron microscope scans, that parts of cells ripped apart by the SARS virus had dangling tendrils similar in appearance to some strains of *Chlamydia*.) Zhong Nanshan and Guan Yi dismissed the finding almost immediately. They had been battling this disease for weeks and knew that antibiotics were totally ineffective, making a bacterial agent very unlikely. Other scientists in Guangdong also disagreed with Hong Tao's findings. Du Lin, the director of the Guangzhou CDC, says that her lab also couldn't find sufficient evidence of a bacterial infection. "It was a big surprise," she says of hearing about Hong Tao's supposed discovery.

Yet in hierarchical China, where seniority and education are venerated, few scientists were publicly willing to challenge the legendary Hong Tao. He had spent more than thirty years studying *Chlamydia* and had also won acclaim for identifying a rotavirus that causes diarrhea. Very quickly, the Chinese scientific establishment rallied around Hong. "As far as they were concerned, that was it," said Henk Bekedam, head of the WHO's Beijing office. "When we met with them, they were very strong about pushing *Chlamydia* as the agent." If securing samples had been difficult before—just as in Hong Kong, scientists in China had been clamoring for samples of this mystery agent—with Hong Tao's announcement, only his lab would receive prized atypical pneumonia samples, while other qualified labs, such as the Beijing Genomics Institute, were cut off from fresh samples. BGI's deputy director, Wang Jian, flew to Guangdong several times in February seeking to secure samples, each time returning empty-handed. "It was pretty hopeless," his colleague, Yu Jun, told *Science*. Chinese researchers who disagreed with Dr. Hong were being shut out of the hunt.

Guan Yi had come back from a second smuggling run, and the samples in his duffel were virtually the only samples in the world not allocated to Hong Tao or to Chinese military laboratories.

Back in Hong Kong, Guan Yi again went to work seeking to culture influenza, while Malik Peiris sought to isolate from those samples some sign of the agent that was causing the disease. Peiris and his assistants chopped up tissue samples with diamond knives, then homogenized the samples inside a laminar airflow cabinet, which whisked away any potentially harmful aerosolized tissue. In the event of a stoppage in the air hose or of a power failure that resulted in the circulation of contaminated air, an alarm would sound. Peiris, whose meticulous, cautious nature is exemplified in his steady hands and methodical approach to working with samples, seldom worked in such a way as to set off any alarms. The work required precision finger work, deft manipulation of matter so small it verged on being invisible to the naked eye—and often actually was, requiring a microscope to complete the task. In its intricacy, this sort of hands-on virology is most similar to building and painting miniature nautical models and then inserting them into tiny bottles. The homogenized samples were thus suspended in saline before being introduced into a range of cell culture lines, each under a hood of sterilizing ultraviolet light that killed any bacteria that sneaked into the experiment. (When seeking to grow a virus, sometimes a scientist inadvertently introduces bacteria into the cell line, thus contaminating the procedure.)

Various cell lines are used when testing for different agents. The levels of specialization are astonishing, with more than one hundred thousand cell lines available for medical and scientific testing from organizations such as ATCC. The cell lines, for example, range from the fetus of a rhesus monkey to the embryonic lung of a cow to a tumor from the liver of a rat. Certain cell lines are better suited than others for growing certain diseases. Hepatitis A virus, for example, will grow better in certain monkey cells, while polio is grown better with embryonic human cells. The cell containers arrive frozen in dry ice and suspended in a culture medium. They are then transferred to sample trays or to individual, sideways-facing vials with upturned necks.

In his attempt to grow this mystery virus, Malik Peiris used the MDCK, LLC-Mk2, RDE, Hep-2, and MRC-5 cell lines—or, in simpler terms, dog kidney cells, monkey kidney cells, human muscle cells, human larynx tumor cells, and human embryo lung cells. This is a group of cell lines geared toward growing influenza or other respiratory viruses such as adenovirus, swine fever, and enterovirus. "We're basically going fishing," Peiris told his lab partner, K. H.

Chan. "Let's hope we catch some virus." Once the samples were prepared and sealed, they were placed in a slow-rotating incubator and then checked every few hours for signs of cellular damage.

For thousands of years, scientists postulated the theory of infectious agents so tiny they could not be observed—couldn't the "bad humors" and "evil spirits" of medieval times have been nothing more than a collective sense of dread at microorganisms so small they seemed to drift in the ether, as scientists would later discover airborne viruses actually did? By the late nineteenth century, the diverse microbial world of bacteria, protozoa, and fungi was established. But as early as 1840, the German anatomist Jacob Henle suggested the existence of microbes, so tiny they could not be observed even by the light microscope, that caused specific illnesses and diseases. Three men—Louis Pasteur, who disproved the theory of spontaneous generation of organisms with his swan-necked flasks that kept yeasts pure; Robert Koch, who demonstrated that specific bacteria caused anthrax and tuberculosis; and Joseph Lister, who developed a technique for obtaining purer cultures of organisms—were instrumental in pushing the study of microorganisms into the modern age.

Most curious, however, was the simple filter to purify water perfected by Charles Chamberland, a collaborator of Louis Pasteur, which allowed for the development of modern virology. Bacteriologists in the 1880s developed a so-called operational concept of sterility, which meant that if liquid was pushed through a filter that was known to hold back bacteria, then that liquid was considered sterile. The filters took the form of hollow, unglazed porcelain cylinders with microscopic pores. The cylinders would be inserted into a tube connected to a hand pump; the entire contraption looked something like a miniature gasoline pump from the 1920s. As these unglazed porcelain filters became even more refined, it was established that any infectious agent that filtered through them and reinfected another organism was deemed to be what at the turn of the century were called "filterable agents." (The term "virus" was then a generic term for any infectious agent.)

Numerous types of filterable agents were soon found. Friedrich

Loeffler and Paul Frosch discovered the first filterable animal virus, that which caused foot-and-mouth disease; and in Cuba, Walter Reed discovered the first human virus when he demonstrated that yellow fever was transmitted by mosquitoes. Still, in both these cases, the only way to study a virus was to grow it in a whole animal, which made experimentation and manipulation very difficult. Virology had to move from animal husbandry to the laboratory.

The idea to grow a virus in individual cells was first realized in 1948 by Elberg Sanford in bacteriophages—viruses that specifically infect bacteria. Soon, scientists at Johns Hopkins University were able to passage—that is, grow—human cells for the first time outside the body, developing a line of cancerous cells from a cervical carcinoma. (Cancer cells are easier to passage than almost any other cells, wild and unruly cell proliferation being the nature of the disease.) By 1950, J. F. Enders showed that *Poliovirus* could replicate in a cell line of embryonic tissue—in a lab, not in a sick child. The underpinnings of modern virology were now in place.

The tests that Malik Peiris and Guan Yi were conducting in their labs on the fourth floor of the University of Hong Kong Medical Faculty Building were not dissimilar from those that Enders had done over fifty years prior, and they employed techniques refined from those Pasteur and Chamberland had used 150 years earlier.

In Geneva, the WHO's influenza chief, Klaus Stöhr, was in daily contact with the University of Hong Kong scientists and increasingly frustrated by the inability of anyone to identify the causative agent they were facing. "At this point, we had to pull in everyone who could contribute," says Stöhr. "We knew we had sick patients showing up in Hong Kong hospitals. We knew there was a serious outbreak in southern China. If you can't establish a pathogen, you can't make a diagnosis. You can't treat anyone. I was having daily conversations with Margaret Chan regarding samples. She had tried her utmost to collaborate with the Chinese. We spoke with Malik and Guan Yi to see if there were any professional relationships and inroads into southern China to help find out what was going on." As for the *Chlamydia* claim? "We were skeptical of that from the very beginning. Now we were beginning to worry that this was an entirely new agent."

At the same time that the WHO's Hitoshi Oshitani was going over his notes on a flight to Beijing, seventy-eight-year-old Toronto resident Sung Min-tsu and her husband were walking down the corridor of the Metropole Hotel when they passed an elderly Chinese man who was coughing severely and struggling to stay upright. He coughed several times in their general direction, and when he looked up, they were struck by his angry, bloodshot eyes. The elderly Chinese man, nephrologist Liu Jianlun, was on his way to Kwong Wah Hospital, where he would admit himself for severe pneumonia. On his way, a few hundred thousand opportunistic virus particles would somehow infect Sung Min-tsu and her husband.

They returned to Toronto on the twenty-third.

The virus had jumped the Pacific.

CHAPTER 26

MEDICAL REPORT

Family Name: Yang Chin

Given Name: Danny

HRN: AAE 3767

DOB: 09/03/54

Nationality: American

Date of admission: 26/2/03

Reason for Hospitalization:

Patient Traveling from Hong Kong

Admitted at the emergency department for fever, dry cough, muscular pain, no appetite for 3 days before admission

On admission: temperature 39.9 (Celsius), BP 130/80 mmHg, mild sore throat, some crepitants in the base of left lung, no dyspnea, SPO2 94%

The patient doesn't have any special medical past history, has not been traveling [sic] in China (Guanzhou [sic] or Fujian province) and has not been to health facilities in Hong Kong.

Initial complementary investigation:

White blood cells 6.6K/ul, thrombopenia 72 K/ul,

CRP slightly elevated 31mg/1

Dengue fever IgM negative, HIV negative

Urine labstix: no leucocytes [sic], no nitrites

Chest Xray: diffuse interstitial bronchitis
Serology: Chlamidia [*sic*] negative, legionella not available
Initial diagnosis:

Severe infection by Influenza B virus

Danny Yang Chin, an American garment manufacturer, was visiting his suppliers in Hanoi after a stopover in Hong Kong and a two-night stay on the ninth floor of the Metropole Hotel. He had landed in Hanoi at 8:00 A.M. on Monday, February 24, arriving at Gilwood Limited's offices at 36 Pham Huoythong, Hanoi, for a meeting with Tina Thuy, an assistant merchandising manager. A tall man, standing at about five ten, Yang Chin wore a yellow button-down shirt, jeans, and sneakers. Because of his firsthand knowledge of Asian manufacturing, he had a reputation in the garment business for being able to source high-quality, cheap denim. Vietnam had increasingly become a low-cost competitor to southern China, especially when it came to a kind of denim treatment that made jeans look naturally aged without the acid-wash tinge that had gone out of fashion. Gilwood Limited had invested money for technicians and washing machines in the Hung Yen Garment Company in Hung Yen province, about thirty miles from Hanoi, and Yang Chin was now eager to meet with local officials and his subcontractors. He and Thuy boarded a Toyota Z-Ace van and drove out to the factory, where the managers had prepared a lunch of spring rolls, mint leaves, minced pork, and steamed chicken. Yang Chin told Thuy after lunch that he could breathe more easily in Hung Yen than he had been able to in Hong Kong or Hanoi. She found that comment curious but ascribed it to the notoriously dense fumes from the motorcycle traffic in Hanoi.

When they returned to Hanoi at 5:00 P.M., Thuy drove them by motorbike to the Ho Market to buy sample trousers and shirts to show to customers back home. "This air is so polluted," Yang Chin said at one point as he stopped walking to catch his breath. (He spoke both English and Chinese, though mostly English, when he was in Vietnam.) Thuy was surprised that this apparently fit forty-nine-year-old was gasping during a routine walk through the market. She drove them both back to Gilwood's offices, where Yang Chin said he would be turning in early.

Gilwood's offices in Hanoi consist of a four-story building with two bedroom suites on the top floor for visitors. From one of them, Yang Chin called his wife in the United States that evening and told her he that felt fine, if a little bit tired from all his traveling.

The next morning Tina Thuy met him in the office, from which they went to a lunch of Korean barbecue at the Daewoo Hotel with the company's quality-control manager. Soon after eating, Yang Chin again complained of feeling out of breath and tired and said he wanted to go buy some paracetamol for his headache. "I just need a nap," he explained.

At 6:00 P.M., when Thuy went upstairs to the Gilwood bedroom suite to check on Yang Chin, he was still sound asleep. He stirred only to say he still felt weak. The following morning, he was scheduled to return to Hung Yen but was unable to get out of bed. One of the company's interpreters saw an ad in the newspaper for a local doctor and called him. This physician was Nguyen Van Thi, a fifty-eight-year-old former field medic of the Vietnamese People's Army. He came quickly and listened to Yang Chin's lungs, took his temperature, and examined his throat. He prescribed a certain kind of *pho*, a soup with noodles, and told him he needed to drink water and eat plenty of fruit. He also sold him a few packets of Odezon, a medicine to ward off dehydration.

Yang Chin was too exhausted to make the drive back out to Hung Yen. It seemed to Thuy that he was lapsing in and out of consciousness. When he was awake, he seemed terribly irritable, as if exasperated at the Gilwood staff. Thuy felt increasingly responsible for the failing health of her boss and decided, at around 6:00 P.M., to seek more professional—or at least more Western—medical care. Thuy and another employee helped Yang Chin up and took him to the Hanoi French Hospital.

Dr. Vien Thue, a forty-three-year-old internal medicine specialist who had studied and trained at Creu Brest Centre University

Hospital, in Brest, France, discussed the new patient's symptoms with Olivier Cattin, the chief physician on duty. Yang Chin had high fever, malaise, headache, and, according to Dr. Thue, "crackles in the lung." A chest X-ray showed slight shadows. Still, the patient, though clearly ill, did not seem to be in acute distress. But because of the high fever and a low platelet count, the two doctors decided to test for dengue fever. Also, in light of the patient's recent travel history to China, both doctors concurred that a test for influenza antibodies would be in order. Cattin raised the possibility that this could be H5N1 and suggested that they contact someone at the World Health Organization. He got in touch with the WHO's Hanoi representative, Pascale Brudon. She immediately called Carlo Urbani, an Italian parasitologist who had been working for the WHO since 1993.

Parasitologists are, even by the standards of eccentric, driven epidemiologists, a breed apart. Their passions are protozoa, such as those that cause malaria, toxoplasmosis, or trichinosis; or helminths, such as roundworms, tapeworms, and flukes. And while pursuit and elimination of these creatures can seem mundane compared with the cutting-edge science of microbiology—very often the solution is prosaically prophylactic, such as mosquito netting in the case of malaria—the calling is one that probably affects more lives than the work being done in any high-tech lab in Geneva or Baltimore. Nearly 1.7 billion of the earth's inhabitants still suffer from one form or another of debilitating parasite. Schistosomes, to cite just one case, are parasitic flukes that begin life inside snails, grow on the surface of freshwater plants, and then complete their life cycle in human beings, who excrete the schistosomes' eggs via fecal waste back into the water supply, where they are again taken up by snails, and so on. Those who bathe, work, or play in that water will absorb the next generation of schistosome from underwater plants through their skin and into their bloodstream, where, depending on the exact species, the schistosomes will make their way to the liver, spleen, urinary tract, kidney, or colon. The long-term effects of schistosomiasis are a wide range of human illnesses, from mild fatigue to liver and kidney cancer. Humans have been battling this parasite for millennia. The only way to avoid infection is to shun contaminated water, but for hundreds of millions of rural and, recently in China, urban poor, that is much easier said than done.

Carlo Urbani was obsessed with helminths, the family of parasites of which the schistosome fluke is a member. Like many parasitologists, when he found a rare sample of fluke, or worm, in its larval stage, he cleaned it and kept it suspended in a small sample vial filled with a formaldehyde solution. The sometimes brightly colored flukes or worms are preciously traded and examined among parasitologists like jewels—there are so many thousands of species that new parasites are constantly being discovered. Never mind that it might require rooting through a sick child's feces to find the prized sample; for a parasitologist, this is like a music maven's searching through a stack of rare LPs and 78s at a garage sale.

When Urbani met fellow parasitologist Han-jong Rim in a remote village in northern Laos while working to control a schisto outbreak, Han-jong informed him that he had discovered a new soil-borne parasitic worm by running the feces of infected villagers through a sieve until he'd isolated a few samples, which he then suspended in alcohol; these he presented to Urbani like a gift of precious stones. Urbani, one of the few human beings who could appreciate the magnitude of the offering, dropped to one knee and held the sample up to the campfire light. The two scientists spent their first evening comparing worm samples and exchanging tales about where and how each sample had been obtained. Urbani's tireless fieldwork and publishing of articles on helminth had helped to bring the flukes back into the spotlight as a global public health concern. Yet if Urbani was an enthusiast in his fetishistic interest in worms, he was also exuberant about almost every aspect of his life. "Carlo was a sensuous Italian," said Eva Christophel, a fellow WHO parasitologist based in Manila. "He always loved to eat well, and when he was dieting, you could really see he was suffering." Urbani not only enjoyed his food and wine but was also renowned for bringing a guitar along on his worming expeditions. It was that nature that made him approachable, even in villages that had rarely seen a Westerner before, let alone a 220-pound, bearded, round-faced Italian in rubber boots singing choral music as he dug through the village refuse pit.

The morning that Pascale Brudon called him from the Hanoi WHO office, Urbani was in his colonnaded mansion with a view of Hanoi's Westlake district. His second-floor master bedroom looked out over rows of kumquat bushes, peach trees, and, in the distance, the Argentine embassy. In the next room, his two sons and one daughter were still sleeping. Outside, he could hear the neighbors' tan-and-white beagle yapping as a motorbike passed by on the street. Numerous foreigners lived in this lush district, and with the morning breezes coming off the lake carrying the earthy scent of peach blossoms, this neighborhood, at its shimmery morning best, reminded Carlo and his wife, Giuliana, of Italy.

When the phone rang, Urbani was anticipating an upcoming conference in Bangkok that would be bringing together two of the world's feuding factions of parasitologists.

"Carlo," said Brudon, speaking in French, "there appears to be a case of avian flu here at the hospital."

Urbani had heard from the Italian consulate about the atypical pneumonia outbreak in Guangdong. And he knew of the avian flu cases in Hong Kong. He asked Brudon if the patient had a travel history to China or Hong Kong.

Brudon told him to call Olivier Cattin or, better yet, to go to the hospital and check on the status of the patient.

The drive to the hospital took Urbani south of the lake, past the Communist Party headquarters, the Foreign Ministry, and then the austere, stark, almost art deco Ho Chi Minh Mausoleum, where he turned onto Dien Bien Phu Avenue and headed east, past the huge statue of Lenin and then the dozens of open-front stores selling Buddha statues and glazed urns. The streets were choked with motorbikes, as usual. There was no rush hour in Hanoi but rather rush *hours*, and these lasted from 8:00 A.M. to 10:00 P.M., seven days a week. Still, despite the proliferation of traffic and commerce, Hanoi retained the kind of picturesque charm that has been banished from almost every other Southeast Asian capital. With its colonial buildings, three-story shop houses, and hundreds of lakes, the city still seemed scaled for human beings, albeit human beings mounted on Hondas or Yamahas. The rest of Asia's supercities, from Kuala Lumpur to Jakarta to Beijing, at least in their downtown districts, seemed to have been constructed with a race of giants in mind.

Urbani strolled through the front gate of the hospital, past the traffic island and the saggy willow tree, and into the entrance, under a sign in English reading ACCIDENTS AND EMERGENCIES. On the second floor, Urbani looked in at the patient, whose fever still hadn't broken, despite treatment with proparacetamol. Danny Yang Chin's blood oxygen saturation level remained at 94 percent, which meant that he was not in respiratory distress. Still, Urbani confided to Pascale Brudon, this was a curious case. Perhaps they should alert the regional headquarters that they had a possible case of avian flu.

He called WHO's Hitoshi Oshitani on his mobile phone that evening and explained the situation. Oshitani, already in Beijing, briefed Urbani about what he knew of the Guangdong outbreak. The question that occurred to both doctors: Was this the same disease?

Urbani said he would try to establish the patient's exact travel history. Also, he would be taking respiratory samples and serum and send them to WHO collaborating labs.

The next day, March 1, Urbani told the Hanoi French Hospital that they should isolate the patient and wear protective equipment. "You should also post a guard," he told the hospital president. Despite Urbani's rapid response, he was too late. The virus had already begun its burn through Hanoi.

CHAPTER 27

■ **February 27, 2003**

■ **Ministry of Health, Beijing, China**

■ **508 Infected, 51 Dead**

WHO'S HITOSHI OSHITANI AND THE CDC'S KEIJI FUKUDA FOUND themselves frustrated by a Chinese government that didn't seem to understand the potential for global calamity if a new pandemic was hatching in Guangdong. Fukuda, the forty-seven-year-old chief of the epidemiology section for the U.S. CDC's influenza branch, had first heard about the possibility of a Guangdong H5N1 outbreak on February 9, from a Chinese citizen in Guangdong who sent an anonymous e-mail to the U.S. Centers for Disease Control and Prevention in Atlanta. The U.S. CDC hears hundreds of such rumors a week and cannot possibly verify all of them. Influenza in China, was not, at any rate, cause for alarm. However, Fukuda did note the use of the scientific term "H5N1," in English, which indicated that whoever was sending the message had some passing familiarity with influenza nomenclature. The next day, an American businessman traveling in Guangzhou reported a local flu outbreak and asked if the U.S. CDC had any information regarding travel safety in that area. Fukuda had also been apprised by Malik Peiris, Guan Yi, and the Hong Kong Department of Health of the circumstances surrounding the Penfold Park migratory bird outbreak. H5N1, Fukuda reflected, seemed to be in the air, literally.

When the Guangdong government held its February 11 press conference regarding atypical pneumonia, Fukuda had taken notice. Nancy Cox, the director of the CDC's influenza branch, was on a plane for Geneva for the annual Northern Hemisphere vaccine meeting, at which influenza specialists from around the world gather to decide which strains of influenza virus should be included in that year's vaccine. (The Southern Hemisphere holds a separate meeting that coincides with its own seasons.) There, Cox was briefed by Klaus Stöhr and the WHO about the Guangdong outbreak, and when she talked to Fukuda, she instructed him to have Sascha Klimov, the chief of the CDC's Strain Surveillance section, get in touch with his counterparts at the National Influenza Center in Beijing to see if they knew anything more about what was going on in Guangdong. Fukuda was already confiding to Oshitani that this seemed like something that might require an international response in Beijing. Fukuda had just returned from giving a lecture series at Osaka City College, in Japan, and was not looking forward to another round of travel. Influenza investigations tended to be open-ended. This wouldn't be like mopping up an outbreak of gastrointestinal disease on a cruise ship.

But the H5N1 cases in Hong Kong triggered his departure. Klaus Stöhr, in Geneva, had told Nancy Cox about the Fujian cases on February 18. "That was it," Fukuda says. "We had human H5N1 in Hong Kong. This changed everything." He returned to his suburban Atlanta home, kissed his wife and two daughters good-bye, and was on a plane for Beijing on the twenty-second.

Fukuda and Oshitani were a perfect pair of diplomatic scientists. With their wire-frame glasses, short-cropped hair, and wide, eager eyes, both men sent all sorts of nonthreatening signals. They leaned forward when they listened. The regular cast of their facial features was a thoughtless grin. Their soft voices were soothing. Both were masters of the protocol required of scientists who worked for international agencies. This was China, after all, and they were representatives of global institutions that had far more business with China than just the matter at hand. However, first and foremost in their minds was the urgency of this potential pandemic. "We needed information," Fukuda said. "As epidemiologists, we wanted to see the pattern of the outbreak, how many cases there had been, who was getting the disease, and, most important, was it under control?"

They agreed on the need to be patient. "This was China," Fukuda said. "You don't just drop in to see the minister of health. But we really felt it was urgent that we explain why H5N1 is such a big deal."

That feeling of urgency, however, was not mutual. They arrived in Beijing on February 23. They would not meet with Ministry of Health officials until March 3.

CHAPTER 28

■ **February 27, 2003**

■ **Hanoi French Hospital, Hanoi, Vietnam**

■ **509 Infected, 51 Dead**

WORKING THE NIGHT SHIFT AT HANOI FRENCH HOSPITAL WHEN DANNY Yang Chin was admitted, slender nurse Luong Nguyen, forty-seven, assisted Dr. Cattin in his initial examination. She recalled that Yang Chin seemed impatient, somehow angry. She knew he was most likely disappointed to be sick and hospitalized, but he seemed to view the hospital staff as being the cause of his illness. When she asked him to turn over on his side so she could listen to his lungs through a stethoscope, he sighed, as if annoyed. Later, the staff would realize that Yang Chin was probably going through some mild hypoxia, which would have left him irritable. He coughed several times while Luong was in the room with him, a dry, unproductive hack.

Luong had the twenty-seventh off, and that evening she cooked noodles with snails and crab for her husband, Thu Vinh, and their ten-year-old daughter, Thu. On the twenty-eighth, she rode on the back of the family's motorcycle as her husband drove them to their home village of Tu Son, about twenty-five kilometers north of Hanoi. With Thu Vinh's position as a mechanic with the state railway and Luong's job at the hospital, they were able to save some money, and they had their eyes on a parcel of land near Thu Vinh's parents' home. On the way back to Hanoi, as she held on to her husband, Luong told him she felt very cold.

When they arrived in Hanoi, a call came in from Luong's younger sister. She was panicked because she couldn't find their mother. Thu

Vinh and Luong took off on their motorbike to search Luong's mother's apartment building. They found her at her neighbor's, watching a Taiwanese soap opera. When they brought her back to her apartment, Luong felt so tired she had to lie down on her mother's bed, and she asked her husband to go to a pharmacy and buy her some antibiotics, which are sold over the counter in Vietnam. She managed to hang on for the ride back home, but by then the fever and body aches were so severe that she sent her husband back out to buy some tranquilizers so she could rest and recover before her afternoon shift the next day.

However, in the morning, she felt even worse, complaining of severe headaches and chills. Of even more concern to her husband was what he saw as a "lack of spirit. She seemed so weak, and she is usually so active." She could stir herself only to rush to the bathroom to pass watery stools. When Thu Vinh called the hospital to tell them his wife would need at least one more day off, he was told that four other nurses from the Hanoi French Hospital were experiencing similar symptoms. "The fever felt like having red ants in my body," recalls Nguyen Thi Bich, forty-four, a fellow nurse at the hospital.

Luong had lost all appetite and could barely hold down a sip of orange juice. Thu Vinh, fearing that his wife would become dehydrated because of the fluid she was losing every time she rushed to the toilet, decided to bring her to the hospital. At almost the same moment, Dr. Thue called and ordered her admitted. At the hospital, as she waited for a chest X-ray, she told Thu Vinh, "I don't know why, but this fever feels different from anything I've ever had."

She was put into a room on the second floor, near Danny Yang Chin, in an area of the hospital that was becoming known as the "Red Zone." Yang Chin was coughing severely by now, sometimes for forty minutes at a stretch. Nurse Pham Thi Uyen, forty-three, would hold him down and pat his back, trying to help him breathe. Near the end of her shift, she too began to feel feverish and achy. She conscientiously finished her eight-hour shift before signing herself in as a patient.

Luong told her roommate, fellow nurse Thi Bich, that she had never felt so weak. Eventually, she was unable even to talk, and when her husband brought her flowers for International Woman's Day, she could not muster the energy for a smile. She whispered to a fellow nurse at one point, "I never knew that smiling made you tired."

Meanwhile, Danny Yang Chin had gone into acute respiratory distress, with, according to his medical report, "severe hypoxia, confusion and blood oxygen saturation level of 60%."

Carlo Urbani, the gregarious parasitologist who now spent most of his working day at the hospital, had never seen a hospital outbreak like this. As he spoke with the CDC's Hitoshi Oshitani and Keiji Fukuda in Beijing, he told them that he believed there were now eight health care workers infected with what appeared to be the same infection that in just seventy-eight hours had resulted in Danny Yang Chin's declining into critical condition and being put on a respirator. "If this becomes a community outbreak, then I am going to need a great deal of help."

Oshitani and Fukuda were killing time at the Beijing WHO offices, meeting with Henk Bekedam and Alan Schnur and eating take-out pizza from the Pizza Hut down the street. Urbani's distress added to their frustration at not getting any response from the Ministry of Health regarding the Guangdong outbreak. David Heymann, at WHO headquarters, had been brought into the conversation regarding the Hanoi outbreak. "When Vietnam started, we really kicked into action," Heymann recalls. "We still thought this was the flu pandemic we were all worried about."

Klaus Stöhr and Pat Drury, at WHO headquarters in Geneva, and Oshitani, working from Beijing, quickly drafted a confidential "Alert/ Request for Assistance," seeking an "epidemiologist who is available immediately to assist with the investigation and prevention" of "acute respiratory illness" in Hanoi, which they sent out through the Global Outbreak Alert and Response Network of affiliated public health institutions.

Aileen Plant, an epidemiologist from Curtin University, Perth, Australia, would answer the call, arriving in Vietnam just one week later.

Meanwhile, Danny Yang Chin was airlifted back to Hong Kong, where he would expire. By then, Nurse Luong was dead. Seven of her colleagues at Hanoi French Hospital also soon passed away.

CHAPTER 29

■ **March 3, 2003**
■ **Shanxi Province, China**
■ **556 Infected, 57 Dead**

THE UNWILLINGNESS OF THE CHINESE MINISTRY OF HEALTH TO COM-
municate what they knew was just one among many critical delays that
allowed the disease to spread from Guangdong to Hong Kong to Hanoi
and, now, to the rest of mainland China. Later, as I thought about the
virus and its emergence, each spike of infection that emanated from
Guangdong—to Hong Kong, Hanoi, Toronto, Singapore—would seem
inevitable, like a great spider spreading its appendages. During the
Internet boom, business writers had overused the expression "viral
marketing," which meant the organic growth and spread of an idea or
brand; it was the prelude to finding a "sticky application." The concept
was that awareness of a business or company would be passed around
like an infection, with the spreader almost unaware of his or her role in
the viral marketing process. Later, as we tracked the spread of this new
disease, which had yet even to be named, the idea that anyone could
market something as ingeniously as a virus replicates itself would seem
hopelessly naïve. I realized that the spread of a virus was indistinguish-
able from the march of humanity. Our every journey, meal, activity, and
bodily function offers a perch for some virus or other. The voyages of
man from Guangdong to Hong Kong and the rest of the world were
really the voyages of microbes, trillions upon trillions of them, hitching
a ride to infect other humans. As I tracked the journey of the virus into
the heart of China—the unfolding of one more spider's leg of disease—

the idea that humans could ever create anything as insidiously efficient as certain viruses began to seem utterly impossible.

Du Ping, a twenty-seven-year-old businesswoman, would be the virus's vehicle as it hitched a ride from Guangdong all the way to Beijing. Du flew from Taiyuan, the capital of Shanxi, into Shenzhen, the border city next to Hong Kong, where some of the first patients, including Fang Lin, had lived and worked, early in the Year of the Goat. If Guangdong was living through the Era of Wild Flavor, then Shanxi, in central China, seemed to be mired in the Era of Failed Five-Year Plans. The province, just a few hundred miles from Beijing, embodied the Chinese rust belt of abandoned factories, depleted coal mining pits, and collapsing public services. The villages surrounding Taiyuan boasted some of the highest concentrations of topsoil lead of any province in China. Taiyuan tried hard to convey the sort of glitzy prosperity one was accustomed to finding in Chinese regional centers now, but it failed by virtue of a bureaucracy that remained steadfastly committed to Communist inefficiency with capitalist-style corruption.

Du Ping, the daughter of a well-connected Chinese journalist, had actually heard reports of an atypical pneumonia spreading through Guangdong. Her mother had even called a few contacts in the region, who told her there was no cause for alarm. Du made two or three purchasing trips a year for her jewelry business, which she ran out of her home. Shenzhen, especially, with its proximity to the border and its international connections, was a clearinghouse for cut-rate jewelry, both real and counterfeit. Especially popular in Taiyuan were the elaborate jade carvings that seemed to be sold at every other shop in Shenzhen, especially the megamalls near the train station. These same carvings were available from any of a hundred wholesalers who operated in shantytowns like Buji, where Du Ping did most of her buying. The jade was most likely fake, Du probably knew, but it was so intricately carved and well rendered that it still fetched good prices back in Taiyuan. Four suitcases of jade carvings, such as those she had in the baggage compartment beneath the bus she was now riding from Shenzhen to Guangzhou, could fetch 120,000 *kwai*, or $7,000.

As the bus motored up the Guan-Shen Highway, Du began to feel feverish. She suspected it was the air-conditioning that had given her a cold. But the next morning, she had a severe headache and what she would later describe as "a great sense of unease." Her husband persuaded her to fly back to Taiyuan without shopping for mother-of-pearl pieces in Guangzhou, as she had planned.

Upon her return to Taiyuan, she made the rounds of local hospitals, visiting Shanxi 302 Hospital, People's Number One Hospital, Shanxi Number Four Hospital, and University Medical Hospital. The virus could not have found a better vehicle than this anxious patient seeking second, third, and fourth opinions. At each hospital, physicians had Du cough as they listened to her lungs, despite her warning that she might have been infected with the atypical pneumonia in Guangdong. Because the reports on the disease that were gathered and prepared in Guangdong and sent up the bureaucracy to the central government in Beijing had never been disseminated, doctors in Taiyuan were skeptical that the minor disease they had stopped hearing about a few weeks before could have caused this woman to fall ill. "They thought it was a cold and gave her normal antibiotics," Du's husband, Chu, would tell Phillip P. Pan of the *Washington Post*. "The health care workers were very responsible, but it's a pity. They didn't have information about SARS." The husband called a hospital in Guangzhou and was told that his wife's illness could be a chlamydial infection. However, Du had already tested negative for chlamydia.

Her parents and husband did what any Shanxi natives with sufficient means would do: they decided to seek better treatment in the capital, Beijing. By the time they left Shanxi, Du Ping had infected eight family members and five medical staff. Her family drove through the night in two Mitsubishi Pajeros, up the mountainous Highway 6 to Beijing and then down the Third Ring Road to Number 301 Hospital, a famous facility for senior party officials that was now open to paying customers.

Du Ping's husband had expected that physicians in Beijing would be better informed about how to treat this mysterious illness. But he and the rest of the family were surprised that even the doctors at the legendary 301 Hospital had no idea what ailed Du Ping. "The illness appeared in Guangdong last November," Chu would say later. "How could a national-level hospital not receive the necessary notice?"

Finally, three days after Du was admitted, a young doctor ordered her isolated in her own room on the respiratory ward. By then, both of Du's parents were feverish. Her mother was admitted as a patient herself, on March 4; her father would be admitted a day later. Both were transferred to Beijing's 302 Hospital (also known as the People's Liberation Army Hospital for Infectious Diseases). Du's father would descend into acute respiratory distress the evening he was admitted and would require emergency tracheal intubation. During that messy process, according to one physician, as many as ten health care workers were infected. Du's father would die later that day, probably the disease's first Beijing fatality.

The virus had reached Beijing, population twenty million.

CHAPTER 30

■ **March 6, 2003**
■ **National Center for Disease Control Headquarters,
Beijing, China**
■ **571 Infected, 60 Dead**

IN FEBRUARY OF THAT YEAR, THE RUMORS OF THE DISEASE IN
Guangdong were a minor concern for citizens in Beijing, who had
welcomed in the Year of the Goat with a decidedly more conservative
palate than Guangdong's, so enthusiasts of Wild Flavor were reduced
to just a few districts in the capital that catered to those seeking
Cantonese fare. Snake, pangolin, lizard, civet, and other delicacies
were available, but with their city boasting the widest array of
regional cuisines in China, Beijingers generally preferred to leave the
civet cats to their stereotypically hot-tempered cousins to the south.
There was one cultural attribute Beijing shared with Guangdong,
however, and that was the atmosphere of breakneck prosperity. After
winning the right to host the 2008 Olympic Games, the city had
redoubled its decade-long construction spree so that virtually every
hutong—traditional neighborhood of wooden houses built around a
courtyard—in the city seemed earmarked for destruction. Those near
the center of the city, close to the government agencies that comprised much of the area inside the Third Ring Road, had almost all
been eliminated and replaced by high-rise buildings, from which
sprouted satellite dishes and, from all but the most expensive, clothes-
lines with garments flapping in the stiff, almost arctic breeze.

As Du Ping's father was gasping for life, even as he expelled his
last breaths during the afternoon of March 6, just two miles away, at

the headquarters of the China CDC, Hitoshi Oshitani and Keiji Fukuda, along with several WHO officials, were finally sitting down with several members of the China CDC. No one in that room could imagine that the mysterious disease that was burning through southern China had already reached Beijing. The Chinese officials present included Liu Peilong, from the Ministry of Health; Li Liming, chief of the China CDC; Xi Xiaoching, also from the China CDC; and several officials from Guangdong, including Xu Ruiheng from the Guangdong CDC and Xiao Zhenglun from the Guangzhou Institute of Respiratory Diseases—Xiao had led the team that investigated the Heyuan outbreak. Most notably present was Hong Tao, the venerated microbiologist who had discovered what he thought was *Chlamydia* in sample tissue from Guangdong.

For Hitoshi Oshitani and Keiji Fukuda, who had waited for over two weeks to meet with counterpart scientists and physicians regarding this outbreak, the presentation they saw was disappointing. The expression in Chinese for this is *yuan fa yue*, or "a walk in the garden." In other words, these visiting experts were being shown a few sites but little of actual value or import. Several clinicians and CDC officials made a PowerPoint presentation regarding number of cases, symptoms, and the results of PCR tests and screens, which indicated that whatever the disease had been, it was no longer spreading. However, both Fukuda and Oshitani were perplexed that their counterparts had not contact-traced the cases or bothered to establish a chain of transmission. They seemed satisfied to note that the disease, whatever it was, appeared to be receding.

One PowerPoint slide explained that, as of February 27, there were 745 cases and only 26 fatalities. Yet, shocking to Oshitani and Fukuda was the fact that 212 of those infected were health care workers. The epi-curve put up on the screen—a graph representing the onset of new cases—showed a satisfying turn toward the flat line just a few days earlier, which seemed to make the case that this outbreak was a historical curiosity rather than an impending pandemic. The Chinese side stressed that such a chain of infection was consistent with what appeared to be an outbreak of chlamydial pneumonia.

Fukuda and Oshitani exchanged glances. The incomplete information the Chinese officials found so reassuring raised even more concern for the two epidemiologists. As the lights came up, Hong Tao, the microbiologist, stood and explained that he had found *Chlamydia* in two lung autopsies and believed the cause of the outbreak to be this bacterial agent.

For Fukuda and Oshitani, this did not look like a *Chlamydia*-type outbreak. Health care workers would not be succumbing at this rate to a bacterial infection, nor would treatment have proven quite so challenging It was also highly unlikely that such a disease could prove as fatal as this one was.

"Have you considered that this could be a viral agent?" Fukuda asked.

Hong Tao did not acknowledge the question after it was translated for him. He instead continued to explain that he had found especially high concentrations of *Chlamydia* in the kidney tissue of deceased patients. As far as the possibility of avian flu was concerned, the Chinese team had so far found just two influenza cases from 162 swab specimens, and PCR tests were also negative. Chest X-rays were also shown to the visiting scientists, and those did not seem consistent with H5N1.

This was convincing data. Despite the lack of follow-up in Fujian, the southern Chinese province from which the avian influenza cases in Hong Kong had emanated, the data indicated that perhaps the first fear of these epidemiologists was unfounded. Also, the high incidence of health care workers infected also indicated that this was probably not influenza. "If this had been influenza," Oshitani explained to me later, "then it should have been more randomly distributed. This disease was pinpointing health care settings."

That night, as the two men discussed the day's meetings over a dinner of Mongolian beef at the Yuyang Hotel, just a few blocks away from the WHO's Beijing headquarters, Oshitani took a call on his cell phone from Carlo Urbani. Urbani explained that Danny Yang Chin, the index patient in Hanoi, had been evacuated to Princess Margaret Hospital, two and half hours away in Hong Kong.

The patient, before he had left, had been in acute respiratory distress and had lapsed into a coma. Meanwhile, there were now thirty-one cases in Hanoi, twenty-five of which were health care workers who had had immediate contact with the index patient. There were already three fatalities and several more patients entering into acute respiratory distress.

"This hospital is now experiencing a major problem due to lack of staff," explained Urbani. "People are basically panicking."

Oshitani, Fukuda, and Urbani discussed the possibility that this was the same disease as that which had infected southern China. "Danny Chin traveled through Hong Kong," Urbani reminded them.

Oshitani and Fukuda told Urbani what they had heard at the China CDC. It certainly sounded consistent: high concentrations of health care workers falling into severe respiratory distress.

"Have they isolated an agent?" Urbani asked.

"*Chlamydia*," Oshitani explained.

There was silence on the line.

"I find that highly unlikely," Urbani said. He had been treating patients with amoxicillin and other antibiotics since the outbreak began and had seen no improvement in those cases. He had received a shipment of amantidine from Hong Kong and was now administering that—to little or no effect on the patients. (Amantadine is an antiviral that tends to be most effective against influenzas; its failure here meant that the agent was unlikely to be an influenza virus.)

"If this is the same disease," asked Urbani, "then what is it?"

The three men were silent.

"And where is it going?" asked Oshitani.

CHAPTER 31

- **March 10, 2003**
- **Prince of Wales Hospital, Hong Kong, China**
- **1,068 Infected, 118 Dead**

HONG KONG'S INCIPIENT SPRING HAZE WAS BEGINNING TO MUDDY the clear winter skies so that Joseph Sung, chief of his hospital's gastroenterology and hepatology division, could now barely make out the Hong Kong Island skyline as he drove his Lexus from his Ma Hong Sam apartment to Prince of Wales Hospital, in Shatin. He parked in the underground garage, and as he waited for the elevator, he reflected on his weekend visit to Seoul, where he had attended a medical conference on intestinal diseases. Seoul had been chilly, and he was grateful to be back in more temperate Hong Kong. Sung was a taciturn man, not prone to outbursts, and if his wife, Rebecca, a gynecologist, had any complaint about her stolid husband, it would be that his emotional intelligence was not on a par with his analytical skills.

Hong Kongers were just now paying attention to such matters, though they did so with an earnestness that undermined the more spontaneous and free-spirited expression such exploration was supposed to encourage. Books on matters of emotional IQ had lately been among the bestsellers at Hong Kong's few bookstores, but Sung preferred to read business manuals, like Rudolph Giuliani's book *Leadership*, along with recent papers on breakthroughs, or lacks thereof, in treating colitis and Crohn's disease. With his deep black hair—colleagues suspected that he might dye it—broad shoulders, and steady, determined walk, he came across as the sort of firm, slightly stubborn clinician an intern or junior medical staffer would willingly follow into microbial battle.

That morning, as he flipped through his messages—his old college classmate, the perpetually worried and anxious K. Y. Yuen, from the University of Hong Kong, had called—he noticed something amiss on his desk. His secretary had placed the duty roster there, next to his messages, but had apparently not bothered to complete it. He called her to check what was going on and was told the duty roster was accurate. There were a lot of sick doctors and nurses.

Eighteen of them, in fact.

Joseph Sung put down the telephone and picked up the book he was reading.

What would Giuliani do?

CHAPTER 32

- March 10, 2003
- World Health Organization China Headquarters, Dongzheminwai, Beijing, China
- 1,069 Infected, 110 Dead

KEIJI FUKUDA AND HITOSHI OSHITANI HAD RUN INTO THE BRICK WALL that too often characterizes the recalcitrant Chinese bureaucracy. "We left those meetings feeling there was a lot we wanted to know and a lot we didn't understand," said Fukuda. "More troubling, we left meetings with the sense the Ministry of Health wasn't really engaged with this as an issue." Concerned at their lack of progress in Beijing and the possibility that the outbreaks in Hanoi and Guangdong were related, Fukuda and Oshitani both decided to depart from Beijing. "If we can't visit Guangdong," Oshitani complained to WHO Beijing chief Bekedam, "there's no point staying in Beijing." Bekedam reluctantly agreed that Oshitani and Fukuda could be more useful in the field. He assured Oshitani that he would continue to seek an audience with China's vice-minister of health.

Within the WHO, there was a divergence of opinion on how much pressure should be brought to bear on the Chinese government—or, indeed, if applying pressure was the right tactic. Oshitani and Bekedam, with Heymann conferenced in, considered bringing Gro Brundtland, the director-general of the World Health Organization, to Beijing to meet directly with Minister of Health Zhang Wenkang. However, it was decided that such grandstanding could end up hurting their cause more than helping it.

The leading figure in promoting a more conciliatory approach was Alan Schnur, the WHO's second-in-command in Beijing and an eleven-year veteran of working with the Chinese government. The Boston-born, Brooklyn College–educated Schnur had great faith in the technical abilities of his Chinese counterparts and had come to the conclusion that, somehow, this outbreak had become elevated into a policy matter. "All we could do was explain to them that this was a global matter, an urgent matter," Schnur said in a later interview, "but we knew they couldn't make any decision."

Schnur had a remarkably long face, like Munch's *Scream* without the hands, and his patience was almost a physical attribute, as tangible and present in any conversation as Oshitani's sighs of agitation or Fukuda's crisply cut-off consonants. Schnur could wait for hours, days if necessary, for the right moment at which to make a suggestion or observation. He had learned from years of working with the Chinese government on tuberculosis and polio outbreaks that you can't bludgeon Chinese bureaucrats into submission. They are the civil servants who are most skilled in the world at warding off unpleasant tasks. At any rate, Schnur knew, the threat to a Chinese bureaucrat from above was of a different magnitude than the threat from an international agency like the WHO. Better to unleash a deadly virus upon the world, a Chinese bureaucrat might reasonably conclude, than become the scapegoat purged for letting the foreigners know about it. Later in the outbreak, reporters covering the World Health Organization press conferences would view Schnur almost as a collaborator with the Chinese government, never realizing the role he had played in building up a relationship that allowed for what little contact there was between the WHO and the Ministry of Health. Still, by March 10, Oshitani had become increasingly frustrated with Schnur's approach.

He was joined in his impatience by the WHO's David Heymann and Klaus Stöhr. As early as March 4, they considered placing a travel advisory upon China, an unprecedented proscription that could set off an international diplomatic row. In a meeting at the Ministry of Health on March 6, Bekedam and Oshitani for the first time actually mentioned to ministry officials the possibility of a travel advisory—such an edict would cause tremendous international embarrassment for the Chinese government and certainly cost the country billions of dollars in lost investment and tourism. "I don't think they understood the consequences at that stage," Oshitani would later tell me. "Some of the officials still thought they could cover up and the disease would be gone in a few weeks."

At least one Ministry of Health official whom I spoke with privately would insist that ministry officials didn't appear to be engaged because they had been told to view this outbreak as a closed matter. "This came from quite high up in our ministry. That's why in our meetings with the World Health Organization, we were emphatic that the outbreak was over but could not give them satisfactory scientific explanations of the etiology, epidemiology, or virology of the outbreak. We did not have that information, and once we were told that the outbreak was officially 'closed,' we could not secure cooperation at the provincial level." The official went on to say that it had become, in effect, a political rather than a medical issue.

When I asked him if a tactical decision had been made to stall and hope that the outbreak would burn itself out, he responded, "We retreated into thinking like that because we had other plans in place. It is not a satisfactory position to take from a public health standpoint. But consider that the information we had from Guangdong indicated that whatever it was, they had reached the narrow part of the epidemiological curve. So the world can't accuse us of malfeasance."

"But the virus was already in Beijing at that point," I pointed out.

"Nobody knew that," he said.

"And what about Hong Kong?" I asked him.

He shook his head. "We never considered Hong Kong. It has to do with old thinking, political thinking as opposed to epidemiological thinking. Once the virus crosses the internal border, it becomes a political matter."

Once it became clear that the WHO would not get down to Guangdong or Fujian from Beijing, it was decided that working through Hong Kong might be more fruitful. Fukuda had worked there during the avian influenza outbreak of 1997 and was well connected with local public health officials and the medical community. As for investigating the possibly related Hanoi outbreak, Oshitani, the holder of a diplomatic passport, could fly to Hanoi without having to secure a visa.

Fukuda boarded a Monday-morning Dragon Air flight and landed in Hong Kong on Monday afternoon, taking a taxi from Chek Lap Kok to the Empire Hotel in Kowloon, just a few hundred meters from the Metropole. When he called his old friend Margaret Chan, Hong Kong director of health, she told him he should plan on staying for a while.

"Keiji, I have two dozen infected health care workers at Prince of Wales Hospital," she explained in her patrician trombone of a voice. "Atypical pneumonia from China. It's here, Keiji, it's here."

THE FIRST RESPONSE TO AN EPIDEMIC IS USUALLY DENIAL. WHEN smallpox appeared in the Roman Empire in A.D. 189, the initial impulse of local prefects was to issue edicts attributing the recent mortalities to an unhappy Jupiter or a vengeful Mars, or to make public pronouncements suggesting that fellow Romans had fallen ill owing to natural calamities, common grippe, or even a poisoned barrel of wine. From the perspective of a head of state, a mayor, a governor, or any ruling body, infectious disease remains among the hardest issues to manage. You can't legislate a virus out of existence, nor can you order a stricken populace to be healthy. There is almost no calamity, save starvation or siege, that can so quickly reduce a city to panic and despair. Yet very often the populace is as guilty of denial as are those ruling it.

In the face of impending doom, there will always be that segment of society that persists in wishing away the problem or pretending it doesn't exist. That was the case in London during the Plague of 1665, in Philadelphia during the great influenza outbreak of 1918, and even in San Francisco's gay community during the earliest days of AIDS, when some homosexuals argued that AIDS was not a new disease but rather a bogeyman conceived by puritanical doctors and conservative politicians seeking to put an end to the promiscuity of the seventies.

Why should China's Mandarins have behaved any differently? When confronted with an infectious disease, they initially downplayed the danger and assumed a tacit policy of wishing the microbe back into whatever species from which it had jumped. What did they really have to go on, a few hundred cases in a nation of over a billion? "Human lives just aren't as valuable in China," says Guan Yi. "They don't see these matters the same way a politician might in the West." Indeed, with infectious-disease outbreaks a far more common occurrence in China than in, say, the United States, it is on one level understandable how China's minister of health, Zhang Wenkang, could have initially downplayed the threat posed by a respiratory infection thousands of miles from the capital. And if it hadn't jumped international borders, then the outbreak would indeed have remained a minor medical curiosity, a matter that respiratory specialists in Guangdong would reminisce about when they finally retired.

At the beginning, this was a very small outbreak. "They had never dreamed it possible that our little town should be chosen out for the scene of such grotesque happenings," wrote Albert Camus in *The Plague*. "A pestilence isn't a thing made to man's measure; therefore we tell ourselves that pestilence is a mere bogey of the mind, a bad dream that will pass away." What China's public health Mandarins failed to recognize was that all disease outbreaks start small. In fact, they tend to start with just one case or a coeval cluster of cases. In denying the severity of the threat, and even the existence of the disease itself in Beijing, China's governing officials were acting in character, both national and political. For that season was even more unsettling than usual, as a political transition was taking place in Beijing on March 15, with Jiang Zemin handing over the title of party secretary to Hu Jintao at the National People's Congress. In a country that has never had a peaceful succession since the Communists took over in 1949, such an event required that a docile media present a united, peaceful, tranquil, and *healthy* China. "Why didn't the government act faster?" asked Zhong Nanshan. "Look at what was going on in Beijing on March fifteenth."

Even if there hadn't been a new leader being appointed, I wonder if such crisis management was even in the arsenal of a Communist Party devoted primarily to its own preservation. How does admitting that there is a new killer disease in the provinces contribute to the state's paramount goal of stability? As long as the state controls the means and modes of the message, then why tell the story of a killer disease? Because an infectious disease, especially one that has achieved human-to-human transmission, will not, as a rule, just go away.

There were those in the Chinese medical hierarchy, such as Zhong Nanshan, who were arguing for greater disclosure and insisting that information be more widely broadcast. As the world would soon discover, if everything that was known inside China had been shared with the world just a few weeks earlier, hundreds of lives would have been saved.

Denial kills, as surely as any microbe.

CHAPTER 33

■ **March 10, 2003**
■ **Prince of Wales Hospital, Shatin, Hong Kong, China**
■ **1,070 Infected, 120 Dead**

DR. HENRY CHAN, THIRTY-FOUR, WAS LEADING A TEAM OF MEDICAL students through Hong Kong's Prince of Wales Hospital on Friday, evaluating them as they made their rounds through the wards—four bays, each housing ten patients. At Ward 8A, they stopped in front of bed 11, where a patient was breathing through a nebulizer, a device that looks almost like a giant kazoo and which aerosolizes a dilating solution that encourages the patient to cough up mucus. As a by-product of facilitating this exchange, the nebulizer disseminated aerosolized droplets as the patient exhaled. For Chan, who himself had graduated from the Chinese University of Hong Kong, the medical school for which Prince of Wales served as a teaching hospital, making the rounds with medical students had become a routine banality.

As they stood at the foot of bed 11, Chan listened as his students discussed this case of community-acquired pneumonia. The patient, J. A. Kwok, a twenty-six-year-old airport freight handler, had first sought treatment on February 28, in the hospital's emergency room. He was diagnosed as having an upper respiratory tract infection and sent home. He returned on March 4, complaining of the same symptoms along with a high fever, and was admitted to the hospital. After being treated with antibiotics, he showed some slow signs of improvement. On March 6, treatment with the nebulizer had begun. As Chan listened to his students' review of what sounded like a common case of pneumonia, he stood in the nebulizer's stream of aerosolized droplet particles.

J. A. Kwok, it would turn out, had visited the ninth floor of Hong Kong's Metropole Hotel for a mah-jongg game the night Liu Jianlun was visiting from Guangzhou. Kwok had probably taken the virus aboard there. By the time he reached Ward 8A, he was shedding virus. The nebulizer would prove to be the most efficient virus spreader imaginable, short of suctioning mucus from Kwok's lungs and injecting it directly into someone else's.

Chan wrapped up his rounds and headed home. He had a strenuous workout at his gym on Saturday and then played badminton and a game of pickup basketball with some friends that evening. On Sunday morning, he attributed his muscle pains to pushing himself a little harder than perhaps he should have. By Monday morning, however, the muscle aches had increased, and he felt them "on an almost orthopedic level." In addition, he noticed he was running a high fever. He would later say he "felt cold from the inside." When he called in sick the next day, he was told to report immediately to Prince of Wales Hospital.

He was one of at least thirty-six other health care workers at Prince of Wales who had come down with the same symptoms.

Before driving himself there, Henry Chan wanted to make sure his affairs were in order. That included writing his will.

DR. DONALD JAMES LYON AND THE PRINCE OF WALES HOSPITAL Infection Control Team had already established that the doctors and nurses suffering from flulike symptoms had all worked on or passed through Ward 8A. Lyon, a consultant in microbiology, ordered those nurses and clinicians who had to work on 8A to take precautions, including the wearing of gowns and masks at all times.

In a meeting convened in a wood-paneled conference room next to Joseph Sung's office and attended by the Prince of Wales Hospital's senior staff, Lyon explained that he could not deduce which patient had caused the outbreak. The illness was already widespread, and the presentation, so far at least, was hard to define. However, since whatever it was was clearly contagious, the hospital had to take special infection-control measures.

Sung had never heard of an infectious disease cutting through a hospital so quickly. In the antibiotic era, diseases did not just strike down health care workers en masse. There was a science-fiction quality to this, Sung reflected, like *Invasion of the Body Snatchers.* He ordered Ward 8A closed to admission, discharge, and visiting, starting that afternoon. Masks and gowns had to be worn at all times, and health care workers who had worked on 8A, and patients discharged from 8A in the last few days, had to be traced.

Considering how contagious this disease seemed to be, Sung had to make some plans to deal with the fifty-plus health care workers and patients who would return to the hospital for examinations and, possibly, treatment. They could not let them turn up in the standard emergency rooms; such a surge would flood the emergency-response capability.

He ordered that a conference room on the ground floor be converted into a screening station. Computers and monitors were installed, as were blood-work machines and improvised dividers separating the various inspection stations. A back stairwell to the radiology department was commandeered so that the returning doctors and nurses could get chest X-rays without sharing elevators with other co-workers or patients. That afternoon, twenty-three health care workers, including Henry Chan, were admitted immediately after their fever checks.

More upsetting, however, was the reopening of Ward 8A. At another meeting the evening before, chaired by Dr. Philip Li, the issue of quarantining was raised, and there was a discussion about the risk of patients signing themselves out against medical advice if they were denied visitors. Legally, the hospital had no way to stop them. But the discharge of infected patients would risk spreading the infection throughout the community. As a compromise, a restricted visiting plan was put in place. Remarkably, there would be civilians walking in and out of Ward 8A even as the medical staff was trying to ascertain the cause of the infection and who was the index case.

By the evening of March 11, there were thirty-six infected health care workers. The day after that, there would be fifty. Prince of Wales Hospital was crashing.

CHAPTER 34

- **March 11, 2003**
- **World Health Organization Headquarters, Geneva, Switzerland**
- **1,091 Infected, 123 Dead**

IN HOSPITALS AROUND ASIA AND NORTH AMERICA, THE PATTERN THAT had played out in Guangdong was repeating itself: index case arrives and seeds outbreak among health care workers that practically paralyzes the hospital. At Tan Tok Seng, in Singapore, the experience was identical to what happened at Sun Yat-sen Number Three Hospital in Guangzhou, Hanoi French Hospital in Vietnam, and Prince of Wales Hospital in Hong Kong. And now, word was coming in to the Global Outbreak Alert and Response team in the WHO war room in Geneva that an agent with remarkably similar etiology was burning through Scarborough Grace Hospital in Toronto. Whatever this virus was, it was not influenza, concluded the team seated around that veneer table on the first floor of the WHO annex building. Also, there were now enough samples collected from outside of China and analyzed by first-rate labs in Japan and the United States, and none of those had turned up any novel influenza viruses.

This was something new: a previously unknown virus that appeared to have achieved widespread human-to-human transmission. In fact, if the numbers sent by the WHO office in Hanoi were correct, then the attack rate of this disease—i.e., the number of health care workers who had come into contact with the index patient and then themselves become infected—was 56 percent. Even more troubling, the percentage of cases that progressed to acute respiratory distress syndrome was "a minimum of 69%." The report gave a case fatality rate of 3 percent but warned that, since "several of the health care workers are critically ill, . . . this will likely increase greatly over the next week[s]."

The reports from China, Hong Kong, Singapore, and Toronto were not much better. "This really looked like an Ebola outbreak," the WHO's Heymann would say later, referring to the epidemiology of the outbreak rather than the agent. "Health care workers always get it first."

"The good news was that it wasn't flu," says the WHO's Klaus Stöhr. "The bad news, then . . . well, what was it?"

The group knew there was very little cumulative experience in the world, especially at the governmental level, in responding to outbreaks of unknown origin. The U.S. CDC had done an exemplary job responding to and identifying an outbreak of hantavirus in the American Southwest in 1993, but that virus had never achieved human-to-human transmission or spread globally. Compounding these problems was the fact that the Chinese government was still denying that there was a link between what had happened in southern China and the outbreaks elsewhere. "Most other governments would have invited other international scientists in," Stöhr believes. "China effectively excluded the international scientific community."

Stöhr, along with several of his colleagues on the Global Outbreak Alert and Response team, decided that they could not wait for one local laboratory to get lucky and find the viral agent. It had taken two years to find the virus that causes AIDS. If it took that long to isolate this agent, then the world could already be living through a new dark age as hundreds of millions were stricken by a novel virus for which they had no natural defenses. Even more alarming, a quick back-of-the-envelope calculation of the number of cases out there put the total known infected at close to three hundred. It would take AIDS nearly six years to infect as many patients in the West; this new virus had gotten to that figure, as far they knew, in about six weeks. How could the usually competitive process of isolating a new agent—the sort of discovery that can make a scientist world famous and open up

avenues for all the grants and studies he or she might ever desire—
be somehow overcome so that they could fast-forward to the discovery part without the obstacle of competition?

Microbiologists are notoriously competitive. By and large, they
are the A students from their respective communities, the brightest
and best analytical minds drawn from around the world. And if
they've made it to the laboratories on the cutting edge of virology,
then that selection process has been even more refined. Virologists,
therefore, tend to be among the most competitively macho of the
microbiologists. Not only do they frequently handle lethal submicro-
scopic matter and seek to get it to replicate, they often do so without
the benefit of prior knowledge of exactly how dangerous the agent is.
In the case of this new virus, early reports indicated that it might be
as highly contagious as measles, which is usually handled only by vaccinated scientists in Biosafety Level Three (BSL-3) laboratories.

As Stöhr considered the challenge, he saw four hurdles—three
material and one metaphysical. The most pressing necessity, as in
most disease outbreaks, was securing samples. With China unwilling
to share or provide samples, Stöhr was left with those labs outside of
the Middle Kingdom, but he knew how hard it was to encourage the
sharing of samples between rival laboratories and scientists. He
quickly made a list of the labs that had samples from sick patients.
He added to that list those laboratories that had lab monkeys, which
would be necessary for testing whatever agent was finally isolated in
fulfilling Koch's postulate. (Devised by legendary nineteenth-century
German microbe hunter Robert Koch, the postulate states that a
microbe can be considered the causative agent of a disease only
when it has been isolated from a diseased host, cultured outside the
host, and then introduced via those cultures to cause disease in
another host. Finally, the agent must be reisolated from this newly
infected host. This is the generally accepted four-step checklist for
establishing a new pathogen.)

Stöhr and his colleagues in the WHO Global Influenza Surveillance Network identified eleven laboratories in Hong Kong,
Japan, Singapore, Canada, France, Germany, the Netherlands, the

United Kingdom, and the United States that had access either to samples or to monkeys. All the labs were invited to join the research project under the condition they agree to work according to a set of rules specifying that all data and information shared among the members of the project be used only to advance the project rather than to pursue other projects or publish papers furthering the interests of only the member lab or scientist. To prevent conflicts between national public health agencies and laboratory requirements, data on cases and samples would be treated separately from epidemiological information and would be kept confidential. The collaboration would be facilitated via daily teleconferences and the use of a secure WHO website to post electron microscopy pictures of candidate viruses, protocol for testing, phylogenetic trees, PCR primer sequences, and the results of various diagnostic tests. This was an unprecedented and astonishing development in modern medicine, and one made possible only by the international nature and structure of the World Health Organization—and the imminent threat posed by this novel pathogen.

A. B. Oosterhaus, the chief of virology at Erasmus University's National Influenza Center, in Rotterdam, one of the member labs, told me, "We had had informal arrangements before, of course, especially with influenza. But this was quite an amazing setup. Scientists are a jealous bunch, as you know, and for us to work together . . . this was an extraordinary situation because we all knew the stakes were so high."

One drawback of the network, however, was that for its architect, Klaus Stöhr, it would mean being awake and on the job around the clock, monitoring the website and hosting the teleconferences that would link laboratories in virtually every time zone on earth.

CHAPTER 35

- **March 12, 2003**
- **Chinese University of Hong Kong,**
  Shatin, Hong Kong, China
- **1,110 Infected, 125 Dead**

HONG KONG, AS THE HARDEST-HIT CITY SO FAR OUTSIDE OF MAINLAND
China, had three network laboratories: the University of Hong
Kong's microbiology department, where Malik Peiris, Guan Yi, and
K. Y. Yuen plied their trade; the Government Virus Unit, where
Wilina Lim was the chief virologist; and the Chinese University
Hong Kong's Prince of Wales Hospital, where John Tam was
head of clinical virology. For these labs, the benefits of working
the international network were less clear cut: In some instances
the cases at the Chinese University of Hong Kong an
Government Virus Unit, samples were piling up faster than
tested, with the previous scarcity giving way, tragically, to
dance as more and more Hong Kongers took sick. But to
out of the network would have been seen as a show of
terms of the international medical community.

Within Hong Kong, the competition between
University of Hong Kong and the University of
remained as intense as ever. Hong Kong College
founded in 1887 by legendary microbe hunter Patric
would go on to found the London Institute of Tro
was the older, more established university and
international attention for its work in fighting
University, founded in 1963, employed Manda

instruction, making it seem more relevant to Hong Kong's future than the University of Hong Kong, which still conducted its curriculum in English. As a result, there was a sense that Chinese University had lately taken over as Hong Kong's most lustrous center for higher learning, and had increasingly become the more powerful and popular university as more and more of the city's elite, or at least those who couldn't afford to study at Ivy League or Oxbridge institutions, chose to matriculate there. Both universities maintained spectacularly high standards and admitted fewer than 5 percent of applicants. The gown-and-gown rivalry between the two had grown bitter over the past few years, especially between the rival medical faculties, with Chinese University's John Tam labeling K. Y. Yuen of the University of Hong Kong as "a great self-promoter." Yuen, for his part, would respond that Chinese University never, in the course of the whole outbreak, shared samples with the University of Hong Kong.

For the two microbiology departments, and the universities they represented, the race to be the first to identify the causative agent of this atypical pneumonia was thus a matter of intense local pride and heated dispute. Though they were both nominally participants in what was being called the WHO Laboratory Network, there was still heavy speculation, within Hong Kong, over which of the two local universities would emerge on top.

CHAPTER 36

- March 12, 2003
- United States Centers for Disease Control and Prevention, Atlanta, Georgia, USA
- 1,112 Infected, 126 Dead

WHEREAS HONG KONG'S MICROBIOLOGY LABS ARE PIGGYBACKED onto hospitals, the U.S. CDC occupies a fifty-acre campus in Clifton, a suburb of Atlanta. Just across from Emory University and its wooded acres of oaks, hickories, and loblollies, the CDC conveys, at first, a reassuring semblance of a college campus. But beyond the sylvan exterior is a bulwark of American science—and defense—as the CDC serves as the front line against a host of biological threats, natural and terrorist, to American and global public health. Jim LeDuc, who headed emerging viruses for the WHO, once described the Infectious Disease Center of the CDC as "the public health service laboratory for the world." Since the CDC was the acknowledged gold-standard lab, in particular for virological testing for new pathogens, it was only natural that it be brought into Klaus Stöhr's ad hoc lab network. The rivalry between the WHO and CDC is complex; in many ways, it is the scientific community's version of the political rift between the United Nations and the United States. At times, the institutions are inextricably bound by the demands of public health; at other moments, each seems offended by the mere existence of the other. This, however, was deemed by Julie Gerberding, the director of the CDC, as a case where "U.S. interests were exactly in step with everyone else: we wanted to stop this thing."

The Special Pathogens branch of the Clifton campus is reached by a steel-mesh catwalk, along which are electron microscopy images of the various killer viruses first isolated in the neighboring BSL-3 and BSL-4 labs. The pictures are a literal murderers' row: Marburg, Machupo, Lassa, Nipah, Ebola, Sin Nombre, HIV. All of them emerging viruses. Most of them first isolated right here. The Special Pathogens branch of the Infectious Disease Center was so vastly greater in scale and scope that it could spend in one day the University of Hong Kong's annual budget for virology. The University of Hong Kong's microbiology labs employed about a dozen scientists; the CDC had hundreds. The CDC had armories of sequencers, electron microscopes, laminar airflow cabinets, and available cell lines. The University of Hong Kong had just one sequencer and one electron microscope. For veteran CDC virologist Thomas Ksiazek and pathologist Sherif Zaki, both of whom had been glorified in numerous books and magazine articles, whatever was breaking out of China and infecting the world posed not only a great challenge but also a chance to add more gilt to their already multipage résumés. They viewed the lab network as a necessary courtesy rather than a genuine avenue to acquire scientific knowledge. During the SARS outbreak, they would scoff at the initial pronouncements of other labs around the world—a reminder that while other labs made mistakes and rushed to announce erroneous findings, the CDC got it right.

The first samples they had obtained, from Hanoi French Hospital, had arrived soggy and badly deteriorated. DHL, for some inexplicable reason, had held the package in transit for several days before Pascale Brudon, of the WHO office in Hanoi, finally screamed for the service to deliver the packages to the United States and Japan. By the time Ksiazek and his team sought to culture whatever had been infecting those bits of tissue, the virus had probably become inactive. (It would later be discovered that the SARS virus could live about twenty-four hours in open air.) A few days later, however, they received a fresh sample, from someone named Carlo Urbani, who had fallen sick in Bangkok.

The arrival of a sample of an emerging virus at a virology lab is a little bit like Jennifer Lopez showing up for the first day of produc-

tion on a motion picture. You know she is dangerous, difficult to handle, and potentially toxic. But at the same time, you can't help but
stare at the newly arrived star—for, without her, there would be no
project at all. These star samples arrive packed in dry ice in
Styrofoam boxes or, worse, on swabs broken off in test tubes or as
bits of biopsied tissue floating in a clear medium in a Corning plastic
bottle wrapped in cellophane. As the assistants unwrapped this
material in a CDC BSL-3 laboratory, holding all the tape, packing
material, and bubble wrap under the laminar airflow hood, Ksiazek,
standing beyond the glass, watched his junior staffers work. He
understood that whatever was in there could be the agent that was
infecting and killing hundreds. Dire, but exciting. He scribbled in
Magic Marker on a piece of scrap paper and held the paper up to the
window. A smile behind an N-95 mask and goggles looks like a flex-
ing of cheek muscle, nothing more, and that's what these two staffers
exhibited as they read Ksiazek's sign: YOU'RE HAVING TOO MUCH FUN
IN THERE.

CHAPTER 37

- **March 12, 2003**
- **Prince of Wales Hospital, Shatin, Hong Kong, China**
- **1,120 Infected, 126 Dead**

KEIJI FUKUDA, THE CDC'S INFLUENZA EXPERT, VISITED PRINCE OF Wales Hospital and saw firsthand what this new disease was doing to hospital wards. By now, Joseph Sung estimated that he had more than fifty infected health care workers. There was almost no protocol in place for treatment. The WHO had no recommendation, and the conventional treatments for severe pneumonia included using a nebulizer, which, as Sung and his staff were now determining, served only to spread the disease. As Fukuda toured the epicenter of the outbreak, Ward 8A, he was struck by how close together the beds were and by the lack of ventilation. There were no fans in the ward, and on humid days, such as today, the walls themselves seemed to be sweating. There were rows of windows looking out on a parking lot on one side and Tate's Cairn, a vertiginous mountain, on the other. The floors were linoleum. There was one nurses' station, which was located at the crossbar of the H-shaped ward. The index case, most likely, had been in bed 11, an east-facing bed. Every patient in the same bay as he, the top-left side of the H-shaped ward, had also been infected. Almost every regular duty nurse on the ward had fallen ill. And most of the physicians who had done rounds on the ward were sick. Even an electrical engineer who had come up to fix the air-conditioning unit fell ill with the mystery illness.

Almost half the patients already had visible consolidations in their chest X-rays to go along with their high, swinging fevers. If medical staff continued to become infected at the present rate, then Sung would soon run out of doctors and nurses. At a meeting called in the paneled conference room next to his office, he explained everything he knew about the disease: the high percentage of cases that would graduate to acute respiratory distress. He estimated, based on preliminary reports he had seen from the WHO regarding the Hanoi outbreak, that about half of the current cases would end up in the ICU. "Our colleagues," he warned, "will start to deteriorate, rapidly. Most of them will have to be intubated. Some of them will die. Those of us who treat them will be at risk of joining them in the ICU."

Sung proposed the formation of a "Dirty Team," a group of doctors and nurses who would treat these atypical pneumonia patients. The other half of the staff would stay off the eighth floor and therefore be at far less risk of infection. He put himself forward as the first volunteer for the Dirty Team. He asked for those with expertise in respiratory and infectious diseases, and he was amazed when over half the staff raised their hands to volunteer.

In hospitals in Hanoi, Guangzhou, Beijing, Hong Kong, Singapore, and Toronto, clinicians were making similarly courageous choices regarding treating these highly contagious and deadly patients. It is remarkable, in all my reporting on SARS, how infrequently I came across a case of a medical professional refusing to treat a suspected patient. Later, when the risks and agent were better known, there would be occasions when doctors and nurses, especially in Taiwan, wouldn't work unless they were given proper protective equipment. But in these early stages of the outbreak, as the tremendous danger was already apparent, it was remarkable that doctors and nurses of hospitals such as Prince of Wales were willing to face the ultimate risk—infection by a potentially fatal viral agent for which there was no known cure—in the line of duty.

Sung himself would never have considered not joining the Dirty Team. To this day, he has not admitted to his wife that he had a choice in the matter. On the third day of the outbreak, he phoned her and told her he would not be coming home, and he asked her to pack a suitcase for him. The hospital had prepared dormitory-style housing for the members of the Dirty Team; Sung stayed instead at his parents' house. His parents, in turn, would move in with his wife, Rebecca, and two children, Winnie, ten, and Joanne, seven. To explain what their father was doing, Rebecca rented the movie *Outbreak*.

"I think that scared them a bit," Sung later recalled.

Keiji Fukuda would return from Prince of Wales to his Wan Chai hotel room and report what he had seen to the Global Outbreak Alert and Response Team in Geneva. The WHO's David Heymann; its director of communications, Dick Thompson; and director of Communicable Disease Surveillance and Response, Guénaël Rodier, that afternoon decided to issue a Global Alert, warning that "since mid-February, WHO has been actively working to confirm reports of outbreaks of a severe form of pneumonia in Viet Nam, Hong Kong Special Administrative Region (SAR), China, and Guangdong province in China. . . . Until more is known about the cause of these outbreaks, WHO recommends patients with atypical pneumonia who may be related to these outbreaks be isolated with barrier nursing techniques."

The alert did make one very important observation: "No link has so far been made between these outbreaks . . . and the outbreak of 'bird flu' (H5N1) in Hong Kong SAR reported on February 19."

CHAPTER 38

■ March 12, 2003
■ Quarry Bay, Hong Kong, China
■ 1,120 Infected, 126 Dead

THOSE OF US LIVING IN HONG KONG WERE JUST NOW BECOMING aware that we were in the eye of a viral storm. That week came the first newspaper stories of a mysterious outbreak at Prince of Wales Hospital, and the beginnings of speculation that this was related to the earlier atypical pneumonia in Guangdong, and somehow connected to a doctor named Carlo Urbani and a hospital in Hanoi—but this was all confusing and, anyway, off the subject. The topic of the day, as anyone could see by turning on the television, was war. At *Time*, we prepared a cover story on what the war would mean for Asia, focusing on how most of the region was either indifferent or opposed to the coming conflict. And then, only at the last minute— and only after the World Health Organization had issued its Global Alert—did we start up a story on the local outbreak.

Almost as soon as we turned our attention to it, however, it seemed that whatever was going on at Prince of Wales Hospital— and in Hanoi, Guangdong, and, by the end of the week, Indonesia, the Philippines, Singapore, Thailand, and Canada—was, as Bryan Walsh would write that week, "Everyone's worst nightmare: a highly contagious, potentially fatal disease of unknown genetic makeup and for which there is currently no antidote or vaccine." We had a quick meeting in my office to talk about the outbreak and how we might cover it going forward, and joked that we needed one of those computer programs they deploy in movies, where the red patch—signifying trouble of some sort—keeps expanding outward until, gradually, the whole globe is infected. Still, we all reassured ourselves that whatever this thing was, it could not be that bad. Remember, we were all healthy and fully vaccinated children of the antibiotics era. Disease outbreaks were not the type of news we were trained to cover.

The next morning, my wife and I had brunch with Ray Bashford and Angela McKay, two fellow journalists, and their children. Their youngest son was listless and feverish and sat on the couch most of the morning, playing with his Yu-Gi-Oh! cards. At that point, this spreading Asian contagion was cause for a bout of speculative teasing that perhaps the boy had this new mystery disease.

During brunch, I recall being the only one at the table who kept bringing the conversation back to the disease and wondering aloud what this thing was. No one else was that eager to engage me on the topic, and I began to wonder if perhaps I was some sort of impending-epidemic hypochondriac. The more I thought about it, the more I questioned if perhaps it was the cinematic quality of the whole affair that had taken me in. What did we really have? A few sick doctors and nurses. And the WHO had even said it probably wasn't avian flu.

And as long as it was contained in one hospital in Hong Kong, then how bad could this virus really get? According to Chinese officials, the atypical pneumonia had burned itself out in China. If this were the agent of a deadly new pandemic, then surely this disease would have infected all of China rather than just remained a local problem in Guangdong. However, as far as we knew, Beijing and the interior had basically been spared. The Chinese Ministry of Health had assured us on CCTV that there was nothing to worry about. Yet, when we dispatched several correspondents to Guangzhou, we discovered that the situation in Guangdong was still critical and that the Chinese government was not eager to share details and information about the outbreak. The World Health Organization and many news organizations were in a similar situation; we were stumbling around in the darkness of Chinese bureaucracy. But at least we had our reporters on the ground, digging up information. The WHO, constrained by being a United Nations agency, could not visit hospitals or laboratories without official permission, which was still not forthcoming. It had been over a month since the WHO had asked for clearance to visit Guangdong, and the Chinese government was still not responding.

As for our reporters, Neil Gough and Bu Hua found atypical pneumonia patients in several Guangdong hospitals, and doctors reported that these were new arrivals, which meant that the disease was still burning. Meanwhile, as far afield as Nanjing and Beijing, doctors were whispering that the disease was popping up in their hospitals—contradicting officials' claims that the illness was confined to Guangdong. The government was sticking to its old number of 305 cases, unchanged since February 11, and that number seemed to be a line in the sand. With the quadrennial National Party Congress coming up, at which formal leadership was expected to be handed from Jiang Zemin to Hu Jintao, Chinese officials were loath to admit that a disease outbreak of unknown severity was blazing through the country. "We wanted clear skies," says one former Chinese intelligence official. "No one had to come out and order party secretaries not to talk about this disease. This is one of those instances where party officials know how to think correctly. China hasn't changed that much."

Other medical officials say the pressure was more overt. One Shanghai doctor reported that local hospitals were warned by municipal officials not to speak to any media, even to the state-controlled Xinhua news agency. Why the secrecy? "Look at what was going on in the country at that time," says one prominent physician from Beijing. "What mattered more, the Party Congress or a few doctors?"

CHAPTER 39

■ **March 13, 2003**
■ **Pok Fu Lam, Hong Kong, China**
■ **1,144 Infected, 127 Dead**

MALIK PEIRIS COULD HEAR HIS SON'S XBOX FROM WHERE HE SAT IN
the living room. He still had an old turntable on which he could play
vinyl albums to drown out the bleeps and bursts of the video game,
but he hadn't been in the mood for music in what seemed like
months. His schedule had become a blur of lab work punctuated by
teleconferences and the briefest snatches of sleep. Despite repeated
attempts to culture something from his samples, he had failed, and
was now finding himself despondent and, for the first time, questioning his own abilities as a virologist. They had gone through the forty-
eight samples Guan Yi had brought down from Guangdong, and had
yet to come up with a causative agent.

What was it they weren't seeing? Peiris agreed with his colleagues
on the WHO Laboratory Network that this was not influenza, but
that was a little like saying a person wasn't Bolivian. There were still
tens of thousands of possible viral agents, and that was only if you
were counting those species already familiar to man. Ecologists estimate that there are about thirty million different species of animals
on earth. If you assume that each is host to at least one species of
virus (and probably dozens), that makes at least thirty million species
of virus, any one of which could conceivably find a route by which to
infect humans. Of the 4,000 identified species of virus, about 150
were known to infect humans, and it was those on which the WHO
Laboratory Network was now concentrating.

Peiris stood up, slipped his loafers on, and walked out the door. He was so distracted he forgot to say good-bye to his wife. Downstairs, he started up his BMW and paused for a moment to let a slightly queasy feeling pass over him. He had been feverish for a few days already and was now regretting not having gotten a flu shot. It was hypocritical of him to urge the shots upon his family and friends while he himself, who worked in a laboratory crawling with influenza, couldn't be bothered to walk down the hall and roll up his sleeve. Now, perhaps, it was too late.

When he arrived at work, he stopped in to see K. H. Chan, his laboratory partner and fellow University of Hong Kong microbiologist. They talked about the new samples that had arrived, both picked up as part of the Heightened Pneumonia Surveillance Network. Chan was reluctant to continue employing the same cell lines; those had so far proved fruitless, and he wondered about the possibility of testing with others.

"What else do we have?" Peiris asked.

Chan mentioned a kidney cell line from fetal rhesus monkeys, most commonly used for hepatitis A tests. (Because of the large amount of shellfish consumed there, Hong Kong has frequent hepatitis outbreaks.) Malik knew this was an unlikely cell line in which to grow a respiratory virus and said he would think about it.

When he arrived in his office and took his seat behind his desk, he received a call from K. Y. Yuen, from the University of Hong Kong's microbiology department.

"Let's try some new cell lines," Yuen suggested.

They had a tremendous advantage over virtually every other lab in the network because of the samples Guan Yi had smuggled down. For political and legal reasons, they could not publicly say that they had these samples, nor could they explain why they had decided to use new cell lines. But at the same time as other labs in the world were getting their first batches of samples, the University of Hong Kong team was already working on their fiftieth attempt at isolating the virus.

Peiris agreed that it was time to try some new cell lines. Dominic

Tsang, an epidemiologist at Queen Mary Hospital, who was well connected with Wilson Yee, a physician at Kwong Wah Hospital, had written to Peiris a few days earlier telling him, "Dr. Yee is seriously considering the option of an open-lung biopsy to obtain tissue for our investigation. Pending patient's family's consent, such procedure would proceed and lung tissue would be delivered to your laboratories for processing." Peiris and his colleagues had just received these fresh samples from Kwong Wah Hospital. The patient was the brother-in-law of the late Dr. Liu Jianlun, the nephrologist who had come down from China and checked in to Hong Kong's Metropole Hotel. The brother-in-law had also passed away from the disease; the virus had to be there.

Guan Yi was back in Guangdong, working with Zhong Nanshan on setting up a virology lab in the Guangzhou Institute of Respiratory Diseases. Peiris didn't bother to tell Guan Yi about these new samples, or the decision to introduce them to new cell lines. After all, Peiris was the chief of human research in the Pandemic Preparedness group, while Guan Yi ran the animal side. At this point, Peiris felt, the disease was a matter for his lab and not Guan Yi's.

CHAPTER 40

■ **March 13, 2003**
■ **Guangzhou Institute of Respiratory Diseases, Guangzhou, China**
■ **1,149 Infected, 128 Dead**

GUAN YI HAD PROBABLY SPENT MORE OF HIS ADULT LIFE WALKING around Chinese poultry farms, wet markets, and wild animal stalls than any other virologist on earth. He knew the stench and filth too well, and during his research in animal influenzas, he liked to sometimes imagine himself as a virus. It was a trope he used in his talks, especially with mainland counterparts. He often explained interspecies transmission by describing himself as a virus.

"If I mutate, if I become bigger, say, or have different needs," Guan Yi would say, "maybe I become too big for my apartment. Then I need to leave and go find another place to live. A virus is the same way. It mutates, one virus particle out of a billion has suddenly mutated and is no longer suited to its environment, so it leaves; it looks for a new home. Of course, most of the time, it can't find a suitable environment, and it dies. But occasionally, very infrequently, it can find a new home, another animal in which it is comfortable."

Guan Yi would then look around, as if he were a virus settling into a new host species. "Oh, I like it here in my new house. I can take over this cell. I can reproduce. I'm a happy mutated virus."

He saw in these wild animal markets thousands of potential new homes for any of the thousands of viruses stacked atop one another in these cages of distressed animals waiting to be slaughtered. The animals shed billions of virus particles in their feces, blood, urine, and tears, any one of which could be a lucky mutation that might find a new host in which to thrive.

Guan Yi often warned his Chinese counterparts that since inter-species virus hopping was inevitable, they had to try to minimize the likelihood. The animal markets, however, did just the opposite. They maximized the possibility that a virus could jump species.

"Say I am a virus," Guan Yi would explain to a bewildered Department of Forestry official. "I look around the market and I see so many possible homes."

At the beginning of the outbreak, when Guan Yi was smuggling his samples back from China, the University of Hong Kong pathology staff had sat in the basement conference room of the Department of Clinical Pathology. Written on the whiteboard was:

Disease
Microbe
Vector
Amplifier

To Guan Yi, those remained the four central issues. Malaria, as an example, would be the disease. The malaria microbe was a type of plasmodium protozoan. The vector was the mosquito, and the amplifier would be the marsh or standing water in which mosquitoes might breed.

So far, when it came to SARS, none of these issues had yet been cleared up.

Guan Yi was working closely with Zhong Nanshan, helping the latter to set up a virology lab in the institute. In the evening, he would take the train back to Hong Kong or crash in the White Swan Hotel. On those mornings, he marveled at how empty the breakfast buffet had become. Usually, the riverfront coffee shop of this hotel was filled with American mothers and fathers having their first meals with their newly adopted Chinese children. Guangzhou often functions as a way station for American families as they acquire U.S. passports for their newest family members. It can be a bewildering sight at first, families whose ethnic composition perfectly splits into Caucasian parents and Asian children. I once asked a Chinese friend of mine what she thought of this.

"It looks like we are selling babies," she said.

The disease outbreak in Guangdong had severely curtailed the adoption business, and as families were reluctant to come to China, the U.S. consulate was evacuating nonessential staff. Now, Guan Yi dined in private, watching the tugs and barges chug their way up the Pearl River.

He was out of touch with his colleagues in Hong Kong and did not even know about the Global Laboratory Network established by Klaus Stöhr. No Chinese labs had been invited to join at the first stages. And Guan Yi had been spending more time in China than in Hong Kong since the outbreak had begun.

What is it?

Where does it come from?

What does it do?

How do you kill it?

He believed that the answer to those questions was somehow in those animal markets.

I am a virus, he said to himself as he sipped his breakfast congee. Where do I live?

CHAPTER 41

■ **March 14, 2003**

■ **New York City, USA**

■ **1,159 Infected, 130 Dead**

THE PATTERN ON THE CARPETING WAS VAGUELY ASIAN—COULD IT BE Persian?—and as Dr. Leong Hoe Man waited for the elevator, he found himself becoming dizzy gazing at the intricate floral pattern. It had been a busy few days, attending a conference on infectious diseases and sitting in on panels hosted by some of the foremost infectious-disease experts in the United States. These affairs were both a perquisite and a burden. He enjoyed meeting with his colleagues and taking a break from hospital rounds; however, for specialists from Singapore such as Dr. Leong, the journey was a backbreaker. In fact, he believed he had caught some sort of cold on the flight over, and since arriving, he had steadily begun to feel worse. The body aches had become more severe, and as he walked down Fifth Avenue earlier that day, hacking into a handkerchief, he once had to stop to catch his breath.

Some combination of jet lag and a common cold virus, he suspected. Yet that afternoon, while he sat in the audience of an emerging-diseases discussion, his muscle soreness had increased in severity, and he now had a debilitating headache. He was also now sure he had a fever, despite the ibuprofen he had been ingesting. Just to be on the safe side, he phoned a fellow physician he knew in New York and scheduled an appointment.

The waiting room had been crowded—a mother and her two children, an old man and a young woman. He took care to cover his mouth as he coughed. It's probably nothing, Dr. Leong reminded himself, but a second opinion and some antiviral medication might be a good idea.

The doctor, a specialist in internal medicine, listened to Leong's lungs, took his temperature, and drew some blood but concurred with Leong that whatever it was, it was probably just a bad cold, some nasty but still-common virus. Leong did not feel it was relevant that before departing Singapore, at Tan Tok Seng Hospital, he had treated two patients with a mysterious new type of rapid-onset pneumonia.

The doctor told Leong that he was fit to fly.

Leong shrugged. The conference was nearly over. He would be flying home in two days.

On his return to his hotel, as he stood before the elevator banks with a few fellow guests, he stared at the carpet and, for a moment, lost his balance and almost fell over. He steadied himself by the mirrored wall above the call buttons and let an elevator come and go before he felt stable enough to board one himself. It didn't occur to him that whatever it was he had caught could be so contagious that sharing an elevator with him could be a death sentence.

CHAPTER 42

■ **March 15, 2003**
■ **World Health Organization Headquarters, Geneva, Switzerland**
■ **1,199 Infected, 131 Dead**

IN ONE WEEK, THE WHO HAD GONE FROM WONDERING IF THERE REALLY was a disease outbreak afoot to knowing it was in six countries and suspecting it in a dozen more. There were hundreds of sick patients in Canada, China, Hong Kong, Vietnam, Singapore, and Indonesia. A huge percentage of those infected were doctors and nurses, each of whom represented a two-person swing in a hospital's battle against the bug: one more patient often meant one less doctor, literally, as some of the world's best health care systems seemed to be imploding.

Now, somewhere over the Atlantic, there was an infected doctor on a plane, flying from New York to Singapore via Frankfurt. Chew Suok Kai, deputy director of medical services at the Singapore Ministry of Health, had informed Mike Ryan, deputy chief of the Global Outbreak Alert and Response Network, of this development. The doctor in question was infected with this atypical pneumonia, and if he was as highly contagious as those patients who had turned up in Singapore's Tan Tok Seng Hospital, then the WHO had to get him off that plane and make sure to quarantine him and everyone else on board. It was 3:00 A.M., still pitch-dark outside, yet Ryan snapped into a frosty state of alertness as he realized the implications of what he was hearing. When that plane touched down, hundreds of passengers would scatter throughout the global transportation system—to connecting flights, to trains, taxis, buses—to carry this bug even farther afield. This Singaporean doctor hadn't even known the disease existed before he left Singapore, and now he just might be the world's greatest spreader.

Ryan spent the next four hours tracking down German health authorities. When Dr. Leong's plane landed for its stopover in Frankfurt, German emergency medical technicians in orange hazardous material suits were waiting at the J-class exits to pick him up. Contact information and itineraries for the rest of the shaken passengers were gathered.

The bewildered doctor was put in isolation. The remaining passengers were transferred to another plane.

"We have to brand this," concluded David Heymann that morning, after the Frankfurt incident, realizing that the earlier labels—"atypical pneumonia of unknown origins," "atypical pneumonia without diagnoses"—not only were maddeningly prosaic but also failed to convey the seriousness of the situation. It was no longer politically correct to name diseases by their geographical point of emergence—Ebola, Rift Valley, Machupo—as that was seen to unfairly stigmatize certain locales. What they needed in this instance, Heymann felt, was an acronym, like AIDS, that rolled off the tongue and conveyed alarm.

Heymann, who had been camping in the Swiss Alps with his son's Boy Scout troop, quickly drove back down the mountain to Geneva to reach the office. At a morning meeting convened on the first floor of the L building at the WHO headquarters, Heymann, Klaus Stöhr, Dick Thompson, and their colleague Dennis Aitken tossed names and labels back and forth, trying to find an acronym that would be both descriptive and easy to say and would convey the seriousness of this new disease. It may have been Thompson, a former *Time* magazine correspondent and now the WHO's director of communications, who first coined the acronym SARS. Heymann doesn't recall who said it, and Guénaël Rodier, who was teleconferenced in, also can't remember who first put the letters together. Immediately, however, there was some disagreement as to whether "severe acute respiratory syndrome" was redundant. "Why 'severe' and 'acute'?" Thompson himself asked.

Heymann believed that "severe" referred to the infectiousness of the disease, while "acute" meant rapid and sudden onset. What wasn't considered, at that moment, was that the acronym was very similar to SAR, or "Special Administrative Region," the Chinese bureaucratic appellation for Hong Kong. Despite seeking to avoid any geographical stigmatization, the WHO had inadvertently done precisely that.

The team then decided to issue an unprecedented Global Alert, which would, for the first time, give the new disease its name.

But they still didn't know what caused it.

CHAPTER 43

■ March 18, 2003

■ University of Hong Kong, Pok Fu Lam, Hong Kong, China

■ 1,331 Infected, 134 Dead

MALIK PEIRIS COULDN'T BELIEVE WHAT HE WAS HEARING. JOHN TAM, over at the Chinese University of Hong Kong, was reporting on the laboratory network conference call that he had positive PCR results for MPV, or human metapneumovirus, in samples from patients who had developed SARS after contact with Hong Kong's index case. He had the virus, he was saying. Sequences of the primers that were needed to test other samples would soon be available electronically via the network's secure website.

MPV is a single-strand RNA virus, meaning it is not self-checking when it replicates. It appears under an electron microscope as an irregular sphere with a (relatively) thick lipid coat and spikes around the outside. It has just eleven base proteins that, between them, provide all the impulse and aggression of each virion as it methodically goes about cannibalizing and destroying respiratory cells. That process, Tam had observed, appears under a microscope like a speeded-up film of a raw steak putrefying. The cell is gradually eaten away as thousands of virus particles gather in the center of the host cell in an orgiastic colony, efficiently reprocessing the host's protein and DNA into more virus.

The mood on the Global Laboratory Network conference calls was almost celebratory. A German lab, working with samples from the doctor pulled off the Singapore Air flight, was also announcing that they had found paramyxovirus—the same family as MPV—under an electron microscope. A day later, they said they found evidence of the virus in infected blood samples. A Canadian lab reported similar findings, as did the pathology department at Singapore General Hospital. All these labs were basing their findings on seeing paramyxo-like viruses under their microscopes. Only Tam's lab had positive PCR tests.

Chinese University had the bug. Or did they? David Heymann of the WHO seemed to agree, announcing, "We are now closer to the reality that *Paramyxoviridae* has caused this."

THE UNIVERSITY OF HONG KONG TEAM GATHERED IN THE DOWNSTAIRS conference room in the basement of the clinical pathology department. They had done PCR tests in the samples smuggled down from Guangdong and hadn't come up with any positives. "This doesn't make sense," K. Y. Yuen told Malik Peiris.

"Something's not fitting," Peiris agreed. Early on, in a few samples, they had found adenovirus, another virus that causes respiratory illness in children, but then, as a precaution, they took nasal swabs from twenty healthy hospital staffers and found adenovirus among those as well, making that an unlikely suspect to produce such severe sickness. When they ran their own PCR tests for MPV on their samples, they came out with negatives. Peiris wouldn't say, definitively, that John Tam was wrong, but he had what he called "funny things" growing that didn't seem to be metapneumovirus.

K. Y. Yuen had written on a whiteboard several possible agents:

Influenza
RSV
Parainfluenza
Rotavirus
Chlamydia
Mycoplasma

Influenza, of course, had proven the greatest red herring, and repeated attempts to isolate that virus had failed. Furthermore, antigen staining for all the above mentioned viruses had proved negative, as had PCR tests for *Chlamydia* and mycoplasma.

Finally, Yuen added one more microbe to the list: coronavirus.

"Unlikely." Yuen shrugged. "But who knows?"

They had something cooking in their own lab, Peiris and Yuen knew. A fetal monkey liver cell line was being decimated, almost certainly by a virus. K. H. Chan had first detected some of the cells dying a few days earlier. After waiting forty-eight hours to make sure of his observations, he reported his findings to Peiris. Could it be MPV? Peiris dismissed the suggestion immediately: MPV didn't grow on that cell line. This was a cell line they had in stock for use in growing hepatitis A; it was infrequently used to grow respiratory viruses.

Yuen and Peiris debated whether they should speak up during the daily laboratory network conference call and question the MPV findings, before finally agreeing to wait and see what they had growing upstairs.

Now, as he looked at the samples under an optical microscope, Malik Peiris could see that something was indeed destroying the cell lines. Healthy cells in a sample tray appear as a row of pleasingly regular bubbles, proverbially fat and happy as they feast on protein medium and replicate. When a cytopathic agent is present, however, the cells present an image of microbial apocalypse as metaphysically compelling as a still photo of a gory car crash. What Peiris was seeing beneath his lens was a landscape of savaged cells, bunched up in self-defense, some with their cytoplasm torn, mitochondria turned to pulp, their decaying nuclei dangling tendrils of shorn protein.

Peiris and K. H. Chan extracted fluid from the dying cell lines and introduced it into a fresh line, to make sure that the cells weren't dying of some contaminating bacteria. Nothing makes a virologist happier than witnessing the death of healthy cells. "I don't know if a fireman looks at a fire as an opportunity," Peiris recalls, "but, having said that, it certainly is one."

Still, they had to establish that whatever was killing these cells was the same agent that was causing SARS. He and Chan gathered serum from patients who were known to have had the mystery ailment but who had recovered, and they conducted a hasty immunofluorescence test of the infected cells, seeking specific antibodies that would indicate the presence of the virus.

IN THE STRUGGLE OF ANY MULTICELLED ORGANISM TO SURVIVE AND reproduce, disease forms perhaps the greatest and most formidable challenge. For complex animals to survive, they have to devote a considerable percentage of their genetic information and biochemical energies to erecting defenses to neutralize and destroy invading microorganisms. Skin is an excellent physical barrier for most microorganisms; similarly, human beings have certain enzymes in our digestive and respiratory tracts that neutralize and degrade bacteria—tears, for example, contain an enzyme that can protect the eyes from certain harmful agents. Our greatest defense, however, is our immune response. Without it, we would never have emerged to range over the huge expanse of ecosystems and climates that we now do; we would not, in other words, dominate the planet. We would have remained, in all likelihood, diminutive tree-borne primates confined to a small niche of sub-Saharan Africa. If our immune systems had not proven to be adaptable and to have a remarkable memory, each time our ancestors ventured into a new territory, indigenous microbes would have wiped us out.

When a pathogen enters the human body, the immune system has several microbial responses that include, during the initial phases of infection, the production and circulation of macrophages, or white blood cells, which attack and ingest, or otherwise neutralize, infected cells, bacteria, and viral particles. There are also helper cells, or T cells, which recognize foreign proteins on the surface of infected cells and destroy those cells. But our most powerful defense against any microbe that has already entered our cardiovascular system is the production by certain lymphocytes—special white blood cells—of specific proteins called antibodies that respond and bind to certain sites on the surface of a virus. The production of these antibodies begins a few days after a virus's introduction to the immune system, and it is the body's memory of this process that allows for immunity to many viruses after initial infection. These antibodies consist of four protein chains that resemble bent pipe cleaners. The chains are too small to refract any light and, therefore, like viruses, have no color, and they are specifically targeted at those viruses possessing distinct surface amino acid sequences. An antibody that is produced to bind with and neutralize an influenza virus, for example, cannot bind with any other viral particle. This biochemical process is one of the wonders of evolution—you could not design a better system if you had to—and when it is compromised, as it is with immunosuppressive diseases such as AIDS, the consequences are devastating.

An otherwise healthy patient who has been infected with a virus will produce antibodies that will be present in high concentrations in the serum drawn from infected blood samples—the serum is the clear liquid at the top of the test tube after the red cells have settled. Those antibodies can be used for any number of medicinal or diagnostic purposes. In some cases, they may be used as a treatment for other infected patients; Prince of Wales Hospital was already trying this as an emergency treatment for SARS patients. Or they may be used as part of an immunofluoresence test, as Peiris's team was now conducting. Such a test seeks to identify the binding of those antibodies with their targeted antigens. The serum would be stained with a certain protein dye and then introduced to a slide containing infected cells. The dye would glow an apple green under a fluorescence microscope only when the antibody successfully bonded with its targeted antigen. Serum from SARS patients would react only when introduced to the virus that had caused the disease.

As they prepared the slides for examination—two sets, one of damaged cells and another of healthy cells, each stained by the serum mixture—Peiris, Chan, and Yuen wanted desperately to see that apple green to confirm that they had indeed found the causative agent.

Now it was time to look. Chan went into the closet-size room off the fourth-floor virology lab, closing the door behind him. Setting the slide under the microscope, he switched off the light. The room had to be in total darkness in order for him to detect the faint glow of a positive result. He waited a moment for his eyes to adjust to the darkness as he wiped his glasses on a handkerchief. He bent too quickly toward the binocular eyepieces of the microscope, feeling the tap of his eyeglasses against the microscope. He squinted at the two circles visible through the eyepiece and then adjusted the focus to bring them together into one image. It was like gazing at a topographical map of a vast and complex archipelago. There were islands, inlets, channels, and estuaries; it was a landscape of damaged cells, but was it glowing?

Yes.

He pulled his head back and then stared down again. Of course there was faint glow. That was the result of ambient light and intense magnification. But was it really glowing? He looked again.

Yes.

This was a positive.

For SARS.

Wait, he thought. Now he checked the control samples, those slides prepared with serum and noninfected cells. These should not glow but instead should appear as faint, distant cellular landscapes, rows of distinctly nonglowing, regular and healthy cells.

The healthy cells did not glow.

Definitely.

He called Peiris downstairs.

They had the virus.

STILL, BEFORE THEY ANNOUNCED THE UNIVERSITY OF HONG KONG'S findings to the world, Peiris wanted to conduct a blind test, running the same immunofluorescence tests on healthy and infected cells to ensure that the results correlated with the presence of this new agent. Researchers sometimes see what they want to see; samples can become tainted; sleep deprivation can cause virologists to make mistakes and rush to judgment. (Isn't that what may have happened with their colleague, John Tam, over at Chinese University?) Peiris decided to conduct the blind test, bringing in an outside virologist to record the results. He contacted Wilina Lim, the chief virologist at Hong Kong's Public Health Center.

Lim had been watching the questions about this outbreak mount much faster than any of the answers. Samples were flooding into her labs, long rows of vials with broken-off swabs sealed inside clear or bloody fluid. If this was a respiratory disease, she asked herself, then why wasn't it spreading in the community? If it was infecting hundreds of hospital patients, then why couldn't she grow it on any of the dozens of cell cultures she had slowly rotating in her incubators? Actually, she thought, many of these questions boiled down to just one question: What is it?

When Peiris called on the morning of March 20, he excitedly told her he had found something in a cell line, and had successfully cultured it in yet another line. He explained that he needed convalescent serum to do an immunofluorescence test, as well as some serum from regular pneumonia patients. To be sure, Peiris asked Lim to provide his lab with three positive serum samples and three negative, all of them unlabeled. He would then introduce the serum to the infected cells and ask Lim to identify which were positive. Dr. Lim would not be told which serum samples were from convalescing patients and which were from blood samples drawn months before the disease had appeared in Hong Kong.

On March 20, on the fifth floor of the Queen Mary Hospital microbiology lab, Dr. Lim determined that three of the samples had the correct tint to indicate the presence of the agent. The other three, she decided, were negative. In a blind test, she had correctly identified the three samples into which convalescent serum from SARS patients had been introduced; whatever was killing those cells was the same virus that was killing patients around the world.

Meanwhile, downstairs in the University of Hong Kong microbiology department, Dr. John Nichols had embedded into paraffin a cluster of Peiris's infected cell lines. In a cutting machine, he mounted a diamond knife so sharp that if it were to make contact with your skin, it would slice to the bone instantly. It is, literally, the sharpest object on earth, capable of splitting hairs, cells, and even viruses. Gazing through a microscope, he shaved infinitesimally thin slices from the speck of paraffin-encased cells, each slice literally ten nanometers across, or one-twentieth the thickness of this page. The slices fell into a tiny basin of purified water and there floated like water lilies. Taking up a chopstick that had one of his assistant's eyelashes glued to the end, he guided one slice into a minute grid of copper. The grid, itself barely bigger than the period at the end of this sentence, was then removed and carried in a sample box to a room next door, where it was removed and locked into a sample holder that looked like part of an industrial lathe and made up the business end of the vast green-and-black steel electron microscope. Nichols switched off the light and turned on the machine, waiting as the samples were bombarded with electrons and an image was constructed. He dialed up the picture and almost immediately thought, That's strange.

The cell had been torn apart. He'd expected that. But what surprised him was the extent of the damage. The landscape visible under an electron microscope is even more complex and variegated than that under an optical microscope. The images have been flattened, sliced away so that you are getting a two-dimensional cut. A speck of damaged cellular protein, under these tens of thousands of powers of magnification, can look like an influenza virus, a chlamydial agent, anything. Nichols was used to gazing at battered cells and searching for days for the viral culprit, trying to pick the hairy influenza virus or the hexagonal adenovirus from the impossibly intricate nucleus of a cell. But here he was seeing an awful lot of viral activity. They were multiplying primarily in the cytoplasm, the interior of the cell, which didn't conform with the behavior of a paramyxovirus. The cells had obviously been ravaged, and virus particles appeared to be budding along the external cell membranes. Nichols tried to focus in on the virus particles themselves. They were round and surrounded by protuberances—spikes perhaps—and they were more regularly shaped than any influenza viruses. He found another viral particle and focused in, taking pictures of the cross sections. Holding the image in the viewer, he flipped through a book of electron microscopy photos of virus particles, seeking a similar pattern. He turned on the light. Strange, he thought, stopping at an image of a round virus with a halo of dots, labeled a coronavirus. This looks similar.

He called Peiris.

Peiris rushed down. They took a look at the killer.

"We were initially elated," says Nichols of that moment. "But pretty quickly it dawns on us. So it's a coronavirus. Now what do we do?"

PEIRIS IMMEDIATELY DRAFTED A NOTE TO THE GLOBAL LABORATORY network. He had missed that day's morning conference call because of their feverish lab work, but now he had to let his partner labs know what was going on. He sent a group e-mail that read,

> The findings to date are as follows:
>
> (a) We have isolated a CPE (cytopathic effect) producing agent from two clinical specimens from patients with typical SARS syndrome. One isolate is from a lung biopsy and the other is from a nasopharyngeal aspirate. The isolations were made in a continuous RhMK cell line. The initial CPE took around 4 days but on second passage, CPE is very quick, within 24 hours.
>
> (b) The virus infected cells do not react with routine immunofluorescent detection reagents we use to diagnose influenza A, B, RSV, parainfluenza types 1, 2, 3 or adenovirus. Virus infected cell nucleic acid extracts [sic] does not react in an RT-PCR assays [sic] for influenza A (M gene) rhinovirus, enterovirus (conserved primers) human metapneumovirus (conserved primers between HMPV and avian pneumovirus), Mycoplasma pneumoniae and Chlamydia pneumoniae.
>
> (c) We used acute and convalescent sera from 5 patients with typical SARS in an immunofluorescent assay using the infected cells as antigen.

The e-mail went on to conclude, "we are confident that the agent in the cell cultures are [sic] associated with the SARS syndrome." That e-mail was dated Friday, March 21, 2003, and was sent at 10:22 P.M. Hong Kong time.

CHAPTER 44

✓ 1/30/12

- **March 20, 2003**
- **Bamrasnaradura Hospital, Bangkok, Thailand**
- **1,400 Infected, 136 Dead**

AS THE CASES MOUNTED AT HANOI FRENCH HOSPITAL, CARLO URBANI had been a constant presence, rallying beleaguered hospital staff, consulting with physicians and comforting patients. He had gathered some of the first clinical SARS samples in the world and sent them as far afield as Japan and the United States. His vigilance had given the planet a heads-up that might have saved tens of thousands of lives. "There is something strange about these cases," he kept telling Hitoshi Oshitani, even after he arrived in Hanoi. "It's not like any disease we've seen."

He had considered canceling his trip to Bangkok to attend a parasitology conference because of yet another surge in SARS cases. Dr. Hoang Thuy Long, who had been appointed the head of the Vietnamese government's SARS task force, had confided to Urbani that he feared the outbreak might already have spread out of the hospitals. They were lucky, both men agreed, that the disease had first shown up in a private hospital, where foreign doctors were free to act without consulting any government bureaucracies. They had closed the hospital; that had kept the disease in check. So far.

Still, the outbreak was a long way from contained, and Urbani was reluctant to leave Hanoi. This bug could be beaten, he had told Pascale Brudon, the World Health Organization representative in Vietnam. You have to be rigorous, and you have to observe laboratory-level infection control. He had worked with Brudon in helping to train twenty-four Ministry of Health mobile medical teams to use sterilization sprays and respirators, and on how to receive and treat a suspected SARS case. Now, Brudon assured Urbani that it would be a good idea for him to attend the parasitology conference. "Worms are your first love," she reminded him.

"No." Urbani smiled and patted his belly. "Food. Then worms."

Why not? His family was in good health, and the conference would bridge a schism that had evolved in the parasitology community—the Hashimoto Initiative Training Centre in Bangkok represented the theoretical side of parasitology. Those Japanese parasitologists treated the field almost as a hobby and devoted time and resources to gathering, identifying, and naming new species of helminth. Urbani understood their urge to collect and categorize, but he would always be a field physician first. He became a physician and joined Médecins Sans Frontières and then the WHO in order to help the poor and the disenfranchised. On his way to the hospital in the morning, he would often pull his Mitsubishi over to the side of the road to help one of Vietnam's many motorcycle accident victims; likewise, he viewed parasitology as a means to improving lives. The parasitology conference promised to connect his fieldwork with Japan's money and allow for greater public health throughout Southeast Asia. "If I cannot work in such situations then what am I here for?" he asked his wife, Giuliana, who had expressed concern at the dangerous situations he put himself into.

He would go.

Only when he was on the ninety-minute flight south would he begin to feel feverish. Then he began to cough.

Remarkably, physicians who have been treating a killer virus do not always react rationally when they first present symptoms. They are capable of remarkable denial, as appears to have been the case with Liu Jianlun, the nephrologist who brought SARS to Hong Kong. He was showing symptoms before his bus ride south, yet self-diagnosis is perhaps the most unreliable of medical practices, and the doctor would infect at least a dozen fellow guests at the Metropole Hotel before he would hospitalize himself. His own failure at self-diagnosis would seed the global outbreak.

There are numerous explanations for a winter fever and cough; the vast majority of them do not require hospitalization. As it turned out, one of the particularly deadly attributes of SARS was this ability to infect without quickly inducing symptoms too severe to preclude travel. In the early stages of infection, almost every patient was still well enough to board a plane, train, bus, or boat, thus potentially spreading the disease ever farther afield. Among recent emerging viruses, only AIDS has such an insidious period of asymptomatic infection; but with AIDS, merely being seated next to someone with the virus cannot get you infected. If you were stricken with Ebola or Marburg virus, you were unlikely to make it more than a day or two before showing severe symptoms. With SARS, one of its most sinister aspects was its incubation period, which allowed it to gain a foothold in a dozen cities around the world almost before anyone was even aware it existed. It was precisely this mechanism of contagion that caused panic in afflicted cities. That man next to you on the train, that lady coughing across the aisle—suddenly the means and modes of transit were rife with potential superspreaders. For the unwitting passengers en route from Hanoi to Bangkok, the stocky Italian man with the bright eyes and round face would turn out to be just such a carrier.

It is a tribute to Carlo Urbani that as soon as he felt the onset of symptoms, he determined that he had likely taken aboard this mystery virus, and he took precautions on the flight to cover his mouth when he coughed and even averted his face from his fellow passengers. As soon as the Thai Air flight landed in Bangkok's Dongmen Airport, he warned a colleague, Dr. Scott Dowell from the U.S. CDC, who had come to greet him, to stay away. They sat down across from each other in a waiting area beneath an elevated roadway, waiting ninety minutes for an ambulance crew to assemble protective gear before driving Urbani to the state-run Bamrasnaradura Hospital in Bangkok, where Urbani asked to be put into quarantine, explaining to confused physicians that he believed he may have caught a highly contagious respiratory ailment. "He was so careful," says Pascale Brudon. "We were positive he didn't have SARS. We thought he was just being cautious." His ailment certainly seemed less severe than those cases he had treated in Hanoi French Hospital. And Urbani's mood remained defiantly positive. "I've never been so pursued by women," he joked to Brudon at one point, referring to the bouquets of flowers and get-well cards flooding into his quarantine room. His colleagues believed he was recovering from what he still thought could be a mutated influenza. (The University of Hong Kong had not yet announced its findings. Though Urbani doubted that this was a mutated avian flu, he, like many other doctors treating the disease, still suspected that this could be some other sort of mutated flu virus.)

The fever, after an initial spike, seemed to be breaking—SARS cases, Urbani already had deduced, often had this lull between the initial mild symptoms and the more severe crashing of the respiratory system. "He was scared, and depressed," said Scott Dowell. "In Hanoi, he had seen patients who were well in the morning and very sick in the afternoon, and he feared it would be quick." His wife, Giuliana, sent the children back to Italy and then flew down from Hanoi, warning Carlo over the intercom that after he recovered, there would be no more zero-distance contagious-disease treatment. Urbani agreed, but both knew that this was a promise he could never keep. Whenever a troublesome case showed up at a hospital and he was in the area, said fellow parasitologist Dr. Kevin Palmer, WHO staff would usually say, "Call Carlo," because of Urbani's clinical diagnostic skills and his soothing bedside manner. That was what had brought him to Hanoi French Hospital the morning after Danny Yang Chin was admitted.

Even as Urbani's fever abated, he confessed to his friend Kevin Palmer, "I'm scared."

Soon, the already familiar viral progression was under way. X-rays revealed cloudy patches, and Urbani's blood oxygenation levels began to deteriorate. He started lapsing in and out of consciousness.

"One of our WHO staff is hospitalized in a major Bangkok hospital with SARS, and is in critical condition" wrote Brian Doberstyn, director, Combating Communicable Diseases, Western Pacific

Region of the WHO, to colleagues at the University of Frankfurt. "He will need to be intubated during the next 12 hours. We would greatly appreciate it if you could make available one of your senior physicians to assist in the critical care of our patient and colleague." The faxed letter noted that expenses would be paid by the WHO, and that the WHO hoped that the expert "will be able to take the plane to Bangkok today." The WHO takes care of its own.

But a top German respiratory specialist is no match for a virus that has subverted the genomic machinery of respiratory cells. Urbani's lungs were filling with fluid. Starved for oxygen, his organs began the process of gradually shutting down. First the brain, then the liver, then cardiac arrest. You die organ by organ, cell by cell, aware at some reptilian level that your life force is being leached out of you. We will never know how much it hurts.

CHAPTER 45

IT COMES IN A CONTAINER THAT MEETS INTERNATIONAL SHIPPING standards for infectious materials—watertight primary receptacle, secondary packaging, and outer shell. That outer shell is broken down inside a BSL-2 laboratory. The secondary receptacle is usually a UN62-compliant shipping package, and that plastic bag with the sample vial inside—if it was prepared correctly—should be sealed in such a way that you are not handling any contaminated material until you open the primary container beneath a laminar airflow cabinet, in this case in a BSL-3 laboratory.

This sample had come from the Ministry of Health in Thailand. It was a lung biopsy from a doctor who had fallen ill en route to Bangkok, someone named Carlo Urbani. The samples, after staining and preparation, were mounted on Cynthia Goldsmith's electron microscope, and damned if she didn't believe she was looking at a coronavirus. Goldsmith, a soft-spoken woman with a pageboy haircut who has stared down some of the most frightening killers the world has known—Ebola, Sin Nombre, Nipah—couldn't believe her eyes. She quickly e-mailed Sherif Zaki, the chief of infectious-disease pathology at the CDC. Under the subject heading THIN SECTION OF EM ISOLATE, Goldsmith wrote, "I can't believe it, but it looks as though Tom's group has isolated a coronavirus. The cells were 'fried' by the microwave, but I'm pretty certain (90%) that's the result."

Independently, the CDC had arrived at the same conclusion as their counterparts on the other side of the world. What was astonishing to Cynthia Goldsmith was what she was seeing through the electron microscope: a coronavirus was a very unlikely culprit. The name "coronavirus" comes from the spiky formations on the virus membrane. This viral family was better known for causing disease in chickens and pigs, and common colds in human beings. It had never been associated with severe disease in humans. Yet as she and Zaki took turns looking through the optical viewer, they were both convinced that this was a coronavirus. This virus had not even been on their list of possible suspects. No one had ever heard of a fatal coronavirus, and now here they were staring at one isolated from the lung tissue of Carlo Urbani.

Within three days, they completed the immunofluoresence testing and decided that what they had cultured in the lab was probably the agent for the SARS virus. They made their announcement on Monday, the twenty-fourth of March, three days after the University of Hong Kong team.

This in no way detracts from the remarkable work done by the U.S. CDC, which remains, in the opinion of the international virology community, the gold-standard laboratory. However, in this instance, the upstart, jury-rigged, no-budget lab from Hong Kong had somehow managed to culture and grow the virus before the best laboratory in the world, and in the ensuing months of media coverage, this would have some scientists from the U.S. CDC fuming over what they perceived as the University of Hong Kong's taking a disproportionate amount of credit. Despite all the great lip service given to the Global Laboratory Network by Klaus Stöhr and the World Health Organization, for those virology jocks doing the benchwork around the world, what mattered was who got the bug first. And the team at the CDC, despite its stellar work, had come in a very close second. CDC director Julie Gerberding, in her announcement of the breakthrough, did not mention the University of Hong Kong team, instead stating that the National Center for Infectious Disease (part of the U.S. CDC) team had "discovered a novel coronavirus."

During a visit to the CDC's Clifton campus, I recall that Tom Ksiazek, the acting chief of the Special Pathogens branch, became a bit testy when asked about whether the Hong Kong lab had, in fact, made the breakthrough first. "Anyone can find a virus," Ksiazek said. "People could have a lot of viruses that have nothing to do with this disease." He went on to point out that the Chinese labs had been claiming that *Chlamydia* was the agent. Then he pointed to one of the doors leading to a BSL-3 lab. Taped to the glass was a color Xerox photo of John Tam of the Chinese University of Hong Kong, who had mistakenly claimed that a paramyxovirus was the agent. "Those labs were saying all kinds of things," Ksiazek insisted. "We don't come out and say anything until we know we are right."

But the Chinese University of Hong Kong and the University of Hong Kong were two completely different teams, and the University of Hong Kong and Malik Peiris had gotten it right.

And they had it three days earlier.

CHAPTER 46

- **March 22, 2003**
- **Amoy Gardens, Kowloon, Hong Kong, China**
- **1,519 Infected, 146 Dead**

A VIRUS IS AN ABSTRACTION. EVEN WITH A NAME, A GENUS, EVEN with colorized photographs taken through electron microscopes, the notion that there was this killer out there that might or might not be airborne, that had caused schools to close and hospitals to cancel all nonessential surgery, made us feel that we faced an ineluctable opponent. We wore masks, we washed with an antiseptic solution, we swabbed ourselves with alcohol-saturated gauze. "The terror of the unknown is seldom better displayed than by the response of a population to the appearance of an epidemic, particularly when the epidemic strikes without apparent cause," wrote Edward Kass in the *New England Journal of Medicine.* You tell yourself that it remains a statistical unlikelihood you could even come down with this disease. The stricken tend to be the old and infirm. And most of those who become sick recover. It is just the very oldest and weakest who die. At first.

Meanwhile, Hong Kong Secretary for Health, Welfare, and Food E. K. Yeoh continued to assure us that the disease was confined to Prince of Wales Hospital. "Words like 'Hong Kong has been quarantined' are detrimental to Hong Kong," Yeoh said. "There is no reason to panic. The outbreak has been largely contained to the hospital system." While that sounded reassuring, the reports from Prince of Wales Hospital itself sounded catastrophic. Virtually everyone who had set foot in Ward 8A had been infected. And even as E. K. Yeoh was offering his assurances, clusters of new cases were popping up in hospitals throughout the city. Yet Yeoh's were the rationalizations familiar to anyone who has lived through a disease outbreak, and in Hong Kong, as we accommodated ourselves to a spring spent indoors, behind masks, under virtual quarantine, we fell into a somnambulant routine of checking the daily infection numbers, monitoring the Department of Health website, and participating in the office pool on the number of new cases each day.

From my office window, visibility had been reduced to just a few hundred meters; there were days when the admixture of pollution and fog was so thick that I could see nothing at all, just murk. In Hong Kong, year-round, we kept tiny heating elements on in our closets, otherwise our garments would become saturated with mold. During this particularly sticky spring, the fungus overcame one of my sweater closets, leaving a puffy growth like putrid sea foam that was almost impossible to wash away. If a climatic condition could be described as ideal for epidemics, this was it: damp, dank, dirty, a perpetual precipitation that seemed like the condensation drip from a billion rusty air conditioners. Even without the possible scourge of this new virus, there was something unhealthy about the nautical wetness. We were living in sea spray laced with factory fumes from upriver in China. The prevailing winds blow down from the Pearl River Delta; the exhalations and effluvium of tens of millions, the exhaust and aerosolized waste of a million factories putting in triple shifts, all at sub–U.S. EPA standards. Our Era of Wild Flavor.

It was the kind of air on which a virus would just love to hitch a ride.

What was changing? For one thing, you didn't dream of going to China, unless you were a reporter working for *Time*, in which case you were coerced into going and then put into isolation back at your apartment for seventy-two hours. Reporting on SARS was done on an opt-out basis, and we never had one reporter opt out. For civilians, there were myriad inconveniences, ranging from an acquaintance of mine canceling her regular mah-jongg game for virological reasons to my actor friend lamenting the fact that he couldn't go up to Mission Hills, in Shenzhen, to bang his new mistress. "I finally found the perfect one," he explained to me, "exactly the type I like. Not too thin. Perfect face. I mean, I really like this one. And now I can't go up there because of this stupid disease." He shook his head at the absurdity of it. "I guess I'll just have to settle for a rub and a yank," he said, using his expression for a massage with what is known as a happy ending.

SARS is hell.

A PATIENT COMES IN FOR A REGULARLY SCHEDULED PROCEDURE. The thirty-three-year-old Wang Kaixi lives in Shenzhen and routinely has his dialysis done at Prince of Wales Hospital. He stays with his brother on the sixteenth floor of Block E of a nineteen-building housing complex called Amoy Gardens, a typical middle-class development on the Kowloon side of Victoria Harbor: five thousand units, seventeen thousand residents, originally built in 1968.

Wang is admitted to Ward 8A, where another patient suffering from pneumonia is being nebulized so that he might expectorate some of the phlegm doctors fear is curtailing his blood oxygenation. That patient inhales a lemon-scented expectorant; he exhales a lethal viral load in fine, aerosolized droplets.

Nearly every patient and doctor on the ward breathes in infectious loads of virus particles.

How many particles? Remember, a virus doesn't, technically, do anything. It just sits there, borne aloft on a droplet or particle, until by chance it nestles in a respiratory tract or mucous membrane. Perhaps the dialysis patient Wang blinks, and a few dozen, a few hundred, a few tens of thousands of SARS viruses enter his circulatory system. Or, even more likely, a thousand or so particles land on his hand, he rubs his eyes, picks his nose, and there, the virus gains entry. The vast majority of particles cannot find the appropriate cells to subvert, or maybe the genomic machinery of red blood cells is inhospitable; likewise the cells in the ocular nerves or even the upper respiratory tract. These virus particles are passed harmlessly through the body and shed via the gastrointestinal tract.

But a few hundred to a few thousand particles find their way onto the highly variegated surface of the lungs, an area that, if unfolded, stretched, and flattened would constitute a quarter acre. The surface consists of hundreds of millions of minute branches culminating in bronchioles that are the mechanisms by which the body absorbs oxygen molecules. That tissue layer of alveoli and bronchioles is all that separates the human bloodstream from the air we breathe. "All a microbe had to do to gain entry to the human bloodstream was to get past that .64 micron of protection," wrote Laurie Garrett in *The Coming Plague*. The SARS virus, by chance, falls upon the alveoli, comes upon lung tissue teeming with cells that facilitate the exchange of oxygen into the circulatory system, and upon these cells—which it finds by drifting, by falling, by being inadvertently carried and hauled by particles, by other cells, by protein wastes in the body—it lands on the cell membrane and is able to bond through surface peptides and pass into the cell. And there, in this human lung cell, a most unlikely place for this particular virus, it is able to subvert the cellular machinery and convert the DNA proteins and nucleotides to its own RNA sequences of proteins. The cell itself starts producing more virus, and soon the cytoplasm of the cell is teeming with virus, as John Nichols observed through his electron microscope, and the membrane is ready to burst. A hundred thousand virus particles spill out and rely entirely on chance to find another cell to infect. Once the virus has found this molecular perch in the leukocytes (lung cells), the body's own immune system kicks in and there is a massive and rapid infusion of the body's infection-fighting weapons. The airy, almost fluffy lung tissue then becomes increasingly heavy, pulpy, and meaty, as an inflamed mix of cell debris, fluid, enzymes, and scar tissue transforms into what doctors call "consolidation." There is something almost cheerful about the delicate pink of a healthy lung—if you were to slice through it, it would be like taking a straight razor to the meat of a raw sea urchin. A SARS patient's lungs become blue or black and inelastic, almost hard to the touch. On an X-ray, this shows up as the "whiteout" effect doctors were noting as SARS cases progressed—when you shoot an X-ray through the dense matter of a pneumonia-stricken lung, it appears as lighter than the rest of the lung. A lung that is in this "whiteout" condition is a lung that has virtually stopped functioning.

It happened in Ward 8A, repeatedly. And in this case, the thirty-three-year-old Wang inhaled, finished his dialysis, and returned to his brother's flat on the sixteenth floor of Amoy Gardens. There, stricken by diarrhea, he used the toilet, repeatedly.

EACH BUILDING OF AMOY GARDENS, VIEWED FROM ABOVE, LOOKS almost like a swastika, with a narrow, two-yard-wide airshaft penetrating the center of each arm. There are eight apartments per floor and thirty-three floors, making for 870 residents divided among 264 households. The hallways are so narrow that one person has to turn sideways if he passes another coming from the opposite direction; the elevators are similarly scaled, and each bank serves eight floors. The building was constructed with the barest of infrastructure—little more than internal wiring for each apartment and water taps in the bathroom and kitchen, leaving the air shafts to serve as the sole conduits for plumbing fixtures, sewage piping, and electrical cables. The pipes are old, and while the management company had recently begun replacing them with newer PVC tubing, Block E still had its old, decaying pipes in place. The air shafts are so narrow, residents liked to say, that they could shake hands across the gap.

It is three floors—about twenty-six feet—from Wang Kaixi's bathroom to Anna Kong's bedroom. Kong, twenty-four, slept in a bedroom the size of a first-class airplane seat, in an apartment the size of a cruise ship compartment. Her years of secondary-school education and her stint at junior college had left her proficient, according to at least one photo of her, at applying a pallorous coating of makeup and teasing her black hair into thick curtains. The five-by-six-foot bathroom in which she prepared herself for work had the usual trio of sewer openings modified to serve as bath and sink drains and toilet. The floor was cheap, chipped enamel tile with grouting stained fecal brown with mildew. There was a fan—which the family switched on every time they used the head—above and to the left of the toilet, next to the tank. Beside the toilet was one additional drain, the covering of which had long ago broken off and been lost.

Anna's mother, Alice, and father, Francis, who slept in the room next to hers, had left the apartment before Anna had risen. The family lived on the sixth floor of Block E of Amoy Gardens, in a flat rented from a family that had purchased the apartment ten years before. The Kongs paid 5,400 Hong Kong dollars, about 800 U.S. dollars, per month rent to live in the crowded conditions that passed for middle class in Kowloon—after Bombay the most densely populated community on earth. The three-hundred-square-foot apartment was simply too small for three human beings in materialistic Hong Kong. As a result, their kitchen/dining room/living room was barely large enough to accommodate a thirty-five-inch television and an overwhelmingly large leather sofa. The family had resorted to storing their out-of-season wardrobe atop the TV and at the far end of the sofa in plastic bins that Alice had purchased specifically for that purpose. The flat was so small that one air conditioner set on low could render the temperature practically arctic. As a result, most of the time the family kept the narrow windows to the air shaft open.

A few futile attempts had been made at decorating the flat— Anna's father kept his collection of old but not-very-precious currency in Lucite blocks up on a rickety plastic shelf mounted next to the refrigerator. There was also a poster of Twins, one of Anna's favorite groups, and an imitation of a fourteenth-century scroll painting of the misty, jagged karsts of Guilin. The flat provided virtually no privacy, and only those who had grown up in similarly close quarters—or who had served on a submarine—could have stood such living conditions. I was always surprised when I visited homes like this, at how many compromises one had to make merely to get through the day. For example, if two people were visiting two residents, there was never space in the living area for all four to sit down simultaneously. One of the hosts would invariably stand or retreat to a bedroom for the duration of the stay. And with so much living being done in such a compromised space, a homemaker's best efforts at hygiene would eventually fall short when she was faced with the sheer amount of detritus and waste created by three or four human beings and one or two pets—Alice had a cat named Mei-mei. One bag of trash, especially the remnants of any dish cooked with the fish oil that is such a common ingredient in Cantonese cooking, could stink up the whole place.

This was, of course, the lifestyle of the vast majority of Hong Kongers. When I visited Amoy Gardens, I realized how lucky I was to live on The Peak, for my children to have a bedroom the size of an entire flat here. And then I would further reality-check myself that even these people were lucky compared with those living just a few miles away, in the truly squalid conditions of places like Chung King Mansions. (Mansion, Garden, Estate, Park, Court—every Hong Kong tenement, no matter how evocative of Upton Sinclair's Chicago, would have a similarly grand name.) These complexes were, as K. Y. Yuen reminded me, "highly vulnerable to disease outbreaks. Population density is the key. A virus jumping from animal to human will be amplified by Hong Kong's housing situation. This city is a perfect disease amplifier."

For Anna, such worries about novel viruses must have seemed as far-fetched as the chances of her father's winning the Mark Six lottery he played every week. She was probably more concerned about her job at Delicious Kicks, in Causeway Bay. The shop, like most of Hong Kong's retail outlets, was being hit hard by the drop in tourism and the general decline in retail trade since the onset of the SARS epidemic. The shop sold primarily branded knockoffs of designer clothes, though the customers might not recognize the floral-print sheer blouses, bell-bottoms, copiously zippered trousers, or zip-up suede boots as being imitations because they wouldn't be aware of the originals. The shop didn't sell to *tai-tais*—socialites—but to girls like those who worked there: Hong Kong's vast working class perhaps knew the labels and brands but could not recognize the real from the fake.

Anna had finished high school and completed two years of junior college at East Island Woman's Preparatory, fecklessly studying Mandarin word processing and becoming just proficient enough to earn a degree but still not competent enough to keep her first job as secretary for a solicitor who helped Chinese immigrants set up new businesses. Her good looks had gotten her hired; her poor skills had gotten her fired.

She was quickly told by her mother, Alice, that she had better find another job: she was still responsible for paying seven hundred Hong Kong dollars, about ninety U.S. dollars, toward her share of room and board.

She found the job at Delicious Kicks the next day, and though her father bemoaned the wasted years and tuition spent at East Island, the work suited her far better than taking down information from haughty, garlic-breathed, tax-dodging mainlanders.

But one morning, after applying her makeup, she told her father that she felt a mild ache in her shoulders and neck; later, as she rode in the minibus to Hung Hom and then caught a double-decker through the tunnel to Central, she must have felt she was coming down with a cold. SARS was a disease that the TV news had been waxing hysterical about but which she didn't take seriously as a threat. Why would it have occurred to her as she went through her day, organizing clothing racks, putting away shoes, and smoking in front of the shop, that she could possibly have caught this disease that was causing pandemonium in local hospitals?

Even when she returned home and sat down to a bowl of soup and turned in early after taking two paracetamols, she doubted that anything was seriously amiss. But in the morning, when she heard her mother preparing tea in the kitchen, she called out and said she was too sick to go to work. Her mother consulted a Chinese herbalist for most of her family's ailments, but she had adopted the pragmatic belief, like many Hong Kongers, that herbalists were best at preventive medicine while Western physicians were better at alleviating sickness. Her daughter, Alice decided, needed symptomatic relief for what seemed like a severe cold.

She helped Anna get dressed in sweats and a T-shirt and took her downstairs, past the McDonald's on the complex's ground level and out onto Prince Edward Street, to Dr. Drew Lau's office. Lau had been the family physician on and off since the Kongs had moved into

Amoy Gardens, and as he examined Anna, he asked her to describe her symptoms.

She relayed a list: headache, body ache, fever, and "I'm very, very tired."

"And she has no appetite," added her mother.

Dr. Lau told her to get plenty of rest and prescribed an antiviral medication. "Let's see how you feel in a day or two."

Back in her room, beneath her coats and leather jackets hanging above her bed on plastic hooks, Anna must have drifted into the waking nightmares familiar to SARS patients. She would then rouse herself with a start and charge across the hall every few minutes to emit a viscous diarrhea; the act of defecating and wiping would be so strenuous that by the time she returned to her bed, she would find herself out of breath and gasping. She would then have to lie still until her next bout at the toilet; the merest shifting in bed left her so tired that she feared she wouldn't make it back to the bathroom. The aches were such that she didn't sleep so much as lie still and hurt. Her mother attempted to get her to drink some soup and then, failing that, convinced her to drink a glass of orange juice. Anna wanted to drink the juice; she just couldn't muster the energy to sit up in bed. While she rested in the afternoon, her mother tried to put in a half day at her job doing the bookkeeping for a private financier of car loans. When she returned home, she found a note stuffed into her front door. It had been left by a nurse from the Health Department warning there had been an outbreak of atypical pneumonia cases in the area and that anyone with the following symptoms should see a doctor: fever, body ache, coughing, no appetite, listlessness. It did not mention diarrhea.

That afternoon, Anna's mother brought her back to Dr. Lau, who performed a chest X-ray. The X-ray revealed two spots on her lungs. Lau took one look at the X-ray and ordered Anna admitted to United Christian Hospital.

CHAPTER 47

■ March 27, 2003
■ United Christian Hospital, Kowloon, Hong Kong, China
■ 1,712 Infected, 168 Dead

"WHAT HAPPENS IF A DEADLY VIRUS FOR WHICH WE HAVE NO TREAT-ment or cure explodes into the middle of a major city?" C. J. Peters, former chief of Special Pathogens at the U.S. Centers for Disease Control had asked a few years earlier. The answer: system failure.

Chan Chi Ming, thirty-nine, a chest physician at United Christian Hospital in Kowloon, had never seen a twenty-four-hour surge like this. While all of Hong Kong's hospitals had been admitting new SARS patients at a rate of about one hundred per day, UCH was going to clock in half that many by itself in one night. The first family to show up had arrived a little after midnight; then came an elderly man, a young woman, another few families, and finally the twenty-four-year-old Anna Kong. They were all admitted to the four wards that were taking SARS patients, and when Dr. Chan arrived at 9:00 A.M. to begin treating them, he was struck by the fact that the first two he treated had virtually the same address, the housing estate called Amoy Gardens.

He asked the admissions nurse to go through the addresses of other admitted patients. When she did, she found an alarming trend: the cases were clustered in the same housing estate.

A day earlier, across the harbor in the Department of Health's Communicable Disease Division, Thomas Tsang, the department's top epidemiologist, had also begun to notice a clustering of cases at the same location, Amoy Gardens. There had been fifteen Amoy Gardens cases on March 26, twenty-seven on March 27, and thirty-four on March 28. Eventually, more than sixty new SARS cases per day were being hospitalized from Amoy Gardens, the vast majority of them from Block E, which would finally label the disease a "community outbreak" rather than one confined to the hospital system.

While in one Hong Kong Island hospital, just a few miles away, virologists were determining what agent had caused SARS, in this housing estate, which was much more representative of Hong Kong, the disease was now mounting its most brutal and deadly assault. In total, 326 residents of this one housing estate, including Anna's mother, Alice, would become infected. What made this all the more terrifying for most Hong Kong residents was the banality of Amoy Gardens. The estate, as it was shown on RTHK television news, looked like any of a thousand other housing complexes. This was no southern Chinese boomtown where hayseed peasants were feasting on snakes and rats: these were buildings just like home, with a 7-Eleven and McDonald's and Park 'n' Shop downstairs. If SARS had struck there, it could surely come here as well. The kudzu-like spreading of the virus in Amoy Gardens was ominous; it had infected residents on the first through twenty-ninth floors of Block E, and had even infected residents in blocks B, C, and D.

The key question health officials and doctors were now asking was, "Has it gone airborne?" Previously, the consensus on SARS was that while it could be spread by particles borne aloft, the virus itself was not airborne. This was a crucial difference, for an airborne virus, like influenza, is far more infectious than one that requires droplets—such as mucus expelled via a sneeze or a cough—in order to spread. You can catch the flu by standing in the same room with someone who has it, or even entering an elevator a few hours after an infected person has left it. While some alarming tests were already showing that the SARS virus remained infectious on a surface for seventy-two hours, as long as the disease required extended direct contact—such as a shared hospital ward—then there was a chance for containment.

While Canada, Singapore, China, and several other countries still wrestled with this novel virus, the world had received its first small virological victory over SARS when Malik Peiris and crew, shortly followed by the team at the U.S. CDC, isolated the virus. And in terms of public health, there was a glimmer of hope in Vietnam. The first flurry of cases at Hanoi French Hospital had been every bit as overwhelming as those in Guangdong, Toronto, or Hong Kong, but after a brutal wave of two dozen cases, and several fatalities, including Carlo Urbani, the disease now appeared to be in check. It certainly hadn't spread widely in the community. This meant that impeccable infection control, combined with first-rate contact tracing and immediate isolation of suspected cases, could curtail SARS. The WHO, because of Urbani and his colleagues' tremendous work in Hanoi, had a blueprint for how to slow down, if not subdue, a hospital outbreak. In Singapore, the government reacted with its usual heavy hand, proving that an authoritarian impulse channeled into public health could show positive results. "We approached fever detection with religious devotion," says Lim Suet Wun, the chief physician of Tan Tok Seng Hospital. "We had to maintain one hundred percent compliance. The first sign of fever from anyone, staff or patient, had to be identified and the person isolated." N-95 masks at Tan Tok Seng were fitted and then tested with an odoriferous gas that, if detected by the person in the mask, meant that the mask had to be refitted for a tighter seal. A former malaria quarantine station was converted by the Singapore Ministry of Health into an emergency SARS hospital, complete with more than one hundred isolation wards. The city-state closed down wholesale markets, cabbies were ordered to pass out receipts to make it easier for the government to track down passengers, doctors were forbidden to move between hospitals. Tan Tok Seng, designated as a SARS hospital on March 22, suspended regular admissions and referred all emergency cases to other hospitals. "We had a zero-tolerance policy for breaking quarantine," says Balaji Sadasivan, Singapore's minister of state for health and transport. By the end of March, nearly one thousand citizens were quarantined at home, their front doors monitored by video camera.

Hong Kong, however, with its greater civil liberties and laissez-faire tradition, could not implement a similarly strict quarantine. Secretary of Health, Education, and Food E. K. Yeoh and Director of Health Margaret Chan consulted with Hong Kong's chief executive, Tung Chee-hwa, daily as to what measures were legal and what would be viewed as a violation of civil rights. As a result, Hong Kong's precautions seemed almost comical compared with Singapore's bold response. For example, the Department of Health ordered suspected SARS cases to report once a day to designated hospitals, therefore guaranteeing that several hundred fever patients would be traveling on crowded mass transit daily. Also, Hong Kong's Hospital Authority compounded the problem by not shutting down infected hospitals to non-SARS cases until well after the epidemic had broken out, and also by failing to have in place a contingency plan for an infectious-disease outbreak prior to SARS. "This SARS is very, very new and very, very elusive," Dr. C. H. Leong, chairman of the Hospital Authority, later told me. "It took a lot of us by surprise. I had never seen anything like this in forty years as a doctor. This was a military situation. This was war. Sometimes in war you can't see so clearly what is going on." Further confusing matters was the March 23 hospitalization with SARS of Hospital Authority chief executive William Ho. Though he would eventually recover, his absence during those crucial weeks complicated the command structure and may have, according to a government report, "compounded the difficulties faced by the Hospital Authority in its handling of the outbreak."

Keeping the emergency wards open virtually assured that the disease would go from the hospitals into the community. "Hong Kong has a very hospital-centered health care system," says Lo Wing Lok, the Legislative Council member representing the medical profession. "Hong Kongers go to the emergency room for almost any minor issue." This was because treatment costs for most hospital patients were often capped at nine U.S. dollars. Unfortunately, until mid-March of 2003, those emergency rooms were also the best way to spread SARS. Only the Hospital Authority, whose budget dwarfed the Hong Kong Department of Health's budget (4.3 billion dollars compared with about 500 million dollars), could implement protocol changes in Hong Kong's public hospitals. "We were waiting for the HA [Hospital Authority]," says Joseph Sung. "That's how the system works."

Yet E. K. Yeoh, who had served at the Hospital Authority before becoming secretary of health, would continue to insist that this outbreak was primarily a hospital rather than a community issue. "His messages to the public at the briefings . . . were confusing and misleading, and they also gave the impression he was trying to downplay the severity of the outbreak," concluded the Legislative Council's report on the handling of SARS.

Had Yeoh and the Hospital Authority acted faster, then Wang Kaixi, the dialysis patient, would not have been admitted for routine treatment to Prince of Wales Hospital after the virus was already burning through Ward 8A, and he would never have brought the virus back to Amoy Gardens. That explained how the virus got from the hospital to Amoy Gardens. But how did it spread once it was there?

Sometimes, as I walked around Amoy Gardens and rode the elevator up in Block E to visit SARS survivors and family members, I would play Guan Yi's game of pretending to be a virus. How would I get from the eighth floor to Anna's apartment on the fifth floor? I would stick my head out of an eighth-floor bathroom window and look down, past the intricate web of pipes and wires, and try to pick out Anna's window a few feet below. A virus particle could simply fall, of course, size providing no exemptions from gravity. But then how would it actually gain entry to Anna's flat and then into her respiratory system? If I were a virus, I could easily go from Wang's eyes or mouth or even anus to Wang's hand and then out the door with him or his brother, to the elevator's ground-floor button—everyone pushes that button—and then Anna comes in and she pushes the same button, and I am then on her index finger and then in her eye. And I am in.

But what about all those infected in Block E who didn't share an elevator bank with Wang Kaixi, what about those hundreds in the other blocks? This complex spans a square kilometer; residents here share the same dry cleaner, the Cantonese restaurant on the main floor, the McDonald's. A virus could be hiding out in any of a number of nooks or crannies. But then, why wouldn't everyone in the complex have become infected?

This had stumped even the best epidemiologists in the world. I sat in Anna's flat a few weeks later and listened to her grieving mother talk about how when she came back out of a coma and was deemed sufficiently healthy, her husband, Francis, had tearfully told her that their daughter hadn't made it. Anna's bed had been wheeled to an observation station, where her father waved to her from behind glass. He couldn't tell if she recognized him—she was already intubated, and her oxygenation level had dipped to the point where she was suffering from severe hypoxia and was probably demented. Most distressing for Francis had been his feeling that his daughter was in pain. She couldn't speak, nor could she move, but there was in her eyes and in the twitching of her nose a tremble that Francis knew reflected deep, unsoothable hurt. Morphine would further hinder the respiratory system, he had been told. So she was without pain medication.

He had brought flowers, for some reason, because he felt he had to bring something. They seemed pitifully inappropriate, yet they were precisely what the nurse chose to focus on as she shouted into Anna's ear while Francis watched from behind the glass.

"What lovely tulips!" the nurse said. "You have a very caring father!"

Why are you yelling? Francis wanted to ask. She's not deaf.

Alice was relating these stories to me as Francis sat on the sofa, watching the news on television. While Anna was hospitalized, SARS had become the text crawl across the bottom of the screen. While tanks rumbled toward Baghdad, a virus was storming Hong Kong.

It embarrasses me, but I would sometimes tune out Alice's lamentations of her daughter's missed opportunities and gaze around the apartment and wonder at how a virus got in. Was there virus in there now? Was Alice exhaling particulates as she spoke? She had been healed, was no longer shedding virus, the doctors had told her. Still, the virus had somehow gone at least twenty-seven feet, from one of Wang Kaixi's orifices to one of Anna's. But how?

One theory postulated that rats were the vector. The same *Rattus rattus* that had brought plague to Hong Kong one hundred years before would later be shown to be carrying SARS coronavirus on its feet. The apartment complex, despite regular exterminations, was swarming with rats. Stephen Ng would write in *The Lancet* his hypothesis that "the epidemic could have been started on March 14 by a rat from Block E going into the apartment visited by the index patient and being infected by contaminated material, such as used tissue paper, leftover food or excreta." The rat then becomes contagious, and "secretions from infected rats, such as urine, droppings or saliva, contain large amounts of virus." When in doubt, blame vermin.

"Have you ever seen any rats?" I blurted out.

"We have a cat," Alice reminded me. "Rats are not a problem for us."

She had a point. If not via rats, then how? Myriad theories were posed by an increasingly hysterical Hong Kong press. Newspapers were running banner headlines every day replete with misinformed "expert" opinion on which medicines might be most effective for preventing SARS. Various masks were reviewed. The *Hong Kong Standard* ran an advertisement on its front page for a product called Lamp Bergere, a scented humidifier that was being touted for its putative antiviral properties. The *Apple Daily* ran a remarkable diagram positing one infected construction worker as the cause of hundreds of cases at Amoy Gardens. The cartoon figure stood atop a construction site across the street from Amoy Gardens and was pictured urinating into a diffusing wind that spread his presumably virus-saturated urine in a fine mist over the whole housing estate.

That was fantasy. The reality looked like this: Anna in an isolation ward with two windows, both looking out onto the street. In the evenings, when the white fluorescent lights were on in Anna's room, Francis would stand in the street and talk to his daughter by phone. He would stay in the cone of a streetlight, pacing back and forth or just leaning against the cold metal. The first few nights, Anna hadn't yet been intubated. She could hold up a mobile phone to hear and listen to her father, though she lacked the energy to respond. She would sometimes sit up in bed and wave, and Francis would wave back. After her condition turned even more critical and she was intubated, a nurse held the phone up to her ear for her. The nurse would come on the phone after fifteen minutes and tell Francis that she had to continue her shift, but that Anna's eyes had moved as she listened, and she believed that Anna understood what Francis had been saying. Francis would thank her, and then the nurse would put the phone to Anna's ear one more time so Francis could say good-bye and tell her to keep fighting. He told Anna that Alice was also sick but was already getting better, trying to inspire Anna to do the same.

Then, after he hung up, he would stand under the streetlamp for a moment, gazing at the lighted ward windows.

One morning, he was told that Anna had died. A doctor whose name he didn't remember called saying she had passed away overnight. Her body was being held in the hospital crematorium and would not be released until the Department of Health decided what to do with SARS victims. The doctor sounded as contrite as a very busy man could.

That evening, for some reason, Francis thought about heading up to the streetlamp before stopping himself.

CHAPTER 48

- **April 1, 2003**
- **Amoy Gardens, Kowloon, Hong Kong, China**
- **2,119 Infected, 243 Dead**

"ONE OF THE MOST STRIKING CONSEQUENCES OF THE CLOSING OF the gates," wrote Albert Camus in *The Plague,* "was, in fact, this sudden deprivation befalling people who were completely unprepared for it." An isolation order for the residents of Block E had been in effect for ten days, given by Dr. Margaret Chan and enforced, in theory, by the Hong Kong police. The reality, however, was that many residents continued to come and go. As to why the residents were not officially quarantined, Margaret Chan would later explain that authorities were worried that panicked residents of Block E would go underground. "As it was, we stopped using the lifts, started using the less frequently traveled exits," according to resident Alice Yuen.

Meanwhile, cleaning crews in biosafety suits arrived and began swabbing every surface of the building—floors, walls, and ceilings—with a bleach solution, forcing the residents to wait in the common areas as they did so. "We had to pay for that ourselves," said Aswin Chan, of the management company that ran Amoy Gardens. "The government wouldn't even take responsibility."

Finally, as public health officials became aware of the environmental risk posed by Amoy Gardens—whatever had caused the rapid rate of infection was most likely related to something in the buildings themselves—they ordered an evacuation. For Block E, the notice of evacuation was made at 6:00 P.M., when the newscasters on Jade TV announced the residents had three hours to pack up and go. "By the time it happened, we were actually grateful," said Ivan Yuen. "We knew there were at least eighty cases. And since we lived on the twentieth floor, we felt like we were watching the disease go up each floor every day. There would be cases on the eighth floor, then the ninth, then the twelfth. It seemed like a matter of time." Public health officials would come in for harsh criticism for forcing hundreds of residents to flee by giving them three hours' notice. Yet the other option—a quick and sudden midnight roundup—would have smacked even more of authoritarianism, and was possibly illegal.

Residents were assured that they would be bused to a "safe, hygienic environment." Indeed, the camps that the Department of Health had prepared for the evacuees were the bucolically named Lady MacLehose Holiday Village and Lei Yue Mun Park and Holiday Village, vacation spots constructed during the colonial era to provide low-budget holidays for civil servants. Despite the promise of vacations in the sun, hundreds of residents took the opportunity to break quarantine and flee, leaving police to ask their neighbors for any possible contact information. The impression of those potential SARS cases slipping through the net contributed to Hong Kong's mounting sense of panic.

As the ten sixty-seat minibuses were loaded, camera crews and newspaper reporters gathered to watch the embarkation as if these were adventurers embarking on some sort of expedition. Cantonese reporters shouted across barricade lines to the bewildered evacuees pulling roller bags. Relatives sought a glimpse of their kin. Once loaded on the buses, the evacuees had to wait for several hours as police closed the roads along the intended route. This seemed the first of what were sure to be more drastic measures.

WHILE I COULD SEE AMOY GARDENS FROM MY OFFICE WINDOW, I could not make it out through the haze from my apartment up on The Peak. The gulf between those of us at this elevation and those living in Amoy Gardens was economic, social, and now, it seemed, virological. Those in the housing estates on the Kowloon side were far more likely to catch SARS than those of us living in tony neighborhoods like The Peak. It was statistical fact. All the apartment complexes where there were clusters of cases—Telford Garden, Lower Ngau Tau Kok Estate—were in Kowloon. Cases on Hong Kong Island were scattered and infrequent, and tended to affect health care workers who had a clear source of infection in a hospital.

To cling to the notion of class or wealth as somehow providing a cordon sanitaire from SARS would be as preposterous as London's medieval nobility believing the poor were more likely to suffer from plague. It seemed that the crowded Amoy Gardens was merely the logical first wave of virological attack. The disease would strike there first because of the higher concentration of susceptible immune systems; but a contagious disease should eventually find its way to every fetid alley and gilded parlor of this city. I thought of that scene at the end of the 1968 black-and-white film *Night of the Living Dead* in which the few surviving uninfected humans are holed up in a house atop a hill. The scourge was coming; it had to be. The only way to keep it away, one fellow journalist announced, was to lock your doors, load a shotgun, and shoot anyone who tried to get in. From up here on The Peak, I imagined I could watch the spread of the infection from Amoy Gardens, down Jordan to Tsim Sha Tsui, across Victoria Harbor by ferry, where hundreds of thousands disembarked in Central. From there, the virus was a taxi or tram ride up the hill and down Barker Road and into my kitchen, and then my children's bedroom.

It always comes down to them, doesn't it? The children. We project our fears onto them. Our worry is not for ourselves, for our own lungs, but for the precious children. Thus far, statistics coming out of Hong Kong's hospitals were noting a remarkably low rate of infection among children. It was postulated that their underdeveloped immune systems did not respond to the virus as aggressively as an adult's immune system, with the result that an infected child might show no symptoms at all and never fall ill, though he or she might still be contagious. But in China, anecdotal evidence had presented a different picture. There, the virus had been a more equal-opportunity killer. Though the elderly and decrepit had fallen ill at a greater rate than the youthful and healthy, there had been tragic cases of pre-teens, teens, and twentysomethings passing away.

My daughters refused to wear the face masks we had adopted as almost good-luck totems to ward off the disease. Public health officials would privately acknowledge that the masks, as they were applied by most citizens, offered little or no protection. Yet in press conferences, these officials were reluctant to strip away this last line of psychological defense.

Air traffic into and out of Hong Kong had already come to a virtual standstill. The 301 flights per day that usually left the territory had already been cut by two-thirds, primarily because there was so little demand for flights into Hong Kong that airlines didn't have the planes to fly out. The projected damage to Hong Kong's economy was five billion U.S. dollars, and growth rates for 2003 were already being revised downward by one to two percentage points. The city itself was remarkably desolate. Shops were forlorn. You could walk into any restaurant in town and be seated at a table of your choosing. I recall dining one night at Vong, the usually crowded fusion restaurant on the top floor of the Mandarin Hotel, and finding myself the only patron in the whole joint; the service was impeccable. Gradually, most restaurants were shuttered up as the owners found no reason to stay open and empty every day. "Hong Kong now looks like a city under threat of a biological weapons attack," wrote Jim Erickson in *Time*. It was eerily quiet, and I could make the drive from my apartment to my office—usually a twenty-minute ride—in about seven minutes.

If I was concerned about the continued good health of my family, I also had to consider the well-being of my staff. We really didn't have any plan for running the office through an infectious-disease outbreak—who does? What if the building were quarantined? We had reporters constantly going in and out of hospital wards and infected provinces in China. One of them could easily return from a reporting trip and then, a few days later, after passing the virus along to a few co-workers, start to show symptoms. That would result in the quarantining of my staff, the closing of our offices, and, effectively, the shutting down of *Time Asia*. That had already happened to one floor of PCCW, the local phone company. We needed a contingency plan.

What precautions were my colleagues at other Hong Kong publications taking? I called John Bussey, who oversaw both the *Asian Wall Street Journal* and the *Far Eastern Economic Review*. He relayed to me the remarkable measures Dow Jones was implementing to ensure that they had backup systems in place in the event of a viral cataclysm. They had decided to keep over half the staff at home, working in isolation, so that there would be no risk of the whole staff being quarantined. "It's all about backup," Bussey explained. "We now have it set up so that we could do the entire magazine by remote, if necessary."

He went on to say that he was looking into producing the newspaper out of a satellite office. "You know," he added, "particles from a human sneeze travel eighteen feet." Then he asked, "What have you guys done?"

"Um . . ." I felt pathetically inadequate. "Um, all that stuff you guys have done. Plus, we're looking into a few more contingency plans."

"It's all about backup," Bussey said. "Let me know if you have any good ideas."

Perhaps, I thought, there was someone higher up the Time Warner organizational chart in Asia who had formulated a plan. The only person I could think of was Steve Marcopoto, the president of CNN in Hong Kong. It turned out that he and his family were back in the United States and would stay there for most of that SARS season. I called one of my counterparts at CNN and was told that they hadn't really thought much about the matter, but they would be happy to hear what we were doing. It dawned on me that no one at the company had a plan for how to manage through this disease outbreak. "Nothing was more fatal to the inhabitants of this city," Daniel Defoe had written in *Journal of the Plague Year*, "than the supine negligence of the people themselves."

Determined not to repeat the folly of Defoe's Londoners, I did what managers everywhere do when they are out of ideas: I convened a meeting.

We gathered in a conference room on the thirty-sixth floor. A few of us were wearing surgical masks; others ignored the superstition. For those who wore glasses, the masks caused condensation to build up behind the lenses, which could be cleared only by pulling off their masks for a moment or two, thereby rendering totally moot any theoretical benefit. I was reassured that the likelihood that any of us around this table had been exposed to possible SARS cases was minimal. We were the management class, in charge of marketing, finance, human resources, and production; no one here could have the disease, could they?

But a few minutes into the meeting, as I suggested that we send out a companywide memo saying that anyone who had come into contact with a possible SARS case should stay away from the office, it became clear that everyone in the room already knew someone who might be infected. In fact, our circulation manager had dined the evening before in Amoy Gardens, at her father-in-law's. Another woman said that her sister was in quarantine. The meeting quickly grew quiet as we realized that enforcing such an order would mean virtually shutting down the office.

"It doesn't matter, as far as we are concerned," said Ivy Choi, a vice president of advertising sales. "None of our clients want to meet anyway. Everyone is too frightened."

Throughout the region, businesses were shunning face-to-face meetings. There wasn't much reason for our business side to report to work anyway. We decided that anyone who could work from home, should. Furthermore, there would be disinfectant cleaning of the entire office, as well as the procurement of antibacterial soap, surgical masks, gloves, and alcohol swabs.

As soon as I was back in my office, a young female employee came to me and explained that her father-in-law had been hospitalized with SARS, and she had just read in the paper that all family contacts of SARS cases should themselves see a doctor for a health check.

"You should really be wearing a mask," I told her.

My reasoning for employees' wearing masks in the office was not so the person wearing it would be protected, but so that anyone infected would not be so contagious. That was the rationale given for the precaution in Japan, where similar surgical masks are commonly worn in winter by anyone with a head cold.

The employee said she would. I told her she had to go home immediately and see a doctor.

Then another junior Chinese staffer entered and said that she had to go to Guangdong urgently. "I'm worried about my aunt and uncle," she explained. "I've been calling them, and I don't know if they are infected or not."

Perhaps, I suggested, visiting Guangdong, the likely epicenter of the whole outbreak, was not a good idea.

"I can't stay here."

"Why?"

"My apartment building is so dirty," she said. "It looks like those buildings where they have SARS."

"But still, going up to Guangdong right now isn't going to be any safer."

She sat down on my sofa and began sobbing. "That is so racist."

She was crying and accusing me of discrimination. I suspected this was one of those situations that, if handled incorrectly, would have me labeled as a lousy boss.

I had both to console her and somehow deflect her charges; I am also half Asian and thus an unlikely source for anti-Asian bias.

"These are stressful times," I told her. "Very nerve-racking for all of us. We're going to have most of the staff work from home. Why don't you work from home?"

She continued her demur, sniffling, "I told you, I don't feel safe in my flat."

"But you would feel safer in China?"

She nodded.

"But—"

"You think *all* Chinese are infected," she said. "You *gwielos* [foreigners] think we are *all* sick."

"No," I said. "It's just that there is a WHO advisory against visiting Guangdong."

"You think Hong Kong is any safer?" she asked. "The WHO has also advised against visiting Hong Kong."

I couldn't console her. "Just try to take it easy. Don't go running off to China just yet."

I let her sit a moment on my office sofa as I pretended to be busy with some papers on my desk.

"Okay?" I said, trying to indicate that our meeting was over. I had to add some blandishment about how seriously we took the health and safety of our employees. "You know, I care about you and want you to stay healthy."

She looked at me blankly.

"Shouldn't you be wearing a mask?" I added. It seemed such paltry defense against this virological agent, but what else did we have? We were now living in a city that the World Health Organization had advised against visiting.

OUR REPORTERS IN HONG KONG, GUANGZHOU, BEIJING, SHANGHAI, and Singapore who were visiting hospitals had already been told to quarantine themselves in hotel rooms rather than return to their families. They would work by remote, we decided. We kept one art director out of the office. One of our editors happened to be in New Delhi, where I told him to stay until further notice. My deputy editor, William Green, and I decided that he would work from home, thereby ensuring that one of us would escape quarantine if our offices were shut down. A photo editor who was visiting Tokyo was told to stay there. Half the Hong Kong office staff were told to clear out. We also arranged for a system by which the European edition could transmit directly to our printing plants in the event that we had to completely evacuate the Hong Kong offices and somehow prepare a magazine from London.

It all made some sort of sense on paper, but in reality I knew that if the virus really burned through Hong Kong, as it seemed was possible, then putting out a magazine was probably the least of my concerns. As I walked down the street to the Wellcome supermarket to pick up some disinfectant soap, I was surprised to see a throng around the electronic doors beneath the red-and-yellow sign. There was a crowd filing into the market and an even denser line backed up behind the registers, with shopping carts piled high with cooking oil, rice, toilet paper, and virtually anything else that could be deemed a necessity. Dodging shopping carts, I walked up and down the aisles, noticing that the rice was sold out, as were the eggs, cooking oil, canned soups, ramen noodles, toilet paper, most of the chili paste, oyster sauce, and even the garlic. There was still some cereal, although the Weetabix and cornflakes were in short supply. Count Chocula, for some reason, had survived the run on the supermarket.

Soap had been cleaned out as well, along with most over-the-counter medications and, for some reason, Band-Aids and gauze bandages. What was going on here?

"Quarantine," one woman said to me.

"Infected port," said another.

I was astonished. It had finally happened. Hong Kong had been declared an infected port. There had been rumors to this effect for days. The WHO, or the Chinese government, would declare the city an infected port and ban all travel in and out, therefore implementing a blockade and curtailing trade. There would be shortages. We would be trapped in a starving city. I looked around. There was still plenty of wine for sale.

I took out my mobile phone to call the office. There was a text message on my phone from PCCW, my service provider. The Hong Kong government was assuring all mobile phone subscribers that Hong Kong had not been quarantined and was not an infected port. Then why the run on the market? It turned out that a fourteen-year-old schoolboy had posted a bogus Internet article about SARS under a fake banner from the Chinese-language newspaper *Mingpao*, claiming that Hong Kong had been declared an infected port, that border checkpoints had been closed, and that citizens had been ordered to stay home. Hong Kong's chief executive, Tung, the hacker's message went on to say, would soon be appearing at an emergency press conference to announce the restrictions in light of the fact that "the outbreak is out of control."

The Hang Sang stock index had plunged one hundred points upon release of the news, and in addition to the run on supermarkets and pharmacies, ATMs and bank branches had run out of cash. It took several hours for the government to figure out what was going on and to get a televised statement from Director of Health Margaret Chan onto the air, reassuring Hong Kong that this had been an April Fool's hoax.

In the midst of this tangled urban tragedy, actor Leslie Cheung, the openly gay and very popular Canto-crooner and star of such films as *Farewell My Concubine* and *Happy Together*, killed himself by leaping from the twenty-fourth floor of the Mandarin Oriental Hotel. Leslie Cheung had been a beloved Hong Kong presence, his cheerful, perpetual boyishness making him a mascot of the city's nightlife. Just in case we needed any more signs, Leslie Cheung's suicide was confirmation for many of us that the good times were officially over.

REMARKABLY, IN THE RAREFIED WORLD OF THE PEAK, NO ONE HAD heard of the hoax. Yet the news of Amoy Gardens and the sense that the virus was now at large in the community had finally compelled most of our friends to leave Hong Kong. Wives and children were fleeing en masse, and those husbands who could were going to work in offices in Tokyo or Sydney. The outbreak now seemed to be entering a fraught new phase. The hoax had vividly brought home to me the rattled nerves of this population and the often irrational responses that disease outbreaks elicit. The first phase of a disease, I had read, is denial. The second is panic. The third, if we are lucky, is rational response. Perhaps we were now in the second phase. Who knew when the blessedly scientific third phase would kick in?

My wife had been determined to stay in Hong Kong, not wanting to squander our precious home-leave allowance on a trip of this nature. But the prior evening, Jim Kelly, my immediate boss in New York and the managing editor of *Time* in the United States, had told me to to send my family and those of my staff who desired back to their home countries. (I would expand this directive to include local employees who wanted to send their husbands, wives, or children out of Hong Kong, even though Hong Kong was their home.) So, with the company picking up the plane tickets, I convinced my wife, Silka, to leave Hong Kong with the kids. She logged on to the Internet and began seeking possible destinations, not wanting to return to her native Netherlands for an indefinite period of time. Seeking to make the best of a bad situation, she decided to visit Sri Lanka, and even persuaded a friend of ours, Juliette, to join her with her two children. Silka arranged to rent a villa for two weeks just outside Galle. Sri Lanka, I was pleased to note, was one of the few countries in Asia that had yet to report a SARS case.

That evening, I took my wife and daughters to the airport, where we met Juliette, her husband Peter, and their two sons. The cavernous Norman Foster–designed Chek Lap Kok Airport was all but abandoned; I would later recall the absurdity of its monumental scale when just a half dozen passengers were using the entire Superdome-size Departures area. There were so few people waiting to check in at the counters that the usual pylons delineating the queues had been removed. There had been much public conjecture regarding the safety of air travel during this outbreak. Most airplanes recycled as much as four-fifths of the air on long flights, and though they supposedly filtered the air, the process would not screen virus particles. This meant that if someone on a plane had SARS, then you would almost definitely inhale some portion of air that that person had exhaled. Already, the WHO had identified flights in which there appeared to have been multiple transmissions of SARS. On a March 15 Air China flight from Hong Kong to Beijing, twenty-three passengers had been infected by a seventy-two-year-old man. Remarkably, the WHO would learn, some of the victims had sat as far as seven rows away from him. That one infected flight had set off a chain reaction of infection, with two flight attendants ending up hospitalized in Inner Mongolia and two trade officials flying to Bangkok infected, one of them ending up hospitalized. When that patient flew back to Beijing a week later, he infected a Finnish man who would himself pass away. Air China claimed to have contacted most of the patients on these flights, but the two stewardesses, two of the easiest patients to trace, told Indira Lakshmanan of the *Boston Globe* that they had not been warned. In fact, Air China would continue to inform passengers of flight CA 112 that it was "good news . . . the SARS patient did not fly on our airline." Yet it would be flight CA 112 that would seed Inner Mongolia's SARS outbreak of more than three hundred cases.

The only passengers at the airport that day were other wives and children being sent abroad by their husbands. We stood at the entrance to Passport Control waving as our families disappeared through the sliding doors. After they left, Peter and I drank coffee at the Pacific Coffee stand and watched CNN as U.S. troops rolled on toward Baghdad.

I would not see my wife and children again for over a month.

CHAPTER 49

- **April 1, 2003**
- **World Health Organization Headquarters, Dongzhimenwai, Beijing, China**
- **2,301 Infected, 255 Dead**

AS UNPRECEDENTED STRIDES WERE BEING MADE IN THE LABORATORY—
it had taken years to identify the causative agent of AIDS; SARS had
been isolated in a matter of weeks—there was also the growing awareness that the extent and true damage of this newly identified virus were
still a mystery. If Hong Kong, Toronto, and Singapore, with their modern, state-of-the-art hospitals and first-rate medical treatment, could be
brought to a standstill by SARS, with thousands of cases and scores of
fatalities, then what must the interior of China be like? There, the medical infrastructure in some poorer provinces was still at developing-
world levels. How would the armies of barefoot doctors cope with a viral
killer for which there was no antidote? One barefoot doctor in the village of Shuiqu told *Time* correspondent Hannah Beech that he would
prescribe an herbal tonic to increase a patient's yin in order to counter
an imbalance of yang brought on by SARS. From Hong Kong, we
inferred the disease to be burning through rural clinics as rapidly as it
had through some of Guangdong's best hospitals—faster, even, we suspected, for Guangdong was on the leading edge of China culturally as
well as technologically. The poorer provinces, such as Shanxi or Hebei,
could have been entering into another plague-induced dark age.

The *New England Journal of Medicine* published an editorial
from Julie Gerberding, director of the U.S. CDC, which warned, "If
the virus moves faster than our scientific, communications and con-

trol capacities, we could be in for a long, difficult race." With the outbreak at Amoy Gardens, and the possibility that the virus had gone airborne, it seemed that the virus had far outpaced those capacities. In those confusing days, we wondered if we were living through the first tremors of a dreadful new plague. Global catastrophes begin with such sporadic outbreaks—the seemingly isolated influenza scares in U.S. Army camps in the spring of 1918, for example, would presage the global catastrophe of that winter—or with rumors of disease in faraway, exotic towns. Genoan sailors returned bearing horrible accounts of a plague afflicting traders in Asia Minor in 1347, months before the bubonic plague struck Europe.

But in some ways, it seemed we had already passed through that phase and were into the next, far more dangerous, one: a previously unknown infectious disease was spreading among us, and we had no cure. And it was spreading primarily in hospitals; in fact, it was being amplified in hospitals. Where could one go for treatment? "The stakes are always higher when a mystery disease is spreading in a hospital," wrote former CDC chief of Special Pathogens C. J. Peters. "When a community realizes the hospital is making people sick, the fear factor goes through the roof." Hong Kong had been living through this fear for more than a month; nearly two hundred health care workers had already been infected. Exacerbating that terror was the belief that China was the belly of the beast, and that as long as an epidemic raged there, Hong Kong would never be healthy.

On television, of course, the war in Iraq was receiving twenty-four-hour coverage, as embedded reporters in flak jackets and helmets riding shotgun in Humvees repeatedly told the world how their presence in these vehicles was history-making. U.S. troops were dashing for Baghdad, and after the initial bogging-down south of the Euphrates, their triumph had become a story the networks could not drag their cameras away from. Our story, of this virus and its equally relentless advance, was relegated to a crawl beneath the images of American vehicles driving through Babylonia.

Then, after weeks of ignoring SARS, the American edition of *Time* was now interested in a story about the disease. The angle they chose was the possibility that Hong Kong's annual Rugby Sevens, an investment banker boozefest wrapped around a rugby tournament, might be canceled. When I suggested that the real story of SARS was, at this point, in China, the writer in New York didn't seem interested. I realized later that, to someone sitting in New York, there is virtually no difference between Hong Kong and China. If that were really true, if Hong Kong's scientists had been forced to abide by mainland political constraints on their research, then thousands more would probably have died from SARS. If Hong Kong's government had been as paranoid and secretive as the central government, then the world would have known far less about the disease than it already did. "The biggest unknown is what is going on in China," said Julie Gerberding. "We are desperate to learn more about the scope and magnitude of the problem there because that really will be the biggest predictor of where this will be headed over the next few weeks."

The World Health Organization, fearing that an epidemic *was* raging, had dispatched to Beijing, on March 22, a second team of epidemiologists, doctors, and virologists, whose mission was to assist the Chinese government in curtailing SARS "by making recommendations" as to how to proceed. The team was made up of Wolfgang Prieser of the Institute for Medical Virology, in Frankfurt, Germany; James Maguire of the U.S. CDC; John McKenzie of the University of Queensland, Australia; Rob Breiman, formerly of the U.S. CDC and now with the International Center for Diarrheal Disease Research, in Dhaka, Bangladesh; and Meirion Evans of the National Public Health Service in Wales. The team had arrived in Beijing on March 23, and were almost unanimous in their stated desire to visit the epicenter of the disease: Guangdong. John McKenzie, because of his stature and seniority, was appointed team leader by a group vote. He was surprised when the Chinese Ministry of Health requested a presentation from the WHO team. The Chinese, presumably, had more information than the rest of the world about this disease. Included in the brief from the Chinese government was a request for technical assistance on *Chlamydia*, the bacteriological agent that Chinese scientist Hong Tao had claimed to have found in some early autopsies on SARS patients and that had been discredited as the causative agent by subsequent international research. Still, the Ministry of Health was requesting presentations, so the WHO team set about diligently preparing PowerPoint slides on atypical pneumonia, virology, emerging infectious disease, and epidemiology.

"The whole atmosphere was very strange," recalls Maguire of the team's first meeting at the Ministry of Health, on March 24. Seated in high-backed chairs around a dark mahogany table, beneath a stylized mural of the Great Wall, with microphones lining both sides of the room and women in white gloves pouring tea, the visiting scientists took turns giving college-level lectures to Chinese public health officials, who, the team felt, already knew everything they were being told. "This seemed like a diversion," said Alan Schnur, deputy chief of the WHO Beijing office.

"We had to wonder," said John McKenzie, "about why they were delaying getting down to the nitty-gritty."

IN HONG KONG, WE LOOKED NORTH AND WONDERED HOW GRITTY THAT nitty might be. At *Time*, we had already realigned our reporting to send as many reporters as possible into China. That was where the story had shifted, we believed, and uncovering what was really afoot in the Middle Kingdom became our priority. The Chinese government was still clinging to the impossibly low number of 305 infected and 5 dead, the figures originally released back in February. Their recalcitrance made it seem that the central government was surely dissembling, and what would compel them to lie but a cataclysm unfolding somewhere in the vast interior? How could the world come to grips with this disease and achieve the WHO's goal of containment if the virus was spreading unhindered in China? Could the epidemic be following the old plague route from Central Asia to Eastern Europe, a long swath of nations with poor health care and malnourished inhabitants? If so, then could the 10 to 15 percent death rate for SARS now being spoken about in Hong Kong become a global die-off of biblical proportions?

In Hong Kong, we became steadily more concerned as caseloads mounted: forty-five newly infected on March 29, sixty on March 30, and eighty on March 31, along with fifteen deaths. The first U.S. cases had been in Utah and Washington, D.C., and there were already about thirty suspected cases throughout the country. "We don't know the reason we've been lucky so far," Julie Gerberding said of the United States' having fewer infections than Canada, "but we're not taking any chances." Toronto's outbreak continued to flare, with nearly two hundred cases. There were mounting infections in Singapore, the Philippines, Indonesia, the United Kingdom, Germany, Australia, India, Thailand, Brazil, Sweden, Switzerland, Ireland, Spain, France, Italy, Kuwait, and, of course, Vietnam. Already more than two thousand reported worldwide. It was only a matter of time before the virus found the right circumstances for more urban outbreaks. London could be the next Hong Kong. Or perhaps Chicago or Kuala Lumpur or Sydney.

We did not yet understand the pathogenicity or contiguousness of this new disease. We understood only the damage it did. Global air travel was down over 25 percent from 2002 levels. Singapore was experiencing 61 percent fewer arrivals. The Catholic Church in Singapore also suspended confession in booths, offering instead the less infectious "general forgiveness" to worshippers. In Toronto, worshippers were asked to refrain from kissing icons or dipping their hands in holy water. Flights into Hong Kong were empty—literally in some cases—as airlines couldn't book a single fare into Chek Lap Kok. Personal bankruptcies in Hong Kong were surging 74 percent; hotel occupancy was down 40 percent; retail sales were down, at the peak of the outbreak, 70 percent. The effects were far-reaching and hard to foresee: Australian fishermen based in Perth, whose catch was sold in Hong Kong and Singapore restaurants, saw demand dry up, and call girls in Hong Kong had virtually stopped plying their trade. Economist Stephen Roach of Morgan Stanley was estimating global economic damage from decreased tourism and trade at about thirty billion U. S. dollars. Businesses in Asia wiped the second quarter from their books.

Commerce historically has been among the first casualties of an outbreak. "Thus it came about that oxen, asses, sheep, goats, pigs and chickens and even dogs . . . were driven away and allowed to roam freely through the fields, where the crops lay abandoned," wrote Giovanni Boccaccio in the *Decameron*, his account of story-tellers who retired from all contact with the plague-ridden world of the fourteenth century. I had sent my family away before I knew what the *Decameron* was about, but I had been acting in exactly the spirit of Boccaccio's sequestered noblemen in my desire to shelter my loved ones from what we feared awaited Hong Kong.

We had to find out what was going on in China. TIME's Beijing bureau chief, Matthew Forney, was saying that based on his discussions with health officials and his scanning of China's daily newspapers, the disease seemed to be in control in Beijing. Like all *Time's* correspondents in China, Forney speaks and reads Mandarin, but he is no panda hugger, as those journalists and China watchers who defend Beijing at every turn are called. He tends to be skeptical of official reports and remains studiously cynical when it comes to the party line. However, with SARS, I initially had trouble convincing him that something was wrong with the official story.

"There have to be more cases in Beijing," I insisted.

Chinese officials had admitted to just twelve cases in the capital, a preposterously low number considering the size of the city and the amount of travel between Beijing and Guangdong.

"I don't know," Matt said. "They might be right."

"I don't believe it," I told him.

"You're the boss," Matt said. "I'll have Huang Yong and Susie call some hospitals."

HUANG YONG WAS A THIRTY-FOUR-YEAR-OLD CORRESPONDENT FOR THE magazine. With his shaved head and large, round brown eyes, he came across as so boyish that it was a surprise to some to find that he was on his second marriage and had just fathered his second child, his son Miles. Huang had lived in Oakland, California, throughout his teens, attending high school in a predominantly black and Hispanic area where he had been among the only Asian students. Instead of becoming assimilated into the United States, he had come to despise America for its persistent and hypocritical racism, and, after a few beers, he could become abrasively jingoistic about China. Yet he often echoed a familiar sentiment when he would declaim, "I love China, but I distrust the government." His heartbreak over the massacre at Tiananmen Square, which occurred when he was nineteen years old, newly returned to China, and excited about the future, was total and had left him feeling as if he had no country. He'd spent two weeks at Tiananmen Square, standing alongside his father, earnestly believing that a new China was being born right before his eyes. On the day the tanks rolled into the square, he was home nursing a cold. The young boy standing beside his father that day—who would have been Huang Yong—was killed by a soldier's bullet. Huang was left hating America and feeling betrayed by China.

Huang had been desultorily working at the Beijing bureau for several years. Occasionally, cynicism in a journalist is an attractive trait to an employer, as we feel we can somehow channel that skepticism into a search for the truth. In Huang's case, such efforts had so far failed. "I'm not sure I have a real reason for living," Huang once told me over lunch. "Marriage, divorce, children, remarriage—it's all a search."

"For what?" I asked.

"For what I left at Tiananmen," he said.

SARS would, for a brief moment, shake Huang out of his torpor.

HUANG YONG'S FELLOW CORRESPONDENT, TWENTY-SEVEN-YEAR-OLD Susan Jakes, had curly brown hair that she often wore in a ponytail over squinted brown eyes and a slightly too-wide nose that disrupted her face and was the attribute that stood between Jakes and sheer beauty. She was single and lived in a bright, airy, modern duplex apartment in Beijing's Dongcheng district. Jakes was a native New Yorker, having grown up on Manhattan's Upper East Side. She attended Trinity in Manhattan and graduated from Yale's history department. Fluent in Mandarin and inclined toward a more scholarly approach to reporting and writing, she publicly wavered between pursuing a career in journalism and returning to academia. I was grateful when she accepted our job offer to move to Beijing, as she was one of the few correspondents who could cover cultural as well as political issues in China with equal facility. But her more studious tack sometimes came at the expense of some of the impulsiveness that makes for a good reporter. She labored over her writing, relinquishing drafts only after some prodding and then worrying publicly over what we in Hong Kong thought of her work—which, almost invariably, was intelligent and well written. Producing stories was such a struggle for her that I had begun to wonder whether Susan would ever become the first-rate journalist and writer that her intelligence and analytic and linguistic skills should have allowed; I now suspected that academia might be her eventual destination.

Her personal life tended to the sort of drama endemic to one in her late twenties. As her boss, I occasionally found the management of Susan to become complicated, as I would need to track her down on a business matter when she was with someone whom, perhaps, I wasn't supposed to know about. Still, through this difficult and trying period, she remained someone I enjoyed talking with on the phone—she had a wonderful sense of humor and was a gifted raconteuse—and I always believed, somehow, that if we just figured out a way to unleash her considerable talents, she would write stories that would get her noticed. SARS, as it would turn out, would awaken in her a sense of her talent and value as a journalist.

To Huang Yong and Susan Jakes would fall the burden of uncovering the extent of the SARS crisis in Beijing. Hannah Beech and Bu Hua, our Shanghai team, would seek to find out how the provinces were faring. Neil Gough, Carmen Lee, and Jodi Xu constituted our team in Guangdong, managed by Matthew Forney, our Beijing bureau chief. As an editor, I was blessed with as talented a team of reporters as there were in Asia. And they were now on the most important story of their careers.

It was a phone call from Susan Jakes, to a friend of hers outside Beijing, that would lead, eventually, to the purging of some of the highest officials in China.

CHAPTER 50

■ April 2, 2003

■ People's Hospital Number 301, Beijing, China

■ 2,397 Infected, 259 Dead

I HAVE NO BIOLOGICAL WAY OF EXPLAINING THIS, BUT I BELIEVE THE
SARS virus also triggers memories, reminds those who have lived
through epochs and been washed along the cruel tides of history that
there are opportunities for redress. Stay alive long enough, keep
looking, don't stop thinking, and then suddenly that vast trove of
memory will be unleashed, not as history or fact but as a moment of
bravery and clarity.

He had watched this before, seventy-three-year-old Dr. Jiang
Yanyong recalled, had seen the best and brightest brought down
because of a lie, for the government's prevarications, recalcitrance, and
duplicity. He had been on duty the evening of June 3, 1989, when the
People's Liberation Army (PLA) massacred the students in Tiananmen
Square. It was a pivotal moment in modern Chinese history, as significant and defining as Deng Xiaoping's reforms or Mao Tse-tung's decla-
ration of the founding of the People's Republic. But while those dates
are noted and celebrated on the Chinese calendar—in Hong Kong, as
in the rest of China, the founding of the People's Republic is a public
holiday—the massacre at Tiananmen is seldom discussed and never celebrated, and the protests themselves are always condemned by the government. If you were to write an alternative history of modern China, a
sort of Plot Against America set in eighties and nineties East Asia, you
would start with that gathering on Beijing's main square, and instead of
the Politburo Standing Committee and Deng Xiaoping ordering the

army to clear the square and "put down the counterrevolutionary riot that has erupted in the capital," there would be a brokered, negotiated deal between the students and Zhao Ziyang, the liberal premier who had been the actual architect of the Deng-era reforms. Instead, we are living that dark alternative history. The Politburo Standing Committee and Deng Xiaoping did declare martial law and order the square cleared; thousands of students were massacred; and Zhao Ziyang was kept under house arrest until his death in early 2005. Political reform in China has been relegated to an issue embraced by masochists and exiles as economic reforms have pushed aside concerns such as civil rights and democracy. Yet during the SARS season, the true human cost of that repressiveness once again became apparent as the government's cover-up of the extent of the outbreak was allowing the epidemic to burn through first Guangdong, then Shanxi, and now Beijing.

Dr. Jiang Yanyong had been on duty the night of June 3, 1989, as the director of the 301 Hospital's Department of Routine Surgery. He was resting in his dormitory room when he heard a succession of gunshots coming from the north. A few minutes later, his beeper paged, summoning him to the emergency room. "We are a big-city hospital," he remembers. "We were not used to seeing young people coming in with this level of trauma." Seven students had been dragged in by their compatriots, fellow protesters carrying red banners. Who would kill children? wondered Dr. Jiang. "The army!" the students shouted, their faces a curious mixture of fear and rage that distorted their features to cartoonish levels of expressiveness. "The fascists. It's a bloody massacre."

The protesters, upon hearing that the PLA was approaching Tiananmen with orders to clear the square, had sent a few dozen students out to Muxidi Bridge, near Dr. Jiang's hospital, where they joined with local citizens and workers to build a series of barricades along the main avenue into Beijing. Drivers of electric-powered buses parked their vehicles across the roads, and students and workers began ripping up the roadway and stacking cobblestones in piles in preparation for a clash with the army. The hope among many of those who were on the avenue that night, or in the square, was that the army would not fire on its own people, on students. Yet Deng Xiaoping had made sure that the units rolling into Beijing were not drawn from the capital's recruits but were instead from Shandong and had been kept in the dark regarding the events unfolding in the capital. They had been told that their mission was to suppress a counterrevolutionary rebellion; and being from Shandong, they would be less hesitant to shoot Beijingers, if it came to that.

As the troops arrived in armored cars and trucks, they launched tear gas canisters over the barricades of buses. The protesters responded by throwing rocks and bricks. About fifty riot police wielding batons stormed Muxidi Bridge, where they were repulsed by a steady stream of projectiles. As regular troops were ordered into the battle, they began chanting, "If no one attacks me, I attack no one; but if people attack me, then I must attack them." They turned their weapons on the crowd. "People began crumpling to the ground," wrote Zhang Liang, compiler of *The Tiananmen Papers*. "Each time shots rang out, the citizens hunkered down; but with each lull in the fire they stood up again." At least one hundred unarmed citizens were shot as the army fought its way across the bridge. Most of those were rushed to 301 Hospital.

"Lying before me were my own people," Dr. Jiang would later write, "killed by children of the Chinese people, with weapons given to them by the people." That night, however, Dr. Jiang did not have time to reflect. There was another salvo of gunshots, followed, a few minutes later, by another delivery of the wounded. He was a surgeon who usually performed intricate cancer and tumor procedures, as well as particularly complicated intestinal operations; he had not seen patients as severely injured as these since he had worked with a PLA engineering unit during the construction of the Chengdu-Kunming Railway, during the 1960s. There he had seen soldiers who had been run through by railroad spikes or crushed beneath truck-loads of timber. Yet even those injuries were different in degree of magnitude and violence from what he was seeing here. In 1989, the hospital had eighteen operating theaters. On the date of the massacre, doctors, working in shifts, spent the entire night performing emergency trauma procedures, attempting to save, according to Dr. Jiang, "those who could be saved."

Eighty-nine students would be rushed to 301 Hospital that night. Seven would die.

An on-duty doctor experiences cataclysmic historical events differently from you or me. Dr. Jiang was so busy trying to stop bleeding, reinflate lungs, and transfuse blood that he could not pause to ponder about why he had been called upon to perform these tasks. It was the bloodiest night of his long and storied medical career, and one that would leave him wondering why and how this had happened. Tiananmen Square was no abstraction for Dr. Jiang, nor was it a political struggle or counterrevolution or culmination of an era of reform. It was blood. Up to his elbows, staining his gown, coating his galoshes. Every surface of each article of clothing would be tinged with students' blood. While it was figuratively on Deng Xiaoping's hands, it was literally on Dr. Jiang's.

HE IS SEATED IN HIS DAUGHTER'S APARTMENT IN WESTERN BEIJING. It is already becoming dark, and outside, in the dusty spring warmth, in those paved gaps between the busy traffic and the buildings, badminton rallies are under way, as they are every temperate evening without rain in Beijing. There are courtyards with sooty bushes and a jungle gym between the buildings of a *danwe*, "work unit," built during the 1970s, when two cramped rooms for a family of three or four seemed a reasonable amount of space. These buildings were allocated according to work unit, and if you worked at a hospital in Beijing, as Jiang's son-in-law docs, then you were lucky enough to be granted a place to live in a six-story walk-up with halls that are never cleaned and walls perpetually coated with slick soot. I've encountered this same earthy, oily grime in aged housing developments all over China, and it has begun, for me, to represent China's prereform past. That history, in my uneducated mind, is one of mire and muck; and no one can be bothered to take a long enough look at the mess to clean it all up. Gradually, even these decades-old complexes are being torn down to make way for more modern, expensive buildings that will unlikely house those currently living here. The folks in these old *danwes* will be paid off and then offered the option to buy some lower-priced housing in a distant suburb, beyond some external ring road that has yet even to be numbered.

Dr. Jiang sits on the bed, sipping from a clear mug of green tea. Next to him is a Konka television, and beside that, for some reason, is a toy U-boat mounted on a plastic display. Jiang is tall, thin-boned, and gap-toothed, and has a delicate chin that flattens to a perfectly horizontal line at its base. As he talks, he is animated, energetic, his Mandarin a torrent, and he corrects the interpreter as she relates in English what he has said. For a moment, when you first see him, you think he must be in his fifties—his hair has remained an unnatural crow's black—but there is an aged droop to his eyes, as if the ocular muscles themselves have worn out from squinting in disappointment at how often his naïve assessments have been proven wrong.

The courage and innocence were always there, wrapped around his history like wreaths around a snake. Born to privilege in a wealthy Shanghai banking clan, he was the youngest son in a family of sisters, doted upon in their large house in the French Concession. He was bright, quick, and handsome, and had the revolution never happened, his life would surely have followed a different trajectory, out of China to a university in the West, then a return to live in the gilded splendor of modern Shanghai. An avid golf, basketball, and tennis player as a boy, he devoted little attention to his studies and was forced to repeat the fifth grade. Yet when his beloved aunt passed away from tuberculosis in 1948, he became determined to pursue a career in medicine.

He then surprised even those Communist classmates who labeled him a "reactionary alien" by sitting for the college entrance examinations a year before he had even graduated from secondary school, "just to get some experience of exam hell," and then by gaining admission to Beijing University's Peking Union Medical College, the most prestigious medical school in China. The faculty included Western doctors who taught in English, such as the famed Canadian surgeon Norman Bethune, who encouraged Jiang to specialize in surgery. An adroit surgeon, Jiang won a residency in the surgical department of the General Hospital affiliated with Beijing University. He was notoriously disciplined as a surgeon, and impressed his superiors with his ability to take quick, decisive action that saved many lives. He married and had two children, a son, Jiang Qing, in 1959 and a daughter, Jiang Hui, in 1960. The family moved into an apartment provided by the hospital. Soon, some of his superiors, when they themselves needed an operation, began asking if Jiang would perform the procedure. For the young surgeon with the quiet confidence and steady, delicate hands, the future should have been bright.

But the Cultural Revolution was a malevolent force of (human) nature that did not recognize talent or utility. It is remarkable to visit the Middle Kingdom and speak to Chinese of a certain age: it seems everyone who survived that era has a horror story to tell. Mao's vision of a deurbanized peasant nation heaped the worst opprobrium on anyone with Jiang's background. He was branded a member of the reactionary classes, and, like so many descendants of those who had thrived under the nationalists, he was repeatedly denounced as an "alien class element" and criticized for having a cousin, Jiang Yanshi, who was a famous agronomist in nationalist Taiwan. As a result, Jiang Yanyong was rusticated to Yunnan province as part of a PLA railway engineering corps—that was where he saw his first trauma cases. Then, in November 1967, when he returned to Beijing, he was arrested and locked up in the top floor of his hospital. Because he was so highly valued as a surgeon, the hospital party secretary was persuaded to allow him to continue practicing during the day. At night, he was escorted back to his ward, where he was occasionally bound until time for his breakfast of cold congee and tea. After a few months, the party chief of the hospital relented and allowed Jiang to sleep in a bed in an unlocked room on the top floor. To make sure he wouldn't flee, Jiang had to surrender his shoes and belt every evening. Through the whole ordeal, remarkably, Jiang believed that if he could somehow report to a high-ranking official outside the hos-

pital what was going on—that innocent doctors were being treated
like common criminals—then surely he would be released and the
party secretary of the hospital would be arrested.

On a Thursday night in 1967, Jiang made his break, holding up
his heavy cotton trousers with one hand and charging barefoot down
the stairs and out a side door. It was a cold evening. He picked his
way down the rough macadam until he found a bus stop, where he
hopped an electric bus to Hepingli, where his mother lived. There,
he told her that he had to get in touch with the party leadership in
Zhongnanhai. "I have to tell them the truth," he insisted. "If they
knew what is going on at the hospital, they would put a stop to it."
Jiang asked his sister to tell her husband, a party member, to pass
along his story.

His mother served him some dinner and fetched him a pair of his
father's old shoes. Later, relishing his sudden freedom and eager to
see some of Beijing, Jiang decided he would walk to his sister's
house. There was a park between his mother's and his sister's, and he
cut through it, breathing in the night air and feeling just a bit gleeful
that now he would see his hospital party secretary get upbraided—
and probably arrested—for his criminal acts.

Leaning back on a bench in the park was a stocky figure in a black
padded-cotton coat and a fedora.

"Little Jiang," the man said. "Have a seat."

It was the deputy head of 301 Hospital; he told Jiang that Jiang's
own sister had turned him in.

Those officials who Jiang was sure would be outraged by the
treatment of surgeons at one of Beijing's elite hospitals? Throughout
China, such injustices had become official policy. China had gone
mad. Dr. Jiang, it seemed, was the last person to figure that out.

Upon his return to 301 Hospital, he was subject to an even
harsher course of discipline, ordered to wash his fellow surgeons'
feet, shovel pig shit from a sty behind the hospital. Then, one midnight, he was awakened in his fourth-floor room and taken in his
long underwear to another room on the same floor. Two uniformed
soldiers holding rubber truncheons were waiting for him. They

ordered him to sit in a wooden chair, then began shouting shrilly that he was a counterrevolutionary, that he was from the wrong class. This was the first of his "struggle sessions." By this point in the Cultural Revolution, these struggle sessions had evolved into sadistic spectacles, sometimes attended by just a few officials and soldiers and at other times witnessed by thousands. The denounced, their faces smeared with ink, would be forced to bow for hours while standing on chairs or to wear grotesquely elongated dunce caps and placards with their names crossed out or deliberately distorted. Afterward, in some instances, they would be paraded through town on the backs of flatbed trucks. Some would be imprisoned, like Jiang, only to be exposed to criticism after criticism, denounced hundreds and, in some cases, even thousands of times.

The soldiers shouted at Jiang to admit his class-criminal past. He had heard all that before, though now he realized that his detention and the perverse language of the struggle session were actually national policy. In his first struggle session, the soldiers beat him so hard on the head that he passed out from the loss of blood. Two weeks later, he was pulled from his room again and beaten about the thighs, back, and stomach until he was bleeding profusely from four places. He was dragged back to his room in his long underwear and left on the floor. He couldn't take his clothes off for two weeks because the dried blood had adhered the fabric to his skin.

Today there are tears in his eyes as he talks about those evening sessions. "But I don't hate those soldiers," Jiang says, shaking his head. "They were just following orders."

When he was well enough to travel, he was sent to a horse farm in Gui Nan, in western China, for reeducation. The place was so isolated that authorities did not even need to put a fence around it to keep the political detainees from escaping. There was only one road from the farm, which led through the desert and to a village three days away by foot. The journey was correctly viewed as being impossible to survive. At the farm, Jiang subsisted on a diet of roots, grass soup, and rice. He did not see his family for five years. Remember, all he had done was be born into the wrong class. What is remarkable about his history is not its extreme hardship but rather how typical it was for Chinese of his generation.

TODAY, JIANG, WHO RETURNED TO 301 HOSPITAL IN 1973, HOLDS A military rank equivalent to general because of his title as chief of surgery at the hospital. He lives in a housing compound in western Beijing reserved for high party officials. In the afternoons, he takes walks through the courtyard and occasionally joins other retirees on the pathways or on benches in the afternoon sun.

Throughout March of 2003, Jiang had been spending more time indoors, like many people around the world, watching television for news of the war in Iraq. The U.S. struggle with the Arab world was of interest to many Chinese, who felt that China would benefit by not being a participant in what was bound to a protracted and bloody clash of cultures. By staying out of it, China could continue to modernize and prosper while American resources were bled away by battles with terrorists and rogue states. China, meanwhile, had no sympathy for fundamentalist Muslim aspirations and was happy to use the pretext of the war on terror to crack down on ethnic separatists in far western Xinjiang Uygur. SARS was only a bottom-of-the-screen crawl on CCTV, a glowing proclamation that it was "under control and there has never been a better time to visit Guangdong Province." The SARS outbreak had so far been reported as primarily a Hong Kong problem; the disease, if it was in China at all, had probably been brought in by foreigners, the official Chinese media were reporting.

Among international public health officials, of course, there was increasing consensus that the outbreak in China was far worse than the Chinese government was admitting. On March 31, the State Council Information Office reported that there had been only twelve SARS cases and three fatalities in Beijing. It seemed impossible: there were thousands of cases in Guangdong and Hong Kong, and hundreds in provinces throughout China. How could Beijing have just twelve cases?

Dr. Jiang found that discrepancy curious but had no way of contradicting the government's numbers. But near the end of that month, a good friend of his fell ill with lung cancer, and naturally Jiang was brought in to consult on the case. The patient, a medical professor, was hospitalized at 301 Hospital. Surprisingly, he developed a high fever and a spot was found on his lung. After another specialist was brought in, Jiang's friend was diagnosed with SARS and transferred to the ICU, before being removed and sent to 309 Hospital, dubbed the official SARS Control and Prevention Center for the People's Liberation Army. Checking on the treatment his friend might receive, Dr. Jiang phoned respiratory specialists at 309 who were former students of his from Beijing University Medical College.

"They sounded very upset," Dr. Jiang recalls of those conversations. "I didn't understand why. There were just a few cases, and that was such a big hospital."

There were sixty cases, Jiang was told, dozens of them medical staff themselves. Seven patients had already died of the disease. Now 309 Hospital was reenacting the crash-and-burn of Prince of Wales Hospital in Hong Kong, of Hanoi French Hospital in Hanoi, of Tan Tok Seng in Singapore, and of Scarborough Grace in Toronto. Dr. Jiang had never heard of anything similar happening at the top-tier hospitals. He called other colleagues and found that there were similar outbreaks occurring at 302 Hospital, which had forty cases, and even at his own hospital, Number 301, which had forty-six SARS cases. "This is a terrible disease," one of his colleagues told him. "It acts so quickly. I've never seen any disease progress this fast. You go from breathing normally to intubation in three days. You die in a week."

Why had the health minister, Zhang Wenkang, said there were only twelve cases in all of Beijing when there were sixty in just one hospital? He had even appeared on television and chastised a foreign reporter for wearing a gauze surgical mask, telling him, "You are safe here whether you wear the mask or not. Beijing is perfectly safe to visit for business or pleasure." Watching the press conference on TV, Jiang could not believe what he was hearing. Either the health minister was lying or he didn't have the correct data. Both scenarios were unacceptable, considering the nature of a highly contagious, infectious disease. Later, while taking a late-afternoon walk in the courtyard behind his apartment complex, he came upon two fellow retired People's Liberation Army senior officers. They discussed the fact that there were far more cases than Health Minister Zhang was admitting.

"There must be something wrong with the health information system," one of the officers said.

"You need to talk to the health minister," said the other officer. "He's a doctor. He should know better."

Jiang agreed. If this disease was as deadly as reports from Hong Kong and Guangdong indicated, then the health minister should be making every preparation to ensure the safety of Beijing, rather than appearing on television assuring the citizenry that there was no risk. Then, Jiang says, he thought about the night of June 4, 1989, and reverted back to that young surgeon locked up on the top floor of 301, before his escape, who thought if only the government knew what was happening there, they would put a stop to it. Naïve to this day, he now speculated that if he just told the health minister about these hundreds of cases in Beijing, about this outbreak, then appropriate prophylactic action would be taken. He would come forward and tell the truth; the Communist Party would see the light.

His wife told him, "Don't do it. You will get yourself in trouble. Why are you sticking your neck out?"

And then he cried, tears welling in his eyes, "This is bigger than me, or you, or China. I have to tell the world. This could be a disaster for all of mankind."

CHAPTER 51

- **April 2, 2003**
- **Guangzhou Institute for Respiratory Diseases, Guangzhou, China**
- **2,448 Infected, 269 Dead**

AS GUAN YI WATCHED THE UNIVERSITY OF HONG KONG'S TELEVISED press conference announcing that the university's lab had identified the causative agent for SARS, he chain-smoked and worked himself into a state of furious resentment. Chinese state television had cut into coverage of the Iraq war to announce the breakthrough. There was K. Y. Yuen, standing before the microphone, talking about how they had gathered the samples, consulted with the health officials, and introduced new cell lines. And there was Malik Peiris, stiffly posed like the Oxford don he secretly wished to be, his hands folded behind him as he explained, with what Guan Yi detected as mock humility, that this had been a "fishing expedition." Chinese state television had covered the Hong Kong breakthrough as thoroughly as CNN was covering the war in Iraq, with reporters doing remote stand-ups from the University of Hong Kong's microbiology department. Peiris and Yuen were being lauded as if they were Koch and Pasteur. And there had been not a single mention of Guan Yi. More disturbing to Guan was the fact that his colleagues here in China, especially Zhong Nanshan, who had undergone great political risks to release the samples, were not receiving any acknowledgment for the role they had played in finding the disease's causative agent. He had immediately called Peiris and demanded to know what was going on.

"I couldn't find you," Peiris said calmly.

"That's because I've been working in China," said Guan Yi. *"Getting you samples.* How can I go back to Zhong Nanshan and tell him I didn't know about this?"

"I'm disappointed you feel that way," Peiris answered.

Guan Yi wondered about what else Peiris was keeping from him. Hadn't they been collaborating on this since the beginning, during that first trip to Ah Chau Island, when they were seeking bird flu? He began to see a pattern: Peiris and Yuen had kept him out of Klaus Stöhr's lab network, and so he had never been privy to the information exchanged there. No Chinese labs had yet been invited to join the network. But by bringing the first samples from China, it was Guan Yi who had given the University of Hong Kong lab the head start that had helped it to win the race to find the virus. And it was Zhong Nanshan who had stuck his neck out and let the samples be taken from his hospital. "I knew Malik had sold me out," Guan Yi recalls.

He called Zhong Nanshan, who had already seen the announcement on TV.

"They shouldn't do this," was Zhong's comment.

Guan Yi had promised Zhong that they would share in any discoveries made with their samples. Even more important, he had promised him that Zhong would be able to go to Beijing with data disproving the *Chlamydia* theory that both men believed could cost China many lives.

"What am I going to tell them in Beijing?" Zhong asked.

"I'm sorry," Guan Yi said, knowing he could never make up for Zhong's loss of face. In virology, Zhong knew, the scientist who comes up with the samples will often be rewarded with a prime author's spot on the academic paper that appears as a record of the discovery. These billings are as important to the scientist as credits are to a Hollywood player or location on a *Wall Street Journal* underwriting announcement to an investment banker. Peiris had awarded himself the first spot on the paper, which goes to the lead researcher, and then given the last spot, the second most prestigious, to K. Y.

Yuen, as the coordinating researcher. The second spot, which is also prime real estate, had gone to Stephen Lai, a pathologist from Princess Margaret Hospital. Guan Yi felt he should have at least been in the second spot. But he was more upset that Zhong Nanshan wasn't listed anywhere at all.

Peiris later insisted that his research simply didn't concern Guan: "We did not find any virus in his samples. And since this research involved human viruses, not animal viruses, this was not in Yi's area. This was a matter for the human side."

Other microbiologists, including John Tam from the Chinese University of Hong Kong, alleged that the University of Hong Kong team could not have made the breakthrough they did without the illegally gathered samples brought back by Guan Yi.

Guan Yi, when he was home in Hong Kong, had even called Rob Webster, the mentor he shared with Peiris and one of the world's foremost influenza experts. When he explained to Webster what had happened, how he felt betrayed by Peiris, Webster said in his New Zealand accent, "There is still an awful lot we don't know about this virus." What he didn't have to say to an eager virologist such as Guan Yi was that that meant there were still some important discoveries to be made.

"They underestimated me," Guan Yi told Webster. "They'll regret this."

CHAPTER 52

■ April 3, 2003

■ 302 Hospital, Western Beijing, China

■ 2,486 Infected, 275 Dead

HUANG YONG, THE REPORTER ATTACHED TO THE TIME BEIJING BUREAU, had been sent by his bureau chief, Matthew Forney, to drop in to a few local hospitals and discreetly see if any of the doctors, nurses, or support staff would talk about atypical pneumonia cases, or "A pneumonia," as the disease was being called in Beijing. Huang was often the reporter we would turn to because of his gift for speaking to bureaucrats and blending in with Chinese peasants and proletarians when that was required. Since he spent much of his time at roadside cafés drinking a few beers, Huang would artfully combine that activity with conversation with a few locals to verify rumors. In this case, he walked straight into the hospital—at *Time*, we would come to call this legally and medically risky practice "bombing" hospitals—and strolled up and down the hallways. The physicians and nurses, he immediately noticed, were all wearing N-95 masks. On later bombing runs, he would carry a bouquet of flowers as a prop. For this trip, he simply began strolling around and asking different doctors if they had any of these dreaded "A pneumonia" cases in the wards. Sure, he was told, they had a few. And they had referred a few cases to 309 Hospital as well.

"How many of these patients do you have?" Huang Yong asked.

"About fifteen," a doctor told him.

That was strange. Fifteen cases. In this one hospital? In all of Beijing, according to the Ministry of Health, there were only a dozen cases.

He strolled back out, stopping to talk to the shopkeeper who ran the snack and newspaper kiosk near the entrance gate. He casually asked how many pneumonia patients she had heard about.

She shrugged. "Dozens."

For the generation that had left their lives at Tiananmen, it was no leap to assume the government would lie. Huang Yong's generation believed that every government—of the United States or China—played with the truth. But shouldn't they be more thorough in their prevarication? If a reporter working for a foreign magazine could stumble upon evidence contradictory to what a government minister was, at that exact moment, probably saying at a meeting or press conference somewhere in Beijing, then that meant the lie had to be so big it could not possibly hide the truth. Which meant the truth had to be so terrifying that it necessitated a lie that had no chance of actually succeeding. Huang returned to his little hatchback and pulled out onto the Fourth Ring Road, sipping his beer while he drove. He followed this train of thought to its logical conclusion: the disease was already here, and not only was it here, it was doing so much damage that the leadership was panicking. The bigger the lie, the more horrible the truth. He pulled in to the parking lot at Ditan Hospital, also known as the Beijing Prevention and Treatment Center for Viral Infectious Diseases. He slipped on his gauze mask and walked up to the registration desk and asked, "Do you have a respiratory department here?"

"What's your problem?" the nurse standing behind the counter asked.

"Lung disease," he answered.

She jumped back a full meter and covered her mouth and nose with her hand. "Go to the emergency ward."

Huang laughed and went to the elevator bank. At the emergency ward, he strolled toward the nurses' station in the middle of an open suite of ICU theaters. There were dozens of patients in critical condition. The nurses wore N-95 masks and appeared to be busy behind the counter.

"I think my mother is here," Huang said.

"Who is she?" asked a nurse seated behind a desk.

"We call her Granny Hu," he said. "She has atypical pneumonia."

"She's one of a dozen," the nurse said.

By the time Huang returned to the TIME bureau offices in the Jianguomen diplomatic compound, he had visited one more hospital, in which he had heard of ten more cases. That made at least thirty-seven cases, while the official tally was still only a dozen.

Matt Forney called me to talk about Huang Yong's reporting.

"You may be right about the government hiding cases," he conceded.

Huang got on the phone, and we talked about what he had seen. I was eager to try to get something onto our website or into the magazine on this subject, and I pointed out that if, in the first three hospitals Huang walked into he had found dozens of infected patients, then who knew how many more there were throughout Beijing?

Forney, when he came back on the line, agreed that there was something to this, but then he raised a valid point: Huang Yong was not a doctor or a scientist, and our criteria of what constituted a SARS case could be different from the government's.

It was a legitimate issue. What would happen if we ran a story about Beijing hiding cases, and it turned out these were not SARS cases but some other pneumonia? More likely, if the government came out and said we were wrong, how could we refute them? We didn't have virus growing in cell lines. We didn't have a doctor backing us up.

"We need a whistle-blower," Matt said, making a reference to *Time* magazine's Persons of the Year in 2002—Cynthia Cooper of WorldCom, Sherron Watkins of Enron, and Coleen Rowley of the FBI—who brought to light the big corporate and government scandals of the previous year.

We both knew how unlikely it was that anyone in China would be willing to take that sort of risk.

CHAPTER 53

- **April 3, 2004**
- **Wanshou Road, Beijing, China**
- **2,487 Infected, 276 Dead**

WHAT HE WAS HEARING SOUNDED SO FAR-FETCHED THAT JIANG Yanyong turned up the volume on his television. The minister of health, Zhang Wenkang, seated behind the microphone tree, was proclaiming that the Chinese government was already dealing diligently with the problem of SARS and that the spread of the disease was already under control. Zhang was a graduate of the Number Two People's Liberation Army Medical University, perhaps the most sophisticated medical school in the country. And he was now rejecting the most basic personal principles of the medical profession.

The next morning, Jiang sipped a cup of barley tea and phoned his colleagues at 309 People's Liberation Hospital. They all had seen the minister at the news conference and said they believed his statements were "bullshit." Their hospital, they told Jiang, already had sixty SARS cases and seven dead. "Perhaps you should leave Beijing," one of Jiang's colleagues suggested.

Jiang later told me he thought a long time about what he had just learned. When he had come forward during the Cultural Revolution to denounce the hospital administration, he had painfully paid the terrible price for daring to speak the truth to power. After Tiananmen, he had kept quiet. But then, that had been a national tragedy, and the notion that the government was wrong—while made graphically clear to him by the blood on his own rubber gloves—would be expressed even if he himself did not publicly make that case. But now he was in a different position. He had more stature than he had had during the Cultural Revolution or Tiananmen; he was now a party member. And he was virtually retired, a respected elder who had treated many of the top brass and had saved several important generals' lives. Perhaps his rank, connections, and party affiliation offered him some protection. Or was he as naïve as he had been when he escaped from his hospital, believing that if party officials only knew about the injustice, they would never countenance it? This was an opportunity to make up for the guilt he felt about never coming forward and stating what he saw the night of the massacre at Tiananmen. Perhaps it was the silence of those in positions to speak with absolute authority, like him, that had allowed the government to claim that the demonstration had been a "counterrevolutionary rebellion" and not the mass killing of innocent civilians. By keeping silent now, was he again participating in a great and terrible lie? Only this time, the consequences of the lie were incalculable.

At lunch, he mentioned to his wife the minister's apparent prevarications, a matter not of misinformation but rather disinformation. Hua Zhongwei frowned. Her husband had paid dreadfully for his earlier outbursts, and so had she, forced to raise their children without a father for years, branded the wife of an "alien class representative." At seventy, she was too old to go through another struggle session. She urged him to keep quiet.

Instead, Jiang penned a note, explaining who he was and the facts about the number of SARS cases in 301, 302, and 309 hospitals. "As a doctor who cares about people's lives and health, I have a responsibility to aid international and local efforts to prevent the spread of SARS." He faxed the note to the government-controlled CCTV-4 and Hong Kong's Phoenix TV, two of China's most powerful news disseminators, using the fax number given for viewer comments and suggestions. He assumed that the stations would quickly get in touch with him to check his credentials before airing the information.

It is ironic that the doctor who probably saved more lives than any other physician during the epidemic would never treat a SARS patient.

CHAPTER 54

- **April 3, 2003**
- **World Health Organization Headquarters, Dongzhimenwai, Beijing, China**
- **2,489 Infected, 277 Dead**

AT THE SAME TIME AS THEY WERE BEING DENIED INFORMATION BY their Chinese counterparts, the members of the WHO team in Beijing were increasingly being pressed by the media for information about SARS in China. Their press conferences had become the most reliable SARS-related event for international reporters to cover, particularly since these were Western physicians and officials rather than Chinese government bureaucrats. As it became clear to the assembled Western reporters that the WHO team was not getting a complete epidemiological picture, many Western and Chinese journalists began to suspect a cover-up. There were already widespread rumors, circulated via cell phone text messages and Internet chatter, of hundreds of SARS cases in hospitals throughout the country and thousands already dead. These rumors reached the WHO team, of course, and while they would be discounted, the epidemiologists knew that rumors about diseases could never be ignored.

The WHO team, from the start, had expressed a desire to visit Guangdong, to tour the epicenter of the outbreak. Li Liming, the head of the China CDC, did not want them to go. His objections were couched in bureaucratic terms. "It takes time to get an invitation. You have to go through appropriate channels. You can't just go down to a province."

And Liu Peilong, of the Chinese Ministry of Health, suggested that the team return home, and then, after a few weeks, perhaps another WHO team could visit and the ministry "would consider the request to visit Guangdong."

For the WHO team—all scientists used to butting up against the irrationalities of bureaucracies—this appeared to be precisely the kind of obfuscation that had seeded the terrible rumors about the outbreak. They felt that to leave China without visiting Guangdong would be abnegating their duties as doctors and public health officials. Whatever had happened down there, the course of the outbreak, the curve of the epidemic, the conditions of its emergence, the treatment and management of cases—all that data was crucial in helping the international community deal with this scourge. How could they allow bureaucratic wrangling to get in the way of their Hippocratic duties? Back at headquarters in Geneva, WHO officials were losing patience with Beijing. "I felt in our own internal discussions we were taking too much time, giving too much time to China," said Gro Harlem Brundtland, the director-general of the WHO.

"At that point, we did not even know, for certain, if the atypical pneumonia in Guangdong was the same thing as SARS," said Henk Bekedam of the WHO in Beijing. "We had our suspicions, but we needed to see for ourselves." If this was the same disease, as seemed more and more likely, and it was already endemic in the world's most populated country, then the chances of preventing it from becoming a global pandemic would be almost nil."

The World Health Organization was resolute: they had to get that team to southern China. The team made a formal request on March 28, relayed via Alan Schnur, deputy chief WHO representative, to Liu Peilong of the Ministry of Health.

For the WHO team, already running out of topics for their daily press conferences, it was becoming hard to deflect questions about when they would be heading south. Each day they showed up in front of the press and hedged. They felt, Rob Breiman explained, "that by not saying when we were going to Guangdong, our credibility was being compromised. We were there as experts and weren't able to demonstrate any utility to our presence."

The team met daily in a conference room at the World Health Organization's Beijing headquarters to strategize how they could best pressure the Chinese government. What they now had going for them, they deduced, was that their leaving the country and saying they were dissatisfied with the exchange of information would be viewed as a PR blow to China. The world was now watching, and here were international delegates being effectively blocked from doing their jobs. "I had never seen a situation like this. They were in some ways in a state of almost catatonia," says Breiman of his Chinese colleagues. "I'd never actually seen such a forceful political response, where they decided what their approach was going to be, and it was going to be cover up until they couldn't do it anymore."

Alan Schnur, the WHO's patient bureaucratic warrior, and Henk Bekedam, the WHO China chief, had been asked to leave the room during a team meeting at which they discussed their options. As the WHO team gnawed the bland Pizza Hut pizzas, they reached a consensus that they would take their problems public, a risky and unconventional move for an international mission, especially in China, where the government was famously recalcitrant. When Bekedam and Schnur were informed of the team's decision, there was some reluctance on Bekedam's part to challenge the Chinese government, while Schnur agreed with the team that the press conference might be an effective goad. It was also, Schnur knew, the only weapon at their disposal.

During their April 2 press conference, the team announced that if they weren't given permission to go to Guangdong, they were leaving on the next flight to Bangkok. That evening, local time, the WHO issued its travel advisory recommending cancellation of all nonessential travel to Guangdong.

BOOK 4
How Do You Kill It?

CHAPTER 55

■ April 8, 2003

■ Jianguomen Diplomatic Compound, Beijing, China

■ 2,941 Infected, 293 Dead

AT TIME, WE HAD CONTINUED TO HEAR RUMORS OF MULTIPLE CASES in Beijing hospitals, yet the Chinese press had become maddeningly silent on SARS, with the coverage now constituting a crawl beneath TV shots of the fall of Saddam's statue and reports on the deaths of Iraqi civilians: "Traditional Chinese Medicine Can Cure SARS." The Chinese press was under orders not to report SARS; reporters were tersely reminded at every turn that there was no upside to breaking SARS news. When a *Yangcheng Evening News* reporter asked Huang Huahua, Guangdong's governor, if there was any chance of SARS passing to people from domestic livestock, she was warned, "You will be held accountable for those words of yours!" And those websites that dared to post SARS-related news were quickly visited or phoned by propaganda officials and ordered to delete any references to the disease. On March 9, the Ministry of Health had held a meeting with the heads of Beijing hospitals to emphasize that they were not to report the spread of the new disease to any media outlets. "We did as we were told," the president of Sino-Japan Friendship Hospital would later tell me. On March 15, when the WHO issued its first Global Alert about SARS, the Propaganda Ministry, by order of the Politburo—China's supreme leadership council—did not pass on the warning. "[The Politburo] knew very clearly how fast SARS was spreading," a Beijing newspaper editor would tell John Pomfret of the *Washington Post*. As late as March

26, Foreign Ministry spokesman Kong Quan told reporters, "This isn't a very serious disease."

That the Chinese government wasn't telling the truth didn't strike anyone as far-fetched—most Chinese people I knew took that as a given—but that they would lie to people about a disease that might kill them seemed beyond the pale, even for a dictatorship. In this context of government prevarication and media cowardice, it was not surprising that Dr. Jiang Yanyong's faxes to CCTV-4 and Phoenix TV had been ignored.

OUR BEIJING CORRESPONDENT SUSAN JAKES WAS DELEGATED TO create a file about the general state of the health care system for a story we were preparing about SARS in Guangdong. The WHO team's brinkmanship with the Chinese government had succeeded, and rather than face the embarrassment of the team's leaving Beijing publicly dissatisfied, the Chinese government had allowed them to fly south on April 3. Matt Forney, our Beijing bureau chief, had been with them, determined to file a definitive timeline story on the initial phases of the outbreak. (In phone conversations with me, he kept saying, "You know, like how in Hot Zone it starts with the guy touching the wall of the cave.") In the meantime, as background, he needed Jakes to run the show in Beijing and do the assessment of the Chinese health care system.

Jakes had no contacts in the Ministry of Health. Trying to think of a way into the subject, she decided to call a political source. Before coming to China, Jakes had worked for Wei Jingsheng, a notable Chinese dissident in New York, interpreting for him and driving him around the city. It had been through that role that she had come to TIME's attention, when a team of our reporters had gone from Hong Kong to New York to investigate a North Korean defector who turned out to be a fraud. Jakes's connections in the dissident community had been useful in the past, and sometimes they gave us the inside track on some human rights stories in China. But those connections did not extend into the Chinese medical or scientific communities; hence Susan's desperate call to one of her political connections, Harold, the relative of a party official.

She asked him if he knew anything about SARS in Beijing.

There was silence on the line. "Call me back from a safe phone."

Often, in China, we suspected our landlines and even our cell phones were bugged. Our conversations, if they involved possible government scandals or could put our Chinese sources at risk, would be circumlocutions to the point where we would sometimes confuse ourselves. When we needed to address specifics, Matt or Susan would switch the SIM cards in their phones from a local Beijing number to an international exchange that was billed through a foreign phone company that we believed was far less likely to be tapped. (We took similar precautions in Malaysia, Pakistan, and other countries where we were afraid the government might be interested in speaking to our sources.) Or, even safer, the reporter would find a pay phone—still common in China—and call from there.

Jakes threw on her denim jacket, rode the elevator down, and stepped into the sunlight of a chilly early-spring morning. Just up the road from the bureau was a red pay phone at a cigarette kiosk, next to a furrier.

"I'm going to send you an e-mail," Harold said when she had called him back. "In that e-mail, there will be a URL to a secure website. At that website, you'll need a password. Type in your old Hong Kong phone number, and you will be able to download a Word file. Read that and call me back."

Jakes bought a pack of cigarettes and ran back to the office. She felt it was risky using the office computers, but Harold hadn't mentioned anything about making sure she was using a secure computer. And he was by nature very cautious. She checked her e-mail, copied the URL, pasted it into a Chinese browser window, and typed her old phone number into the dialogue box that appeared.

She downloaded the Word file. At the top, it read, "Jiang Yanyong, Doctor," and said that Jiang was a longtime Chinese Communist Party member. It also gave his phone number. She read the note. There were a few Mandarin words she didn't recognize, and she had the office assistant, Li Cheng, help her translate. The letter began, "The number of patients infected with SARS may be significantly higher than statistics made public by China's Ministry of Health indicate," and went on to say there were at least sixty patients at one Beijing hospital. Most amazingly, the letter was signed by this doctor.

Jakes went back out to the kiosk and called Harold.

"Who is this guy?"

"He is who he says he is. A doctor. A party member."

Jakes was nervous this letter would be difficult to verify. "Can I call this guy? Will he talk to me?"

"Call him," Harold assured her. "He's at home."

Jakes knew what she now had. A big story about a big lie. This was what had fallen into her hands in the form of the doctor's letter.

Still using the pay phone, she called the number on the letter. Dr. Jiang Yanyong answered.

When she identified herself, Dr. Jiang told her, "Everything I want to say is in the letter."

"But I need to ask you some more questions," Jakes pleaded, "to flesh this out a little bit."

He paused for a moment, and then, speaking in a lower voice, he said, "Okay. Let's meet at the teahouse at four o'clock in the Ruicheng Hotel, in the western part of Beijing, near the 301 Military Hospital."

Jakes went back upstairs and e-mailed me that she thought she had something "scoopy" and that we shouldn't wait to get it into the magazine; we should put it on the TIME website as soon as possible. I e-mailed her back, asking her to call me from a safe phone.

She briefed me on what she might have, and we both considered how to verify the contents of the letter. That she was meeting Dr. Jiang that afternoon was a good start; Jakes would get a quick read on whether he was reliable or a crank. (At that point, we had no way of being sure.) But with a story like this in a country like this, the reporter's instincts, while being a very good starting point, would not be enough. Jakes, to her credit, repeatedly reminded me that we had to get this confirmed. I was going to wait and see how her meeting went. Also, I wanted to go back and speak with Huang Yong about his earlier hospital visits. Couldn't we put a story together using both of these elements that would expose the cover-up? We decided we would have Huang Yong go out on another bombing run to the hospitals mentioned in Dr. Jiang's letter, to see if he could verify some or all of the letter's contents.

As soon as Jakes was off the phone with me, she received another call, this one from a lawyer whom she had called the day before asking if he knew anyone who knew anything about SARS.

"Why don't you come to my office right now," he suggested. "I think I might have something you want to hear about."

She took a taxi to his office, on the fourth floor of a modern office building, and when she walked in, after he closed the door, he told her that he had a cousin who was a doctor at the Military Academy of Sciences.

"Will she talk to me?" Jakes asked.

"No," the lawyer explained. "But I can call her, and you can listen while we speak."

Jakes would later realize that this plan had been prearranged by the lawyer and his cousin in order to screen the cousin from any possible accusations of talking to a foreign reporter and violating the gag order that had been handed down on March 7. As for the veracity of the source, we had worked with this lawyer before on several stories for *Time,* and found him to be reliable and honest.

The lawyer dialed his cousin's mobile.

"Tell me again what you told me before," he said, and handed the phone to Jakes.

Jakes listened as the doctor spoke of a situation even more terrifying than that described in Dr. Jiang's letter. She described the first case to come to Beijing—the woman who had driven from Shanxi and seeded the Beijing outbreak. To Jakes's surprise, that had been back in early March, during the National People's Congress. The hospital director at the Military Academy of Sciences had told his staff that there was SARS in Beijing but that no one was to mention a word of this outside the hospital in order that they not interfere with the National People's Congress and the leadership transition. The doctors at the meeting were not allowed to take notes.

Since then, the woman went on, there were numerous cases at several hospitals. Number One and Number Two hospitals each had dozens of cases. "[The hospitals] are practically filled," the woman said. And 309 Hospital, specifically mentioned in Dr. Jiang's letter, had forty new cases in just the last week. Number 301 and 302 hospitals were also being overwhelmed. "*Yuan yuan chao guo*," the woman doctor said.

The official numbers are lies.

After leaving the office, Jakes caught a taxi. She didn't want to phone me because of worries that her phone might be bugged. Instead, in the back of the cab, she began to translate the contents of Jiang's letter, confirming everyone's worst fear: not only was SARS in Beijing, but there had also been a cover-up of that terrifying fact.

Jakes arrived at the Ruicheng Hotel in western Beijing at two forty-five that afternoon. The Ruicheng is a typical Chinese business hotel. The lobby is on the shabby side, and the registration desk tries to sell you various discounted tours of Beijing and dinner tickets for local restaurants. The foot traffic through the lobby tends to be Chinese businessmen carrying their male purses and talking on mobile phones on their way to meetings in one of the restaurants on the second and third floors. Jakes took a seat in the lobby, feeling slightly out of her element as she continued to translate Dr. Jiang's letter with the aid of a Chinese-English dictionary.

With each Chinese businessman entering, Jakes would glance up, wondering if this was Dr. Jiang. When he finally walked in, he paused for a moment and then, seeing Jakes, the only foreigner, he gestured for her with a quick backhanded wave to follow him. She shoved the letter and dictionary into her purse and took off after him as he headed for a corner of the lobby. He led her through a service entrance, up an elevator, and down a hall, where he asked a hotel employee for directions. Jakes realized he didn't know where he was going as he walked into a cafeteria, which, it being midafternoon, was closed. Finally, he found a teahouse and took a seat in a booth partitioned from the restaurant by wooden screens. Around them, they heard the clatter of mah-jongg tiles, which, besides clandestine business meetings, was the primary purpose for these little teahouse rooms.

Jakes's first impression was that Jiang was nervous. But once they ordered and began to chat, he calmed down. He talked about his work as a surgeon, spoke in very clear Chinese, and gave the names of medical procedures in very good English. He was, Jakes quickly deduced, exactly who he said he was in his letter.

Finally, Jakes asked, "Why did you write this?"

He paused. "I live in a *danwei* in a military complex. My colleagues there are also doctors, public health officials, many of them retired. When we run into each other, we talk about what is going on. We had all seen Zhang Wenkang's press conference. We talk about how he is lying. As a doctor, I cannot stand by while there is a terrible disease threatening the people and they are not hearing the truth about it."

Dr. Jiang didn't resemble the dissidents or malcontents whom Jakes had dealt with during her time associated with the dissident movement. Those fellows had had an agitated quality that came across almost as a twisted sense of entitlement. They had been through so much that they no longer felt it necessary to consider the feelings of others who hadn't suffered as they had. If you questioned their reasoning, they would just dismiss you. This fellow gave off a completely different vibe. He was an elder, and like a teacher coming forward because he was disappointed at how his students were behaving. He emphasized that he was a party member, that he felt, as a member of the leadership, he had an obligation to state the truth. Jakes was enjoying his company, and as they pored through his letter sentence by sentence, she felt increasingly comfortable that the material was journalistically sound. Her listening in on the phone call from the female doctor had essentially confirmed everything Dr. Jiang had written.

"If I were an ordinary person in Beijing and came down with fever and a cough," said Dr. Jiang, "then I would probably just go to bed instead of going to a hospital. I would never know that I had contracted a fatal disease because the leadership has not told the people."

"Are you sure you want to sign your name to this?" Jakes asked him.

"Yes."

"Aren't you afraid this will get you into trouble?"

"Everything I have said is true." He nodded. "I can prove it."

"That doesn't mean you won't get in trouble."

"I have constitutional protection."

Jakes shook her head. She can take on an almost haughty demeanor when she feels someone is making a mistake. So far, she had maintained a deferential manner; Dr. Jiang was, after all, an elder accustomed to being treated respectfully. Now, however, she felt she needed to be understood clearly, to pierce Dr. Jiang's naïve belief in China's constitution. Jakes had worked with numerous Chinese who should have been protected by the constitution yet were sentenced to reeducation, put through the psychiatric hospital system that is often used in lieu of prison, or forced to flee. The truth provided no sanctuary.

"The constitution doesn't protect everybody," she said. "It will not protect you."

He smiled. "I am seventy-two years old, and I have lived in this country a long time." He knew exactly what was at stake.

Jakes thought about Dr. Jiang during her taxi ride home. He was taking a bold and groundbreaking step. There had been instances where a Chinese official might reveal something to a reporter anonymously. Yet Jakes could not think of another case in which a high-ranking Chinese official or military officer who was still living in the country would speak on the record about a government lie. Even in trail-blazing work like *The Tiananmen Papers*, edited by Andrew J. Nathan and Perry Link, the compiler of the government documents had remained anonymous. Jakes concluded that, like many heroes, Jiang was almost as misguided as he was courageous. Perhaps it took a combination of such attributes to come forward in a dictatorship.

Jakes called me from the taxi. I was impatient in Hong Kong. We seldom worked late on Tuesdays, and I had kept Tim Morrison, the editor of the website, at the office with me so we could get this story up quickly. Jakes said she had everything she needed and would start writing as soon as she got home. She got off the phone with me and called Huang Yong, asking him to come over and help with the translation.

Huang arrived at Susan's apartment just after she did, with beer and cigarettes. He was wide-eyed. They were both almost giddy as they went over the translation of Jiang's letter, and then Jakes wrote a quick story to go on top of excerpts from the letter.

She sent the story to me at 9:00 P.M. We had it up on our website an hour later.

"A physician at Beijing's Chinese People's Liberation Army General Hospital (No. 301) in a signed statement which he provided TIME, says that at one Beijing hospital alone, 60 SARS patients have been admitted, of whom seven have died," Susan had written. (That day, the official *China Daily* had put the number of SARS infections in Beijing at nineteen.) Jakes's story also detailed the Ministry of Health meeting at which hospital directors were told not to talk about the disease because of the National People's Congress. "Officials were forbidden to publicize what they'd learned about SARS," Dr. Jiang wrote.

Almost immediately, the story was picked up by media all over the world. In China, where our website was blocked, the story was translated and quickly disseminated through chat rooms and text messages. During the first quarter of 2003, Chinese sent 26.5 billion text messages to one another. It had become the preferred means of communication for many Chinese, their written pictograms being particularly suited to communicating efficiently via short messages. "Beijing Doctor: Government Is Lying About Number of Cases," said a typical message. "Foreign reports: SARS has reached Beijing," said another. Initially, with so many rumors flying about SARS, there was some speculation among Chinese about the reliability of the story. Yet *Time*'s name attached to the story added credibility. By the next morning, the sentiment of Henk Bekedam of the WHO— "There seems to be some question over when the central government knew about SARS in Beijing and what they actually knew"—would

seem very generous. At the blog *Peking Duck*, a more typical sentiment was expressed: "The new regime was supposed to be ever so
sensitive to the needs of the people. So what is its first accomplish-
ment? It gives its citizens the finger, concealing a deadly health
threat to make their meaningless People's Congress look pretty."

For Susan Jakes, the story was a signal achievement and made her
reputation among the Beijing press corps—and around the world.
She appeared on CNN and NPR and would spend the next day field-
ing calls from reporters asking how they could reach Dr. Jiang. Even
John Pomfret of the *Washington Post*, whose reporting on SARS—as
it was with almost everything China-related—had been the standard
against which we measured our own coverage, called to ask for the
number.

Disease, of course, is a common and familiar subject in literature.
And even emerging viruses have their share of great accounts—Matt
Forney had mentioned the *Hot Zone* repeatedly as he covered
SARS; another colleague would mention *And the Band Played On* as
we talked about what was happening in China. And as I thought
about that account by Randy Shilts of the early years of AIDS, this
did seem to have a similar arc: disease emerges, no one knows what
it is, there is a heroic effort made to identify the agent, and now, to
complete the similarities, there is a government cover-up that far
exceeded even the Reagan administration's indifference to AIDS.
Only here, in Asia, the whole cycle seemed to have been speeded up,
so that what had taken years to unfold in the case of AIDS—from
emergence to epidemic had been five to ten years—was now occurring in what seemed like a matter of weeks. And perhaps there was
an even more dramatic precedent, the writer Bryan Walsh suggested
to me in Hong Kong: *Jaws*.

"How so?" I asked.

"The virus is the shark in the water," he explained.

"And that mayor who keeps on insisting that the beach is safe?"

Bryan smiled. "That's the Chinese government."

CHAPTER 56

- **April 10, 2003**
- **You'an Hospital, Beijing, China**
- **3,110 Infected, 303 Dead**

HUANG YONG CHUGGED DOWN A BOTTLE OF BEIJING SPRING BEER AND lit a cigarette. He had parked on the street in front of the You'an and decided to finish his brew before "bombing" this croak house. I knew that the *Wall Street Journal*, among other publications, had ordered reporters to steer clear of these hospitals and was having them take extreme precautions in covering this disease. This was something of a moral dilemma that I dealt with by offering our reporters an opt-out clause on all SARS coverage. Huang Yong shrugged when asked whether he thought these "bombing" runs were dangerous. He would pull on his black leather car coat and say, "Oh, I think I'll be fine." And fortified by a few beers, he didn't hesitate to venture into what would turn out to be some of the most dangerous hospital wards in Beijing. While I worried for his safety, I admired him for his courage and desperately wanted to publish his discoveries.

The excitement of Susan Jakes's scoop with Dr. Jiang's letter had inspired Huang and awakened in him some competitive zeal. Also, this was among the first stories he had worked on since joining TIME where he felt that something greater than the reputation of a newsweekly was at stake. If this epidemic had, as the great infectious-disease reporter Laurie Garrett had put it, "only just begun," then the well-being of China—not the government that Huang didn't give a shit about, but the people whom he loved—might be at grave risk.

By getting slightly buzzed and poking around in Beijing hospital wards, Huang was performing, he believed, a great patriotic duty.

Still, nothing could have prepared him for what he found. In an aged and run-down hospital ward, crammed in two to a room, were dozens of patients. "Every single one of us in this building," a nurse infected with SARS named Zhang would tell him, "is a SARS patient." The patients were sprawled on dirty sheets, catatonic, fighting for their lives. "There are at least a hundred patients here, several hundred," said the nurse. "Conditions here are really bad. We're not allowed out of this room. We piss in this room, crap in this room, and eat in this room. At least half the patients here are doctors and nurses from other hospitals."

Huang Yong had expected lies; he had not expected the cover-up to be over a plague severe enough to wipe out Beijing. This was the charnel house that we all had feared was out there, like the vast open-air Atlanta mausoleum of *Gone With the Wind* transplanted to central Beijing.

Huang steeled himself and continued walking through the wards. He regretted not wearing a surgical mask yet felt obligated to continue his journalistic rounds. Another nurse in a surgical mask and gown stopped him. "Look, I'm not pushing you away. I do this for your own good," she explained. "It's too dangerous here. It's really a terrible disease; even we who work here don't know when we'll get it. No place is safe in this hospital. All of these wards are full of SARS patients. There are over one hundred at least. Don't believe the government—they never tell the truth. They say it's a deadly disease with four percent mortality? Are you kidding me? The death rate is at least twenty-five percent. In this hospital alone, there are over ten patients dead already."

"Never believe what the Health Ministry tells you," warned Nurse Zhang as Huang Yong left the hospital.

Huang would crack open another brew in his car and chug down almost half of it before driving back to the bureau. When he arrived, he walked into Susan's office to report what he had seen. He had a smile on his face.

It was all true, he related to her, every one of those awful rumors. He had not only confirmed them; he had seen it all for himself.

And, he would tell me later, he had never felt so alive.

"This is the hospital ward China's Ministry of Health doesn't want you to see," began *Time*'s coverage that week, written by Hannah Beech and reported by Huang Yong and Susan Jakes. "According to the Chinese government, most of these patients—and perhaps hundreds or even thousands of others across the nation—simply do not exist. . . . Numerous reports from local doctors over the past week suggest that the nation's health-care system remains hostage to a government that values power and public order before human lives."

CHAPTER 57

■ April 11, 2003

■ World Health Organization Headquarters,
Dongzhimenwai, Beijing, China

■ 3,130 Infected, 312 Dead

THE WHO TEAM WAS SATISFIED WITH WHAT THEY HAD SEEN DOWN
south. Guangdong had, in part thanks to Zhong Nanshan's aggressive
clinical leadership and a very rapid response by the Guangdong public
health authorities, done an admirable job of containing the outbreak.
"The authorities in Guangdong have carried out a very detailed surveillance of the case, and the numbers reported in Guangdong do in fact
represent reality," said Mike Ryan of the WHO. If Guangdong had
truly beaten the bug, the WHO team suspected, then there was hope
for the rest of the world.

Yet now the team was confronted with Beijing, and the real possibility that SARS was more widespread there than previously acknowledged.

They immediately went to the leadership compound in Zhongnanhai, central Beijing, a rarefied lakeside enclave reserved for
Chinese leaders and visiting dignitaries. Vice-Premier Wu Yi, the most
powerful woman in the Communist Party and a protégée of former
premier Zhu Rongji, greeted them in a formal reception area.

It is hard to attend such a meeting without a sense of awe and privilege. Zhongnanhai is adjacent to the Forbidden City and occupies land
once the retreat of Chinese emperors and their court. It may be called
the Palace of the People, but to visit this hallowed ground today is to
follow in the footsteps of every courtier or envoy who ever had an audi-

ence with the Chinese court. The ranking members of the Communist Party reside within its vermilion walls, similar to those that once enclosed the Forbidden City. To be allowed to reside inside Zhongnanhai is the surest indicator that a bureaucrat or official has reached the pinnacle of Chinese power. Yet among those inside the compound, which is so secretive and carefully constructed that it is impossible to see from the outside, there is intense jockeying for prime position. Mao Tse-tung, for example, lived in one of the exclusive lakeside villas and held court in a sedan chair next to his indoor swimming pool. Deng Xiaoping took one of the villas but preferred to sleep in his immense office. High-ranking officials vie for the walled compounds on the former imperial grounds between the Middle and South lakes, preferably facing south. These former *hutong*-style complexes have been converted into lakeside villas, many with private swimming pools and other perquisites. Security within Zhongnanhai is tight, and even those who are allowed to live on the grounds are required to show specific passes as they go from one section to another. The WHO team was getting a rare glimpse of courtly life as they waited for Wu Yi in a massive reception hall before one of the freshwater lakes stocked with carp and rare tortoises.

Wu Yi coifs her silver-black hair in an immense, intricately curled pile, as if she has borrowed a giant's pompadour. She wears wire-rim glasses and has a wide nose. When she listens, she purses her lips. Before speaking, she parts them once, closes them, and then talks. High-ranking Chinese officials are used to being listened to, unfamiliar with being interrupted, and unaccustomed to being questioned, making it exceedingly difficult for anyone speaking with them to raise a controversial issue.

For the meeting with the WHO team, Wu sat in the traditional seat at the northern side of the room, on her armrest a cup of tea that she did not sip. SARS had been weighing on her mind, she admitted, telling the WHO team that she hadn't been sleeping well since the outbreak had flared up. In fact, she'd had to resort to sleeping pills. She expressed how grateful China was for the WHO's cooperation.

The team was soon impressed with Wu's technical grasp of the science involved. She understood very well what the issues were, and the necessity of such particulars as contact tracing and acquiring samples. She had personally dispatched teams to two provinces to investigate outbreaks and further promised improvements in China's updates on the number of infected. For Henk Bekedam, as the head of the WHO Beijing mission, this meeting would mark the beginning of a fruitful relationship that would allow him, when necessary, to appeal directly to the highest levels of government.

Yet the team was left wondering why, if Wu Yi represented those highest levels, the government was unable to acknowledge SARS as the paramount national issue that Wu Yi seemed to be admitting it was. "What she was saying was that they were aware of the problem; they got it," says WHO team member James Maguire. But there was something off-putting about this admission. "You realized," says Meirion Evans, "that if the central government cared this much and still couldn't cut through the bureaucracy in terms of implementation, then who could?"

Apparently, no one.

On April 7, when China's new premier, Wen Jiabao, visited the national CDC offices in Beijing, the *China Daily* reported that Wen "stressed that China has the SARS epidemic under control." Wen's own comments that day were actually more telling. He said to Li Liming, head of the China CDC, in earshot of other CDC physicians and scientists: "It was wrong that the military was not reporting cases of SARS," adding that the CDC should start telling the truth to the people.

He then pointedly asked the CDC how many people in Beijing had SARS.

"We couldn't tell him," said one person who was in attendance.

But even as the premier was expounding on the need for transparency, his health minister was still covering up. John Pomfret wrote that Bi Shengli, a leading virologist in Beijing, had warned a senior official in Health Minister Zhang Wenkang's office, "We have a disaster in the capital in this new disease. We have got to do something."

Bi was told that the minister already knew about the problem.

"We have to negotiate with other ministries and government departments before anything can be done," the ministry official told him.

"Well," said Bi, "nothing was done."

One Chinese official who was not directly involved with SARS would tell me that the paralysis was the result of his fellow officials' thinking of their own careers first. "It's not a selfish impulse," he explained. "It becomes second nature. Minister Zhang did what came naturally: he wanted to understand the spheres of influence. He knows he can't interfere with the military hospitals. Those are the domain of his boss and protector, Jiang Zemin. So he was paralyzed and could not really ask for that information from military hospitals."

So far, Dr. Jiang Yanyong's revelations concerned primarily cases in the military hospital system, which reported up a separate chain of command from the public hospitals. This failed to explain why the cases in the hospitals Huang Yong was "bombing" had been unreported. These hospitals, it would later emerge, were among those ordered by the Beijing municipal authorities to squelch information on SARS.

The WHO's plan to quietly write and file a report on Guangdong was put on hold. Everyone now conceded that there was virus burning through Beijing. In fact, there had been for well over a month.

After Madame Wu Yi concluded, "This is a big, big problem," WHO's James Maguire decided to call Dr. Jiang Yanyong.

"I would be happy to talk to you," said Dr. Jiang when he came on the phone, "but you need to get the approval of my commanding officer."

The party boss of 301 Hospital and a fellow hospital bureaucrat had visited Dr. Jiang in his apartment on Wanshou Road and told him that he had to stop speaking to the media. From now on, all media requests had to go through the hospital president.

But after prefacing his statement with that proviso, Dr. Jiang told Maguire which hospitals the team should visit.

The WHO would now also be in the business of "bombing" hospitals.

CHAPTER 58

- **April 14, 2004**
- **Quarry Bay, Hong Kong, China**
- **3,750 Infected, 369 Dead**

THERE WERE SIXTY-ONE NEW CASES IN HONG KONG ON THE ELEVENTH and two more deaths, forty-nine newly infected on the twelfth, forty-two on the thirteenth, forty on the fourteenth, and forty-two the next day, with nine fatalities. That day, Ka-Kui Kwok, one of our production staffers, won about two hundred dollars in the office pool by predicting that there would be forty-one newly infected—the office infection pool was structured like *The Price Is Right*: you had to be under the number; one over and you would be out of the money. I was always optimistic and tended to lowball the guess, choosing in the twenties. We never established an office pool on the number of fatalities. The WHO would later report, "the case fatality ratio of SARS ranges from 0% to 50% depending on the group affected, with an overall estimate of case fatality of 14% to 15%."

A 15 percent case fatality ratio is frighteningly high. Marburg, one of the strains of Ebola, among the most lethal viruses ever to cross the species barrier, has a case fatality ratio of 20 percent. Measles kills just .2 percent of those who are infected in the developed world and 2 to 10 percent of children in the developing world. Malaria, which remains one of the globe's worst scourges, killing more than a million people a year, has a fatality rate of under 3 percent. Smallpox killed about 30 percent of those who contracted it. SARS, then, was a world-class killer.

By mid-April, SARS in Hong Kong had taken on a certain

rhythm. Most of those who did not need to stay in Hong Kong had already departed. And many of those who had departed, like my downstairs neighbors, decided to convert their trips into permanent relocations. For those of us who stayed, there was almost nothing to do in the evenings. Nightclubs were closed. The Rolling Stones had canceled their two scheduled shows in Hong Kong, as had DJ Shadow and Santana. My actor friend told me, "even the Russian hookers have all left town." Rather than drive home, I would stop in at the American Club in Tai Tam for dinner. The buffet, of course, had been shuttered. The swimming pool was drained. The children's play area was closed. The staff was masked. It began to look, I realized, as if Hong Kong, and indeed much of the world, was still subscribing to the miasma theory of disease, the Victorian view that "foul air" was somehow to blame for the outbreak.

Indeed, Hong Kong instituted rigorous disinfecting and scrubbing, as if litter-free streets would ward off disease. Bow-tied chief secretary Donald Tsang led "Team Clean," while Betty Tung, wife of beleaguered chief executive Tung Chee-hwa, donned more protective gear than most hospital workers as she toured Lower Ngau Tau Kok housing estate to pass out hygiene kits. Newspapers immediately slammed Tung for wearing the protective gear that ICU doctors were finding in short supply. And what protection did face masks actually provide? One's eyes, perhaps the best conduits for virus particles, were still exposed. Newspapers were filled with advertisements selling "antibiotic incense" and health-giving tonics—"Boost your immunities!" and, my favorite, "For Bountiful Lungness."

I would sit by myself in a booth at the American Club, eating my chicken vindaloo. Yet I was, in my own small way, succumbing to the same hysteria. We are, even at the best of times, bombarded by advertisements for pharmaceuticals that list symptoms to be alleviated. Every day, we read in newspapers about research indicating that too much of this or not enough of that will result in cancer, heart disease, or some other chronic illness. We are warned to vaccinate our children and ourselves, to take vitamins, to eat a balanced diet, to exercise—all in the name of warding off illness. The pharmaceutical-industrial complex encourages mild hypochondria. And the more recent development of the businesses of holistic medicine, New Age pharmacology, health food, organic food, vegan restaurants—it is all, at its most basic, about warding off sickness. Wheatgrass. Folic acid. Echinacea. Vitamins C, D, E, B_{12}. Zinc. Fish oil. We are fortified by a host of supplements intended to keep us fit. The first sniffle, fever, cough, chest pain, ache, or rash, and many of us are off to have it divined by our physician—or healer or masseuse or dietician—as a possible symptom of impending illness. And if he or she does not prescribe something to banish the symptom, we feel as if we have been cheated.

And here, into these vast cities, these communities of hypochondriacs, comes an actual, real disease, a genuine reason to worry and panic. No wonder folks started disinfecting their houses with the government-recommended one part bleach to ninety-nine parts water. They had to do something.

Among my precautions was a completely unscientific decision to eat almost nothing while in or around my office besides Chicken McNuggets. I shied away from using the office kitchen; I had once seen the cleaning lady, a well-meaning woman with a cheerful grin who spoke a dialect that no one seemed to understand, wipe a table, the counter, and the basin around the sink, and then, with the same damp cloth, reach under her shirt and wipe her armpits. Part of our office's "Team Clean" program was to have this woman on duty every day, all day, which didn't inspire confidence.

But McDonald's, I figured, had to be among the most sanitary dining establishments in Hong Kong. And Chicken McNuggets, I reckoned, were probably the most sterile items on their menu. The chickens were fed an antibiotic-rich diet while raised on factory farms; they had to pass inspection by the Hong Kong Department of Agriculture and Fisheries; and they were slaughtered in industrial pens and then chopped, breaded, and bagged on mechanized assembly lines—untouched by any potentially virus-infected human beings. After they were delivered to the franchises, they were then deep-fried in oil heated to over 350 degrees, enough, I guessed, to kill almost any living organism or active virion. The McDonald's staff in Hong Kong wore rubber gloves, and, throughout the process, from slaughter to preparation, the nuggets never come into contact with potentially virus-bearing human flesh. I ate at McDonald's every day for lunch during SARS season: nine McNuggets and a Diet Coke. I lost eight pounds.

The SARS Diet.

I would throw the cardboard McNugget packaging away, light a cigarette, and then, some afternoons, take a pair of binoculars and survey Jordan, Kowloon Bay, and Lower Ngau Tau Kok. There, in the distance, was Amoy Gardens. Local authorities had begun gathering rats and cockroaches from the apartment complex to see if they carried a viral load. Hong Kong's Department of Health determined that of eighty-three rat specimens, six samples from droppings and two samples from throat swabs had been found to be positive for SARS. Rats, the study charged, could serve as mechanical carriers for SARS even if they themselves were not infected. Partly as a result of this investigation, Amoy Gardens became notorious for poor hygiene and pest infestations. Yet how different could that complex be from any of a thousand others in Hong Kong?

The prevailing hypothesis as to what had triggered the outbreak at Amoy Gardens, announced at a Department of Health press conference later that week, would be that virus-laden feces had been aerosolized by a cracked fourth-floor drainage pipe. Feces, of course, are proven viral carriers, as any particles that don't find a cell to infect are passed through the intestinal system; a single gram of human feces may contain up to one billion virus particles. This sewage pipe crack found at Amoy Gardens, according to the Department of Health investigation, "emitted droplets carrying contaminated sewage into the light well every time a toilet was flushed." The droplets would then be sucked into tenants' apartments via "dried-up U-traps"—the same sort of curved pipe that is beneath most sinks. That U-shaped pipe beneath a floor drain, when filled with a barrier of standing water, serves to keep insects, vermin, and viruses from reaching the bathroom. At Amoy Gardens, these drains had long dried out, so that when tenants turned on their bathroom exhaust fans, the fans created a vacuum, hoovering up droplets of virus-laden feces that would either infect the tenant of the apartment or pass through the fan and back into the light well, becoming a finer, more aerosolized virus-feces cocktail. One of the more recently noted symptoms among SARS patients was stomach-cramp-inducing diarrhea, causing up to twenty watery bowel movements a day. Thus, each infected tenant at Amoy Gardens would use the toilet often, flush frequently, and, in doing so, send another viral load down the pipes to be aerosolized and blown back up the light well—and into other tenants' noses, eyes, and mouths.

As I sat in the press conference where E. K. Yeoh announced these findings, I found a few factors confusing. Didn't this mean that the disease had basically gone airborne?

"No," Yeoh insisted. "It had been 'aerosolized.'"

I don't know if anyone in that conference room found this to be a reassuring differentiation. It also did not explain why so many other residents of Amoy Gardens, those who didn't live in Block E but in entirely different buildings, had become infected. Were the virus-laden feces really being blown that far afield?

Yes, we were told, it was possible that the bathroom exhaust fans had created an upward "plume effect," shooting virus-laden feces droplets up out of the top of the building, like a Roman candle of microscopic shit.

The secretary of health assured us that the cracked sewage pipe would be fixed and that U-pipes would be filled with water, preventing any microbial travel through the drains.

ON APRIL 12, THE MICHAEL SMITH GENOME SCIENCES CENTRE IN British Columbia, Canada, had announced that it had sequenced the entire SARS coronavirus genome, a crucial step toward understanding the virus. The SARS coronavirus, Caroline Astell, project leader at the Michael Smith Genome Sciences Centre, would tell me, was "the largest RNA virus we've ever seen." Every organism on earth, besides a few viruses, stores and deploys its reproductive instructions by DNA, or deoxyribonucleic acid, the familiar double-helix "code of life," rendered in Ping-Pong ball and pipe cleaner models in high school biology classrooms around the world. RNA, ribonucleic acid, is generally used as a messenger to carry genetic instructions. A few viruses, such as the SARS coronavirus, instead store their genes as RNA. DNA is far better suited than RNA to serve as a storage medium for genetic information. The "double" part of the helix can, in effect, check on the work of the genetic machinery as it replicates itself, "correcting" most errors, i.e., mutations, as they occur. Viruses with their genetic code stored as RNA instead of DNA lack this proofreading ability to check on their own reproduction. As the nucleotides of a next-generation virus are assembled according to information stored on the viral RNA, there is a far higher chance of an error occurring. Adding volatility to the process is the fact that RNA is a more reactive polymer than DNA, which means that there is more likelihood that any part of the genome may become altered because of the presence of foreign chemicals or proteins.

Though not a reliable storage medium, RNA does provide an interesting survival strategy for a virus. All viruses go through a very rapid duplication cycle. A single RNA virus can produce ten thousand copies of itself in six hours; a human being might reproduce, perhaps, every eighteen months. Hence viruses, even more stable DNA viruses, are evolving much faster than we are.

An RNA virus accelerates that already quickened process. As the RNA replicates one strand of itself, it averages one mistake per ten thousand nucleotides copied. Such instability means that once an RNA virus has entered a cell and successfully begins to create copies of itself, it is producing numerous mutated versions, with differences ranging from just a few nucleotides to dozens or perhaps hundreds. The vast majority of these genetic changes—perhaps 99.9999 percent of them—are inconsequential or unsuccessful, or they create a virion that is not successfully "active." Occasionally, however, because of the tremendous rate of mutation, an evolved virus will emerge, one that it is actually better suited to its cellular environment. It might have different spike proteins or a slight variation of peptides that enables it to bind more efficiently with a host cell, making the disease it causes more infectious or contagious.

The SARS coronavirus, it turned out, had nearly thirty thousand nucleotides—a nucleotide is a specific combination of purine or pyrimidine, a sugar molecule, and phosphate; genes are the sequences spelled out by groups of these nucleotides—making it the largest RNA virus ever recorded. While genome size does not automatically correlate with mutability, it is a depressing fact that RNA viruses do mutate more rapidly than DNA viruses, and that this was a very large RNA virus. "SARS is a big, big boy," said Kathryn Holmes, who literally wrote the book—or at least the relevant chapter of the book *Field's Virology*—on coronaviruses. Viruses from the coronavirus family had previously been known to cause only mild infection in humans; along with rhinoviruses, they are a major culprit for the common cold. They were traditionally so unthreatening that the 2,629-page *Harrison's Principles of Internal Medicine*, the world's best-selling English-language medical textbook, devotes just a page and a half to them. Yet coronaviruses that infect animals can undergo troubling mutations. For example, a coronavirus that caused gastroenteritis among pigs mutated in the late 1980s into porcine respiratory infection. Another coronavirus that was the cause of shipping fever among calves became the reason for dysentery among adult cattle. Perhaps it had been a mutation—a random set of genetic tumblers clicking into place—that had allowed this particular RNA virus to become so infectious among humans at Amoy Gardens? Even more alarming, virologist Stephen Tsoi and a team at the Chinese University of Hong Kong were conducting research indicating that the SARS virus, as it rolled through humanity, was mutating and could become increasingly pathogenic, causing disease of greater severity. His team would take the genomic sequence from the earliest cases in Hong Kong, which were very similar to those in Guangdong, and find that there were already significant changes occurring in later strains taken from Singapore and Toronto. These mutations, he noted, were occurring primarily in the spike protein of the virus, precisely the means by which the virus attaches to a cell receptor. Perhaps what had occurred at Amoy Gardens—the emergence of new symptoms, the apparent greater infectivity—was that sort of mutation.

I thought back on the images I'd viewed through John Nichols's electron microscope at the University of Hong Kong: viruses budding by the hundreds around the membrane of an infected cell; other cells collapsing and breaking apart as the SARS virus finally cannibalized all the proteins and genetic material it could. As the cell literally broke apart and fragmented like a torpedoed ship—the process described in textbooks as "lysis"—thousands of viruses would then be released into the host. Lysis is the virus's crowning achievement, the moment at which it has successfully commandeered everything of value in the cell: all those virus particles, each potentially carrying three mutations (ten thousand nucleotides times three), heading off to infect more cells.

Yet here was E. K. Yeoh, assuring us that *aerosolized* SARS coronavirus was not *airborne* SARS coronavirus.

Oh, how we clung to that fine semantic point. "Airborne" meant that it was everywhere, all the time. That if you rode in an elevator a few minutes or even hours after someone who had it, then you would probably inhale it. "Aerosolized" meant that you had to be in the elevator at the same time as someone who carried it, that you had to inhale his infected mucus particles. "Aerosolized" meant there was hope. If we scrubbed, cleansed, and disinfected, were valuable members of "Team Clean," kept our N-95 masks strapped on tightly, and were diligent about fever checks and avoiding virus-laden feces, then we had a chance. "Airborne" meant we didn't.

I went home. I filled my U-pipe. Six inches of standing water in a U-shaped drainpipe to keep the world's microbes at bay.

CHAPTER 59

■ April 14, 2003
■ Sino-Japan Friendship Hospital, Beijing, China
■ 3,755 Infected, 370 Dead

HE HEIYU, THE DIRECTOR OF BEIJING'S SINO-JAPAN FRIENDSHIP
Hospital, wore a pressed white lab coat over his tailored slacks and
black rubber-soled shoes. His office was a spacious corner suite that
he kept so spotless that it could have served, in a pinch, as an operating theater. This hospital was among Beijing's elite institutions,
renowned for having the latest technologies, and was the emergency
ward of choice for wealthy Beijingers and visiting foreigners. The
two buildings that make up the hospital are modern, the hallways
wide, and the color palette of the interiors soothing pastels. The hospital is the embodiment of a contemporary big-city hospital and one
of the institutions that Beijingers could point to with pride, the equal
of any hospital in Tokyo or New York.

Still, the economic reforms had posed vexing financial challenges
for Director He. Each department had been forced to pay its own
way; staff remuneration was based on each section's achieving certain
earnings targets. One Sino-Japan cardiac specialist I spoke with
would explain that in the past ten years, he had seen the hospital transform, going from considering medical issues as primary to putting
business concerns ahead of all others. He bemoaned the changes,
claiming that the ethical standards of his own profession had probably
declined since the party had opened up the economy. As a senior doctor, he was forced to consider the financial means of each newly admitted patient. His section's bonus was determined by a simple formula:

revenue – costs = salary + bonus. This hospital, like many others, sub-contracted out maintenance and the provision of basic hospital sup-plies. Hence the initial reluctance, as the SARS epidemic emerged, to freely dispense masks and other protective gear; doctors and nurses viewed the cost of each mask as coming out of their own pockets, which, in effect, it was. If a patient was not insured and did not seem likely to have the means to pay, he or she was discharged. If not, then the doctor admitting the patient might have to pick up the tab himself. Sino-Japan Friendship Hospital, despite its sleek exterior, had become more akin to Adam Smith's pin factory than to the Mayo Clinic.

Yet as the paragon of modern medicine in Beijing, this hospital was certainly expected to uphold basic Hippocratic standards. The WHO team had begun a round of Beijing hospital visits, responding to persistent reports that there were far more cases in Beijing than the thirty-seven now acknowledged by the Ministry of Health. They had already dropped in on 309 Hospital, where doctors admitted that they had more than fifty patients who had been transferred from 301 Hospital. "They told us that 301 hadn't been able to handle those patients," says Henk Bekedam. The WHO team toured the wards, went through the hospital's records. So far, 309 alone had had seven SARS deaths. All of Beijing, according to the Ministry of Health, had had only two deaths.

"Why hasn't this been reported?" James Maguire wanted to know.

These doctors' response was that of course the cases and fatalities had been reported—up the military chain of command. And the WHO team could begin to understand that there was a jurisdictional issue that seemed to be getting in the way of accurate case counts. At another military hospital, they would see that while there was a great deal of traffic between the military hospital system and the civilian system—doctors and patients seemed to move freely back and forth, and both systems took on any patients with the ability to pay—the bureaucracies themselves, the Central Military Command and the Ministry of Health—had virtually no formal communication. Those cases in mili-tary hospitals, as data relating to the People's Liberation Army, were considered state secrets.

Yet there was another bureaucratic entity with an interest in keeping a lid on SARS information: the Beijing city government and its own health department.

After a visit to the Beijing CDC, the WHO team made a quick decision to drop in on the Sino-Japan Friendship Hospital, which was rumored to be housing twenty SARS patients.

Hospital director He received a call from the Beijing Municipal Health Bureau a half hour before the WHO team was scheduled to arrive. The bureaucrat on the other end of the line gave He some extraordinary instructions, so remarkable, in fact, that Director He asked the bureaucrat to repeat them to make sure he had heard him correctly. The second time the Beijing Municipal Health Bureau official spoke, Director He had turned on a cassette recorder attached to his phone line. He wanted to make sure he had these instructions on tape.

CHAPTER 60

- **April 15, 2003**
- **Kempinski Hotel, Beijing, China**
- **3,989 Infected, 375 Dead**

TIME'S BEIJING BUREAU CHIEF, MATTHEW FORNEY, HAD RETURNED from Guangdong, where he had written a comprehensive timeline of the earliest phases of the disease and was taking the precaution of quarantining himself in the Kempinski Hotel. He had called Dr. Jiang at home and been told by his wife and then by Dr. Jiang himself that he could no longer talk to *Time*. "I have been so ordered," Jiang said.

The *Time* team in Beijing gathered at the Kempinski to plan that week's SARS coverage. There was now mounting speculation that what was going on in the provinces was the real story, and we were determined to get out to what we believed were the worst-afflicted areas. Susan Jakes and Kaiser Kuo would head to Inner Mongolia. Hannah Beech and Bu Hua were on their way to Shanxi. Neil Gough would continue to report in Guangdong and also head to Guangxi Zhuang.

After the team had received their marching orders, Susan Jakes returned to the TIME bureau in Jianguomen, where she gathered a bottle of Advil and a few surgical masks for her reporting trip to Inner Mongolia.

The office phone rang, and Jakes answered it. A Chinese woman spoke in English, "I am a concerned Beijing citizen, and I have information from the doctor you know. Meet me in Wukesong in a half hour." The woman said she would be wearing a denim jacket.

Wukesong is a subway station at the far end of Beijing. There was no way to get there by car in a half hour. With the SARS situation in Beijing looking increasingly out of control, riding in the close quarters of a subway car had become an unpleasant prospect, but Jakes had no choice. When she arrived at Wukesong Station, she walked back and forth in front of the ticket wicket until she was approached by a Chinese woman in her thirties with short black hair and sharp cheekbones. The woman was slender but broad-shouldered. There was a giddy anxiousness about her that Jakes found a little disturbing, as if this woman, who called herself Wu, were enjoying the clandestine nature and melodrama of this rendezvous. She could, Jakes felt, be the sort of person who took unnecessary risks, who secretly wanted to get caught.

But there was also earnestness in her mien, Jakes noted. She acted as if she were on a mission, and that no matter how much fun the cloak-and-dagger games were, this was still part of a larger cause.

"The WHO has been trying to reach our friend," Wu explained. "He has been told by his hospital that he is not allowed to talk."

She handed Jakes a letter. "Can you take this to the WHO? It's a matter of life and death."

Jakes took the letter, now understanding that it had to be from Dr. Jiang. In the taxi on her way back to the office, she read the Chinese text. It was indeed from Dr. Jiang, and it was a remarkable document. If what he had written was true, then the WHO would never be able to get a handle on this plague: the Chinese government did understand the severity of the issue but was still covering it up, and the world's health could be in jeopardy.

Her responsibilities as a journalist and her obligations to humanity were now blurring. This seemed a matter of larger significance than merely getting a scoop for the magazine. Jakes called Jim Palmer, a public relations consultant working with the WHO, and suggested he meet with her immediately. Then she called Matt Forney, explaining what she had. Forney said he would join her and Palmer in the parking lot outside the WHO's Beijing headquarters.

Jakes arrived before Forney and described to Palmer the nature of the document in her possession before asking him to pass it along to James Maguire, Henk Bekedam, and the rest of the WHO team. Jakes wasn't looking for any quid pro quo, she explained. "I'm here as an emissary from Dr. Jiang. That's it. This is a letter that talks about SARS in Beijing."

Palmer worried that the *Time* correspondent was playing a journalistic game, perhaps seeking confirmation of something from the WHO. As a representative of an international body, Palmer was sensitive about attaching the WHO to a story that could unnecessarily offend the Chinese government.

Matt Forney's arrival in the WHO parking lot further confused matters. Forney believed that before passing the letter along, TIME should at least be assured by the WHO that the magazine would have first crack at publishing the contents of the letter. He also wanted interviews with the WHO team. "I didn't know who we could trust," Forney would tell me later. "So I wanted to make sure that we were covered." Forney found out during the conversation that Palmer was having dinner that night with Charles Hutzler, the *Wall Street Journal*'s Beijing bureau chief and one of the best reporters in Asia. Forney had worked for the *Journal* before joining *Time*; it would kill him to lose this scoop to his old employer.

Forney and Palmer began going back and forth about the terms of TIME giving the WHO the letter, in a conversation that seemed completely beside the point. Then Palmer boasted of having been in the meeting with Wu Yi, the woman Politburo member, and how he and she had had a very frank and honest exchange. He said that the WHO did not necessarily need this letter. And he would not even promise he would pass it along to the WHO investigative team.

While Jakes was again trying to cut through the adversarial tone of the meeting, her phone rang. It was a radio station in Hong Kong that was calling to conduct a previously scheduled interview.

By the time Jakes had concluded the interview, Forney and Palmer had agreed to go inside the WHO offices and make a copy of the letter. The WHO would retain one copy for its internal use, and TIME would keep a copy to use as possible material for a story.

CHAPTER 61

- **April 17, 2003**
- **Inner Mongolia, China**
- **4,087 Infected, 389 Dead**

SUSAN JAKES AND KAISER KUO, A FREELANCER WHO OFTEN WORKED with us, had flown to Hohhot in the windswept Autonomous Region of Inner Mongolia and were now riding in a beat-up Jinbei taxi through the Mongolian steppe, on their way to a remote snowbound village that, as far as they could discern, was named Shuiqu. The huts there were made of a combination of cinder block, stone, and mud. Back in Hohhot, they had had to remove their sweaters because of the arid heat, but here in the snowbound steppe, they found the shepherds wrapped in thick knitted wool coats and seated on *kohngs*—charcoal-heated beds—and smoking brown cigarettes. When Jakes and Kuo were seated in the hut and had managed to make clear the purpose of their visit, the shepherds all started yammering at once in thickly accented Mandarin, which frequently gave way to their local Mongolian dialect. To Jakes's surprise, even these villagers seemed to understand what SARS was and that if anyone had the symptoms—fever, cough, backache—they would take him to see the local barefoot doctor. When Jakes and Kuo then dropped in to see the one-eyed practitioner, they found that his entire dispensary consisted of cotton balls, a bottle of rubbing alcohol, and a few pairs of old shoes.

If any of these people were actually to contract SARS out here, Jakes reflected, they wouldn't have a chance.

Jakes and Kuo already knew there was virus in Hohhot. The outbreak there had been seeded by the two China Air flight attendants who had never been informed by their bosses that among the passengers on their March 15 flight from Hong Kong to Beijing was a highly contagious SARS patient, who would infect twenty-four fellow travelers. Liu Sutao, the deputy director of the region's health department, had told Jakes and Kuo that they already had more than thirty SARS cases in Hohhot. The virus could hop a ride out to the steppe as easily as Jakes and Kuo had hailed a taxi. What hope did a barefoot doctor with cotton swabs have against an emerging virus?

Back in Hohhot, Jakes was sitting in a hotel lobby, waiting for a meeting with another government official, when she received a call from the same "concerned citizen" who had given her the letter from Dr. Jiang. Wu Xi seemed slightly hysterical and was going on about the patients being hidden from the World Health Organization. She said that the nurses at 309 Hospital and Sino-Japan Friendship Hospital were terrified. "You have to tell the world," the woman was saying.

Jakes had to hang up, as a senior health department official had just arrived. She was too excited to eat any of the ten courses served—Chinese officials banquet relentlessly—and was looking for a moment to excuse herself to make a phone call. As a few junior officials began drifting away from the lunch, she slipped out and called Matt Forney.

"Did you ever look into that letter?" she asked him.

"I didn't have time," he said.

When she left the letter with Forney, she had suggested that Huang Yong look into verifying the material in it. Forney hadn't seemed that interested, saying, "We did the cover-up story last week." He was focused on producing a political story about how the Chinese government was responding to SARS and the potential for a leadership shake-up.

Jakes hid her frustration. Now she asked if it was still okay to have Huang Yong look into the matter.

Huang was quarantined at the Hotel Lido. Jakes reluctantly realized that she had no choice but to speak to him over his unsecured hotel phone. She explained to him the nature of Dr. Jiang's letter.

"That's unbelievable," he responded.

"Well, look into it."

"I'll see if I can confirm it," he said.

Jakes didn't see how he could.

She disconnected and went back into the hotel banquet hall. The officials were glad she had returned, and again promised they were doing everything they could to ward off SARS. Still, Jakes wondered, was there anything that could ward off a plague that was already there?

Huang returned to the TIME bureau and took a seat behind his desk, which was piled high, as it usually was, with Chinese newspapers. When he wasn't out working a story, he would skim the local press for possible story ideas and pass these along to Forney or Jakes. Now, however, he had a real story, perhaps the biggest of his career. How could what Dr. Jiang had written really be true?

CHAPTER 62

■ April 17, 2003
■ Sino-Japan Friendship Hospital, Beijing, China
■ 4,088 Infected, 390 Dead

DIRECTOR HE HAD TO CALL A MEETING OF HIS DEPARTMENT HEADS. His staff was furious at how they were being treated, at what they were being asked to do. They were still, after all, doctors, and as word spread of what had transpired, the staff now threatened open rebellion. Director He was a veteran of political turf wars and a survivor of various consolidations and institutional upheavals—he had overseen Sino-Japan's transformation from a state-run to a for-profit hospital and had maintained the institution's reputation as among the best in Beijing. The power wielded by hospital directors varies and has diminished somewhat in the reform era as control has devolved to department heads, especially those in areas such as cosmetic surgery, which bring in the most revenue. Thus, hospital directors, while still holding a title that allowed for Director He's imperious manner, had to be diplomatic in order to ensure their hold on the mandate of their department heads.

The team members now gathered before him were livid. They were marshaled around a black wood conference table in a corner room slightly too small to accommodate them all without some invasion of personal space. Yet even with the room as crammed as it was, the department heads gave Director He plenty of distance. No one, it seemed, wanted to be perceived as being close to Director He, spatially or professionally.

"I had no choice," Director He said, justifying his edict. "I would never elect to compromise the safety and health of my staff or violate sound treatment protocol."

"That was not the act of a doctor," the head of nephrology argued.

"I would never make such a decision on my own," Director He insisted. "These are extraordinary times. SARS is a new disease that is causing the leadership to take extraordinary measures."

"What does that have to with treating patients?" asked another physician.

"Or possibly making more staff sick?" asked a nurse.

"Listen to this," Director He told them. "This is a conversation between myself and the Beijing Municipal Health Bureau. Just so that you understand I was following orders."

On the conference table before Director He was a boom-box-type stereo. Director He pushed Play on the cassette player. His voice could be heard asking the official from the Beijing Municipal Health Bureau to repeat his instructions.

CHAPTER 63

- April 17, 2003
- TIME Bureau, Jianguomen, Beijing, China
- 4,454 Infected, 398 Dead

HUANG YONG SAI SMOKING CIGARETTES, FLICKING THE ASHES INTO a blackened brass ashtray he kept on his desk next to the newspaper piles. The hospital spokesmen had passed him along to the Beijing Municipal Health Bureau. The health bureau referred him to a local English-language SARS information hotline.

He had already tapped out his most obvious contacts. His father had been a scientist who had pioneered China's electron microscopy programs in the 1960s, when it was viewed as a matter of national pride to develop such technology without Western help. Through Huang Senior, Huang Yong was acquainted with a few aged members of the science and medical elite, yet none of them knew specifics about the current outbreak.

Desperate, he began calling friends and asking if they knew anyone who worked in Beijing's hospital or public health sectors, not really expecting to come up with a source. Yet a friend of his suggested a doctor from the Sino-Japan Friendship Hospital whom he vaguely knew and gave Huang Yong the doctor's mobile phone number.

Huang called and quickly explained who he was and that he and the doctor shared a mutual friend. Then he told the doctor what *Time* magazine had learned.

The doctor was silent.

Fearing he would hang up, Huang added that what they already had was going to be published anyway, and this was merely an attempt to make sure they had the facts correct. Huang was gambling that if something was really wrong with the reporting, then this doctor would come out and say it. Still, that wouldn't necessarily confirm the story.

Huang listened as the doctor took a deep breath and sighed. "*Time* magazine has very good information. It's true."

The doctor then recounted to Huang the story of the World Health Organization visit. The hospital had fifty-six SARS patients, thirty-one of whom were doctors, nurses, and other medical workers. A few minutes before the WHO team arrived, a fleet of ambulances had pulled into the horseshoe driveway in front of the hospital and along the road that divides the administration building from the old wing. The hospital director had ordered the stricken health care workers loaded onto gurneys. The staff was scrambling to move these patients onto elevators and down to the waiting ambulances. "The fleet of ambulances," *Time* would write that week, "started their engines . . . and dispersed into Beijing's smog-filled traffic. Inside the vehicles was a deadly secret: 31 coughing, shivering hospital workers who had caught severe acute respiratory syndrome (SARS) from their patients. Riding alongside were nurses who recoiled at each contagious breath their dangerous charges exhaled. As the white vans took a leisurely tour of the Chinese capital, a team of experts from the WHO walked into the China-Japan Friendship Hospital, hoping at last to gain a more accurate sense of the scope of Beijing's mounting SARS crisis. Instead, they were presented with deceptively uplifting news. . . . Especially encouraging, as those ambulances were meandering around Beijing, it appeared that no medical staff were confirmed to have contracted the virus."

The WHO team had arrived in midafternoon and during their tour had spoken to a few health care workers. "We were told that there were health care workers who had gotten sick but who did not meet the criteria for SARS cases," Maguire recalls. The team left the hospital with the impression that there had not been an outbreak among health care workers. Those rumors about dozens of infected health care workers at Sino-Japan had to be mistaken, the team felt.

"But there was an overall impression emerging, from visits at other hospitals and clusters of cases here and there, that something was wrong with the reporting system," says Henk Bekedam. "There seemed to be holes in the ability to detect, report, and take care of patients."

The Sino-Japan doctor was now confirming with Huang Yong that this was more than a "hole"; it was a pattern of deception the scope and scale of which were very hard to imagine unless, like us, you had been tipped off. He confirmed Dr. Jiang's letter and also the telephone call from "the concerned citizen."

Huang Yong asked him, "How could you do this?"

The doctor said softly, "We are ashamed."

After that, calls to 309 Hospital confirmed that before the WHO team had visited, the hospital had moved forty-six SARS patients into a separate long-stay facility, a sort of convalescent hotel, on the hospital grounds. One doctor at 309 Hospital had stated, "I have seen an internal Ministry of Health report which puts the number of confirmed SARS cases in Beijing at between 200 and 300 based on accounts from individual hospitals. Another internal document I have seen says that in the last 10 days there have been more than 100 new cases reported in Beijing." All this while the Ministry of Health maintained that there were only thirty-seven cases in all of Beijing.

CHAPTER 64

■ April 18, 2003
■ Ministry of Health, Beijing, China
■ 4,577 Infected, 400 Dead

IT WAS AN UNUSUAL SUMMONS. THE MINISTER WANTED TO MEET with the World Health Organization team. Alan Schnur, at a press conference the day before, had responded to a question from a *New York Times* reporter by saying that the WHO now believed there were between one hundred and two hundred SARS cases in Beijing, far more than the Ministry of Health had admitted. Henk Bekedam and the WHO team had concluded that the Ministry of Health was either lying or incorrect about the number of SARS cases in the city. Until there was an accurate reckoning of cases, then the most basic epidemiology could not yet commence. Where were these patients? How had they become infected? And whom had they been in contact with? Only by obtaining real numbers and accurate data could public health officials get a handle on the scourge. "Things just weren't getting done the way the Guangdong experience said they should be done," says James Maguire.

Emerging diseases pose questions from the most basic (What is it?) to the most metaphysical (What does it mean to say a virus is neither alive nor dead?). Still, from the asking of questions, the collecting of data, and the review of treatments and protocols comes the beginning of complex answers. The Ministry of Health and the Beijing Municipal Health Department, which reports to the mayor of Beijing, were unwilling to pose the simplest queries. They seemed to confuse suppressing news of the disease with curing it.

Amid this news vacuum, Beijingers were text-messaging one another rumors and SARS aphorisms at the rate of twenty-five million sends a day.

HEALTH MINISTER ZHANG WENKANG, ONCE THE WHO TEAM HAD gathered, took a seat at the top of a U-shaped chair formation and nodded solemnly.

"We have some mistakes," he told the team. "We have gone through the hospitals and found more cases than we had estimated."

Even the WHO reckoning of cases—one hundred to two hundred—was, Zhang had concluded, far too low. "We are going to hold a press conference Sunday, in which we will announce the actual figures."

Then the minister, a taciturn man with rectangular wire-frame glasses and gray hair combed into a wave, neatly breaking from left to right over his wide forehead, took on a tone of voice that caught the WHO team off guard. This was a man who never departed from the official script and who did not hesitate to dress down any reporter or staffer who, he felt, was breaking with protocol. During the early days of SARS, at his very rare public appearances, he had defiantly asserted that SARS was under control and had imperiously dismissed any suggestion even hinting at the contrary. He had presided over a cover-up of a national AIDS crisis, insisting, despite terrifying evidence, that there were only thirty thousand cases in the entire country, before finally admitting there were more than a million. And he had ordered that various outbreaks of food poisoning in which thousands of children had been sickened be kept secret. He believed the primary responsibility of a health minister was the peace of mind that comes with the outward appearance of social stability. Yet his obfuscation and the cover-up were seeding greater panic than a frank admission of the status of the virus. He looked shaken by the realization that he had badly miscalculated.

Zhang Wenkang had inadvertently become a pawn in the power struggle between factions supporting the new president and party secretary, Hu Jintao, and the former party secretary and current central military commander, Jiang Zemin. The paralysis in terms of responding to SARS was in part due to bureaucratic uncertainty as to who was actually running the country. It should be pointed out that for all China's reforms and modern, almost neocon economics, the country remains in the thrall of various cabals of engineers for whom the possession of power often rests on abstractions. (Most members of the Politburo Standing Committee were engineering students.) There is no fundamental difference in ideology between Jiang Zemin and Hu Jintao. Both believe in economic reform, but not at the expense of political control. Possibly, Jiang Zemin is more of an internationalist, proud of his relations with foreign heads of state and CEOs; but given time, Hu Jintao might also see the wisdom in fostering the personal relations with foreigners that come at virtually no expense for China but provide great latitude in allowing China to pursue its own interests. Western politicians and businessmen often make the mistake of believing that because a Chinese plutocrat is someone you can sit down with and talk to, he will also pursue policies and agendas that fit into a Western framework of rational policy. This is true when it comes to economic issues, perhaps. But when it comes to human rights or censorship, Chinese leaders have done a remarkable job of taking those issues off the table. That is often the quid pro quo of being able to sit with a Chinese leader and have that conversation. When *Time* was banned in China, I was told that Gerald Levin, the CEO of AOL–Time Warner at that time, would bring up the topic in a meeting with Jiang Zemin, whom he had introduced at a Fortune Global Forum event as "my good friend." Yet Levin never raised the thorny issue of censorship of his own company's media properties, so pleased was he to be granted an audience with the Supreme Leader of the world's most populated state. There is a pattern to this China exceptionalism, and at almost any international event where Chinese government officials are guests, you will find most of those attending unwilling to mention human rights, censorship, Taiwan, and other controversial issues in the presence of those officials. I've encountered this at the World Economic Forum and other high-profile global events. It is an unstated precondition of Chinese attendance at these conclaves. So eager is the world to meet

China that China is allowed to set the terms of the meeting. Jiang Zemin had played this hand very skillfully, and Hu Jintao, in due time, would perfect the same trick.

First, however, the domestic political rift was one that had paralyzed the government. Hu Jintao was nominally the head of state, but anything that directly threatened Jiang Zemin could bring on a grave crisis. Jiang still had the army and his power base in Shanghai, where, during the 1980s, he had been mayor; the city had prospered under his party secretaryship. Now, just as Hu Jintao was taking office, he was confronted by a national health crisis whose scope was hard to define. To further complicate the issue, the military hospital system was under Jiang Zemin's jurisdiction and would report to him rather than to the Politburo.

Yet the outbreak and the scandals erupting around the ham-fisted cover-up of the disease gave the Hu Jintao faction an unexpected political opening. This could be the issue that gave Hu the populist sheen that would lubricate his ascension to real power. He recognized that SARS was both a grave danger for the people and an opportunity for him.

Perhaps because China is a plutocracy in which information is so tightly held, historians have cited great populist gestures as being the catalyzing moments in epochal transformations in Chinese history. Mao Tse-tung had signaled the start of the Cultural Revolution in 1966 by going to Wuhan and swimming across the Yangtze; one result of his famous crawl stroke was the banishing of Deng Xiaoping to exile and the death of Mao's greatest rival and fellow revolutionary, Liu Shaoqi. Deng Xiaoping, after returning from exile, set the country on the path to reform, demonstrated his willingness to use force to ensure the party's hold on power at Tiananmen Square, and launched his famous "Southern Tour" in 1992, in which he visited Shenzhen, Dongmen, and numerous factories up and down the Pearl River Delta, bypassing the party conservatives and going straight to local cadres and party committees. In Shenzhen, he planted a tree at a local park and repeated his famous statement, "To get rich is glorious." As newspapers began reporting Deng Xiaoping's journey, it was widely taken as a signal that the post-Tiananmen period of austerity was over. He had launched the Era of Wild Flavor.

Now, Hu Jintao embarked on a similar tour of the south, only instead of factories and worker collectives, he visited hospitals and virology labs. For ten days, Hu and Premier Wen Jiabao crisscrossed the country gathering information about SARS. Back on April 7, Wen had visited the Beijing CDC and asked pointed questions. On April 11, Hu flew south to Guanghzou, SARS ground zero, making widely reported appearances in hospitals and laboratories. Meanwhile, Jiang Zemin had returned to his power base of Shanghai, causing Beijingers to text-message one another that the former Paramount Leader was afraid of the virus. Wen Jiabao warned the country, at an emergency meeting of the State Council on April 13, that China's economy, international image, and social stability were at stake, saying of SARS, "the overall situation remains grave." Hu had made a brilliant political calculation; he would personally take over the anti-SARS campaign, making it a centerpiece of his inchoate administration and using it as a wedge issue to consolidate power. After all, if he was the anti-SARS guy, then who would want to obstruct him and be known in a hundred million text messages as the pro-SARS villain?

The day before the WHO was summoned to the Ministry of Health, the Politburo had met in a special session in which Hu and Wen vowed all-out war against the epidemic. That meant that officials had to stop lying. Those orders were the leading story on the state-run news broadcasts that night and were on the front page of every Chinese newspaper the next day. For government officials on the wrong side of this policy change—men such as Zhang—this was a brutal repudiation.

Now, Zhang Wenkang was offering the visiting officials cups of tea and sighing audibly. The conversation took a philosophical turn, with the minister discussing the sacred trust of public health and how he viewed each member of the WHO team as a "dear friend."

Only days later did the WHO team understand why the minister had suddenly become so maudlin. That meeting was his last official act as minister of health.

CHAPTER 65

- **April 20, 2003**
- **Quarry Bay, Hong Kong, China**
- **4,920 Infected, 408 Dead**

THE NUMBER OF NEW CASES IN HONG KONG HAD BEGUN A PLEASING decline, with only thirty on the eighteenth and thirty-one the day after that, with twelve fatalities. On the twentieth, we had only twenty-two new cases, the fewest in four weeks. There hadn't yet been a repeat of Amoy Gardens, and most new patients could be contact-traced to a hospital or health care worker. That meant, for those of us who had stayed out of the hospital system, that the chances of getting the disease had gone from seeming inevitable just a few weeks before to unlikely. Yet Hong Kong remained desolate. There had been 115 deaths in Hong Kong due to SARS, but the city was so abandoned you would have guessed the body count at a thousand times that.

That afternoon, Hong Kong felt a little like those scenes of post-virus London in the film *28 Days Later*—I was the only shopper on the entire fourth floor of the Landmark Building, one of Hong Kong's swellest malls. The surgical-mask-wearing shop attendants leaned on their glass-topped counters and stared vacantly at their racks of on-sale clothes; an employee at Kenzo told me the shop was averaging two customers a day. And she was counting me as one of them.

As I was leaving, my cell phone rang. It was Susan Jakes calling from Beijing. China's minister of health, Zhang Wenkang, and Beijing mayor Men Xuenong had just been ousted. It was the highest-level Chinese government purge since the uprising at Tiananmen Square in 1989.

"It's unbelievable," Jakes said. She described the press conference that morning at which Gao Xian, the deputy minister of health, had revised the number of SARS cases in Beijing upward by a factor of nine, from 37 to 339. There were 402 additional suspected cases in the capital, Gao said. In Shanxi, there were 108 cases, in Guangxi Zhuang 12, Hunan 6, and on and on through each province for which the government had even the sketchiest of statistics. Nearly 2,200 SARS cases in China, up from 350 just the day before. It had been an unprecedented press conference for the Chinese government—nearly two hours of live questioning with international reporters hammering the deputy minister with tough questions. Then the government took the remarkable step of canceling the national full-week spring holiday.

I had to stop. Right there on Queen's Road. The Chinese government had just admitted it was wrong, catastrophically so, to the point where it had just held a public, nationally televised mea culpa. And then they canceled May Day. And this was due, in part, to the work that Susan Jakes, Huang Yong, and the rest of our China team had done. For a moment, standing there next to the Dunhill boutique in my light blue surgical mask with a Nokia pressed to my ear, I was proud.

But my next thought was, What now?

There would be 289 new cases in Beijing the next day, along with 28 dead. The day after that, there would be 105 more in the capital. Throughout the provinces, various regions were reporting similar spikes in new cases. How much of this was due to new infection and how much was actually belated reporting of existing cases was impossible to discern. Either way, the impression these numbers gave was of a massive national outbreak. The WHO was saying that the disease, apparently present in twenty cities, provinces, and regions, could explode across the country if sharp measures were not implemented immediately.

Think of all China's bad impulses, all that machinery that gets cranked and fired up to perpetrate some banal and ultimately pointless government campaign. Does anyone remember the Three Represents? The Three Whatevers? The Three Antis? These had all, at one time or another, been the centerpieces of massive propaganda efforts to invade and subvert citizenry mindshare. Now that Hu Jintao had repositioned the entire government propaganda apparatus behind fighting SARS, China was being inundated with twenty-four-hour-a-day SARS coverage as intensive as the O. J. Simpson Bronco chase and the Iraq war rolled into one. If in the Era of Wild Flavor, Chinese television had drifted slightly away from the omnipresent propagandist imagery of peasant farmers in Mao suits walking boldly into the future, arm in arm with their urban proletariat comrades, the capability to produce such imagery had only been mothballed. Now those peasants and proletarians had been converted into doctors and nurses.

CCTV, on each of its nine channels, began to air nothing but SARS coverage, all either prescriptively informational or hagiographic. Doctors in Guangzhou, Guangxi Zhuang, and Beijing who had given their lives for the nation. Nurses who had manned their posts till the virus claimed them. There were twenty special medals commissioned by the government to be awarded to various "Heroes of SARS." The virus was identified as the great common enemy (a villain whose perfidy took everyone's mind away from the criminal mendacity that had allowed this villain to spread through the country in the first place). Gauzy montages of nurses waving from their windows of quarantined hospital blocks, doctors who could speak with their children only by phone or IP video link, squinty-eyed virologists gazing into test tubes and microscopes. This was a nation suddenly at war with the virus, with a patriotic intensity that made the U.S. networks' coverage of "Shock and Awe" look like measured, objective journalism.

"Our doctors and nurses will never shirk from their duty to defeat the national menace of SARS," announced one CCTV voice-over. "They are working around the clock to make sure that China can continue to prosper in good health. HEALTH CARE WORKERS! KEEP FIGHTING!"

Hu Jintao would then be shown walking through a hospital greeting physicians, and then in a laboratory, nodding and listening to a technician as if Hu himself were working on finding a vaccine, a remedy, the cure. Where was Jiang Zemin? He had vanished. Hu was making a bold political move that by hitching his reputation to the campaign to defeat SARS, he would, almost as soon as he had taken office, become associated with an issue that it was impossible to dispute. Who *likes* SARS? Who is in favor of an infectious disease? It was a brilliant calculation, and one that provided the first glimpse into why, exactly, this stolid former engineer had risen so efficiently through the Chinese political hierarchy. Citizens' committees were formed. Beijing announced the formation of a five-thousand-person SARS task force, whose duties were, presumably, to enforce the imposed quarantine of thousands of citizens and the closure of all movie theaters, discos, Internet cafés, libraries, and churches in the capital. On April 24, elementary and middle schools were closed. Beijing municipal authorities shut down one major twelve-hundred-bed hospital with two thousand employees because of fear of a vertical transmission pattern similar to that at Hong Kong's Amoy Gardens. Meanwhile, text messages were being thumbed saying that martial law would soon be declared in Beijing.

The greatest concern for the Chinese government and the WHO was Beijing's population of migrant workers, next to Guangdong's probably the largest in the country. It had been migrant workers who were among the first affected in Shenzhen and who then spread the disease to Heyuan and, eventually, into Guangzhou. Now the fear was that as hundreds of thousands, possibly a million, migrant workers left Beijing for their home villages, they would bring with them the virus, seeding local outbreaks. These so-called floating people tended to live in more crowded and squalid conditions than native Beijingers or the prosperous and emerging middle class.

In Beijing, it was as if the panic that we had been living with in Hong Kong had been pent up and then unleashed in a day. "I'm very worried about getting on a train with so many people," said a student surnamed Wang as he waited for a poorly ventilated train back to his native Changzhou in Jiangsu province, "but I'll do anything to get out of Beijing. It's simply become too dangerous." The scramble to make up for lost time had only succeeded in spooking the population. Those who weren't fleeing were staying home, indoors. "Beijing, a city whose wide avenues are usually jam-packed with crowded buses, squadrons of bicycles and even the occasional donkey cart," wrote Hannah Beech, "had transformed into a ghost town." Beijing was now joining Hong Kong as a pariah city, and the WHO confirmed the verdict, imposing a travel advisory on Beijing and Shanxi province and recommending the postponement of all nonessential visits. Vietnam announced that it was considering closing its borders with China. Taiwan was flaring up with SARS clusters in Taipei and Kaohsiung. The WHO also slapped a travel advisory on Toronto, the first such proscriptive measure taken outside of Asia. And even more ominously, India reported its first cluster of cases in the Goa coastal region, in the south. "There have been fears," the WHO noted, "that if the SARS virus reached India, it could spread rapidly in the subcontinent's crowded cities and health facilities."

The government's massive anti-SARS campaign was like a national parlor trick that had sawed the population of Beijing in half. The city was emptying out as hundreds of thousands panicked and made for the hinterlands.

CHAPTER 66

- April 24, 2003
- Taiyuan, Shanxi Province, China
- 6,119 Infected, 431 Dead

CLUTCHING TWO WHEELED-SUITCASES FILLED WITH SURGICAL MASKS, rubber gloves, disposable hospital gowns, and a few pairs of goggles, I stood in line behind my fellow passengers. We were waiting to be scanned by a thermal imaging device that gave a digital body temperature reading to the robed, goggled, gloved, and masked personnel poised behind a rickety wooden desk near the arrivals gate. China's anti-SARS campaign comprised a broad array of dubiously effective measures—neighborhood committees that informed on those who had ventured beyond city limits and should therefore have been in fourteen-day quarantine, daily reminders to eat hearty winter greens to build up resistance—yet there was one that was bluntly effective: relentless, constant fever checks as you traveled through the country.

Lacking a fast and accurate diagnostic test, the Ministry of Health decided to use fever as the screening criterion for SARS, and within days it had set up perhaps the most elaborate and most intricate fever-detection system in the history of the world. You could not walk in or out of a bank, government office, train station, or office building without being thermal-scanned. If you were running a fever and it didn't subside after a few minutes of waiting and a second test, this one possibly with a handheld thermometer, then you would be rushed off in a locked ambulance to secure quarantine for up to twenty-one days. Fever checks were a lowest-common-denominator approach to curtailing the spread of SARS, an approach that did not take into account the possibility of asymptomatic carriers and a pre-fever incubation period of up to ten days. However, University of Hong Kong scientists had already concluded that patients were at their most contagious between ten and twenty days after infection, in which case they would almost certainly manifest a fever. "We knew that some patients didn't cough as heavily as others, some had diarrhea and others didn't," said a Ministry of Health official. "The only symptom that presented in one hundred percent of cases was fever." It was hard to argue with the logic, though it meant that an average Beijinger would have his temperature taken a half dozen times a day, and if you were on the road, as I was, traveling through China, you could have your temperature checked twenty-five times a day. It was a crude but surprisingly effective method. (When you were driving, there would also be teams of gowned and masked workers who would spray your tires with a disinfectant. I'm not sure what that had to do with SARS, but it added to an overall vibe of combating microbes, even those hiding in tire treads.)

I had come to Taiyuan bearing medical supplies, with the idea of helping out a local charity by donating protective gear while at the same time trying to set up some plan whereby TIME could contribute money and supplies to help fight SARS in the Chinese hinterlands. It was also a pretext for me as a reporter to get some sense of the scope of the outbreak in Shanxi, China's third-hardest-hit province after Beijing and Guangdong—and one of the poorest. If China's Gold Coast, that cacophonous and polluted stretch running from Shanghai to Guangdong, was the new face of China, then interior provinces like Shanxi were bringing up the rear. In fact, it was those provinces in the middle of the Middle Kingdom, like Shanxi, that had fared the worst through the Era of Wild Flavor and, before that, the great reforms of the eighties and nineties. The small-farm agriculture and subsistence mining that had predominated in the early revolutionary period had given way to larger-scale endeavors that had not succeeded in improving prospects but had instead inflicted remarkable damage on the environment. Every stand of timber south of the capital had long since been clear-cut for fuel to run inefficient smelters, which left their toxic residue on some of the most polluted topsoil in China. Replanted trees did not stand a chance.

Regional Chinese airports, even in cities more prosperous than Taiyuan, can be dreary affairs. In the best of times, there are fewer merchants than storefronts, and those merchants who are selling offer an unappealing array of souvenirs: faded postcard books, crappy calligraphy sets, fake jade jewelry. There is usually a small, forlorn restaurant selling bowls of rice with a glutinous protein on top—duck? pork? It's hard to tell—and maybe unappetizing cakes wrapped in cellophane. If there is a newsstand, it is oversize, with too much shelf space, allowing for the appearance of wide expanses of empty display racks between neatly stacked issues of bland newsweeklies. To Westerners, or those used to Beijing's or Shanghai's crowded newsstands, such cavernous displays give the impression of a province where not that much is happening.

With the advent of the national "Strike SARS Hard" campaign, Taiyuan International Airport—one flight a day to Ulan Bator justified the international label—was even more sparsely populated than unusual. My flight in had also been virtually empty, my only companions a pair of American families from Utah who were here to pick up babies from a Shanxi orphanage.

As soon as I arrived in my hotel, the Shanxi Grand, I noticed that my cell phone did not receive a signal; nor could I connect to my company's e-mail server. When I am in remote cities in China, I often feel lonely. It is a peculiarly desolate feeling, part of it stemming from my being absolutely sure that I do not know anyone within a thousand miles. But it also has to do with feeling isolated in time with the anachronistic Mao-era amenities and conveniences. The hotel service retains a solemn Communist surliness; the food, for the most part, has not yet evolved to Beijing or Shanghai standards; and the government officials remain far more suspicious of foreigners than those back on the Gold Coast. And Shanxi in particular is famous for its indifference to foreigners and its lack of success in attracting foreign capital for joint ventures. It seemed, as I rode around the capital and strolled through its trash-strewn parks, to be a province that had been left behind by the reform era. And the government officials I met seemed stubbornly proud of their resilient indifference toward progress, which they equated, I presumed, with staying true to revolutionary ideals that their corrupt Gold Coast cousins had abandoned—or perhaps, as my local contact, an immense Swede named Tor Helgeson, who managed a local Christian charity that I was considering as a partner, explained to me: they might be making so much money skimming from the status quo that they feared foreigners would cut in on their action.

Yet now, like so much of the country, the province was mobilized in an anti-SARS frenzy. Shanxi had nearly two hundred SARS cases and several clusters right in Taiyuan. It was still one of those remote provinces from which it was notoriously difficult to gather accurate information, and even the central government was bemoaning the inefficiency of the province's SARS reporting system. Furthermore, no matter how often the central government insisted on openness regarding SARS, the reality of modern China was that any chance of implementing this new frankness was in the hands of a hundred thousand or so local bureaucrats who tended to resist any change that was not either immediately profitable or imposed under threat of reeducation. As a result, Shanxi's public health officials were continuing to resist Beijing's efforts to garner information or to give data to the World Health Organization, believing that such information would only embarrass Shanxi. A typical response was that of Dr. Zhang Hanwei, director of the Shanxi Provincial People's Hospital, in Taiyuan, who berated his staff about the high incidence of SARS in his hospital. "The government is very unsatisfied with these numbers," he told them at a staff meeting. "Our provincial leaders are very angry." He then announced a new policy called "The Three Nos," which came straight from the Communist Party Publicity Department: no talking to the media about the nature of SARS, no talking to the public about doctors' personal experience treating the disease, and no communicating with the WHO about anything to do with SARS.

If SARS was raging in the hinterlands, if there were still vast wards of SARS patients who had not yet been counted or acknowledged, if China was in the throes of a new plague that was burning through a poor, malnourished population, then I was sure it would be in Shanxi, which had, among other anti-SARS precautions, taken the very prudent step of banning kite flying.

TOR HELGESON KNEW SEVERAL PUBLIC HEALTH OFFICIALS AND retired CDC doctors through his charity work. They would occasionally come to him with requests for specific, very expensive equipment— usually Mitsubishi Pajero four-wheel-drive vehicles—which he would have to negotiate down into something his charity back home would actually provide, such as pallets of polio vaccine or a dialysis machine. Helgeson had delivered numerous supplies and medicines, and had to put up with disappointed Chinese officials who claimed they needed SUVs to drive to remote villages and clinics, to reach those sick children. Helgeson would explain that he was able to reach those villages just fine in his ten-year-old minivan

And SARS, for many of these officials, was a terrible disappointment in that there didn't seem to be vast amounts of expensive equipment that could be bought to prevent or diagnose the disease. The thermal scanners had been a rich grafting opportunity, but after that, what was there—boring old X-ray machines? The provincial authorities quickly instituted a series of SARS stations in virtually every county and even in some villages. Each was staffed with nurses who were to study a government-provided SARS Fact Book, which they would be tested on; those who passed the test would qualify for a supplementary income bonus, paid for out of the 250 million dollars the central government had earmarked to fight SARS. On my first day in Taiyuan, I visited one of these SARS stations to find a nurse seated in a dirt-floored room studying her Fact Book. I asked her what she would do if a suspected SARS case were to show up?

She quickly flipped through her book and, finding the appropriate page, announced, "Take his temperature."

Very good. And then?

She turned some more pages. "Chest X-ray."

I looked around her dirt-floor nurses' station. A cot. A chair. A filthy screen separating them. There was nothing resembling a radiology theater.

When I inquired about this, she smiled and indicated that I should follow her.

Next door, in another dirt-floored room, beneath a ceiling of white fiberglass material that had the Korn-Ferry logo repeating at regular intervals, was an X-ray machine from which the power cable had long ago been commandeered for another piece of equipment.

"Does it work?" I asked.

"Yes," she lied.

"Show me."

"The technician isn't here," she said.

Still, if SARS was endemic, shouldn't poor, backward Shanxi be swamped by cases? The head of the local Red Cross, Feng Jin Sheng, told me he believed SARS would be more difficult to contain than AIDS, which had set off an epidemic the human toll of which is still being tabulated. But then he made an interesting remark: "It will be easier to control in rural areas. You can just close down the whole village."

So all those millions of peasants now streaming into the countryside, won't that cause a catastrophe? We were already reading reports of migrant workers dying by the side of the road and ambulance crews refusing to collect the bodies. Each village had established barricades to keep out strangers; it would allow in only its native sons. If those returnees were sick, that could be a problem. "Treatment remains our greatest worry," Feng said.

I asked him what we might provide him in the form of donations.

"We need an ambulance," he said. "And respirators. And blood-testing equipment."

My suitcases full of protective gear seemed a pathetically small offering.

TO LEAVE TAIYUAN, YOU HAD TO PASS THROUGH A FEVER STATION, then another to go over a bridge, then another to leave the county.

To get off the highway and enter any village required that the driver and passenger give documentation of who they were and then state their business. Since I was a foreigner, I was allowed to proceed into any village I wished. These were usually dusty mud- and brick-walled and tile-roofed enclaves of anywhere from a dozen to fifty houses inhabited by several clans. The livelihood of a village could be discerned by surveying the local landscape—tea bushes or barley indicated agriculture, while a more mountainous landscape meant mining. The mining towns were the dustiest, and the first one I visited was situated downstream from small, family-operated strip mines. There was presumably a great deal of runoff from these mines, as chemical by-products of purifying the ore, such as mercury, drained into the streams that ran past the main road and served as a sewage system. Children were often playing in these streams. Families had been told not to drink groundwater, so freshwater was rationed from a cistern that was a source of constant worry among village elders, especially now that SARS had put a stop to regular deliveries of any kind. The two men who were staffing the roadblock had dug out their old People's Liberation Army jackets and were attempting to impose on the whole measure a semblance of martial order, but they quickly backed down when my taxi driver shoved a few *kwai* at them.

Yet it was obvious, as I spoke to the local barefoot doctor, that there was no SARS here. Mercury poisoning, perhaps—the children certainly looked a little undersize. "If someone comes to me with a temperature," the doctor said, obviously repeating instructions that had been given to him, "I will send them to the county for quarantine." Chest ailments and lung problems were frequent—this was a mining town—and he would generally prescribe a series of antibiotics or herbal remedies.

Curious, I asked what the life expectancy in this village was.

He nodded as if he had never thought of the question. "The oldest man in the village is fifty."

The villages in the province were poor, and perhaps not the healthiest environments in the best of times, but the barricade and quarantine system was plainly going to keep movement among them to a minimum. At one village, I came upon a barricade composed of cabbages piled in a mound three feet high, with women seated behind them as if they were minding a store—which, in a sense, they were, as these cabbages were also for sale.

Yet despite the best efforts and assiduous inspection of virtually every vehicle on the roads in Shanxi, there were still cases—thirty-seven new infections on April 22, eleven on April 24, twenty-two on April 26, and then twenty-nine two days later. You drive through the villages of Shanxi, look at the farmers sitting by the side of the road in muddy rubber boots, the kids in flip-flops pissing in polluted streams, the women carrying sacks of rice, and you think, All those people, they're just meat for a virus

CHAPTER 67

- April 26, 2003
- Number Two Hospital, Shanxi, China
- 6,226 Infected, 450 Dead

THE VIRUS HAD COME THROUGH HERE, PAUSED A MOMENT, SPREAD its tentacles to a half dozen hospitals, and then continued on its way to Beijing. That had been over a month ago, while Chinese scientists were still insisting that *Chlamydia* was the agent. And now, these doctors and hospital officials were nervous, as if somehow China's supreme leaders in Zhongnanhai would blame their country cousins for sending the disease onward, into the leadership compound itself, where an assistant to the widow of a former top official had come down with SARS and was being treated in 301 Hospital. The virus had come to the very heart of the empire.

As I met with physicians in conference rooms in Shanxi hospitals, they were reluctant to describe any spread, contagion, sickness. They had taken the strictest measures, they assured me, had implemented barrier nursing from the very beginning, had treated the index case, the woman jewelry merchant, as if she were as toxic as a cobra. Their concern was that they did not want to be perceived to be at fault, could not admit to a foreign reporter that their hospital had been the one, that Beijing was stricken because of a screwup at Taiyuan's People's Number One Hospital. Instead, hospital directors in pristine lab coats sat with me, their hospital party secretary beside them, and smiled and lied and then made up urgent appointments they had to keep. The director at Taiyuan Number Two Hospital looked at me with a smile and told me, "I can't talk about this."

"What about the new openness?" I asked. "I thought it was okay."

"Do you really think anything ever changes?" he said.

That reply was perceived by the hospital party secretary as going too far, and she showed her displeasure with a sharp comment in Chinese about how they were late for their committee meeting.

THE SKY WAS BLUE ON TOP AND THEN FADED YELLOW ALL AROUND THE edges. I headed south, along empty expressways, past barricades and fever stations where I had to submit to inspections and pay various fees. My driver had just one cassette, Leon Lai singing in Mandarin, which he played over and over again. The desolation of central China is remarkable in that it is a thoroughly populated desolation. You arrive at a nasty patch of land, of scarred earth, of dust-covered rock, of rusted scrap metal, of animal feces in a mound, and you think, Surely, nobody lives here. But someone does. Right here. There's his house, a shack next to the feces mound. And he is staring at you.

I had trouble figuring out the geography. Was this desert? Chaparral? Mountains? We seemed to be passing through various climates and floras, each denuded and debauched according to the resource it offered. In the best of times, the yellow loess soil that gives the eponymous river its name is very hard to farm, the topsoil already eroded by flash floods washing down from the numerous ravines. Up those ravines, peasants were still living in caves. The desert was former farmland. The brush former forest. The barren crags former alpine slopes. Now it was all wasteland, as far as the eye could see. As the sun receded up a distant ravine and then slid behind the backbone of a narrow range, the rocky earth went from yellow to gray. We were alone on the road, and the dark of the Chinese countryside closed in around us as tightly as an undersize coat. The night wraps you. You follow the cone of your headlights as far along the highway as they go. And you pray that the trucks in front of you have working brake lights. (China has about a hundred million trucks and maybe ten million pairs of working brake lights.) I have seen too many accidents in China where the cause was simply a trucker stopped on the road and the driver of the sedan behind him never saw his truck until his steering column was being jammed into his jaw. I put on my seat belt.

At night, the road to Changzhi winds its way through mysterious-seeming country, and during that season of SARS, I imagined the virus racing along with me, in each of these little villages, snaking in and out of every mud hovel. Out there. Stalking me.

WHAT WAS I DOING? I ASKED MYSELF WHEN I REACHED THE CITY OF Changzhi. I was here, I realized, to see the front lines of the war between man and virus. Hong Kong may have been the hardest-hit city in the world, but SARS would be won or lost out here, on the Chinese steppe, or whatever this was, where barefoot physicians would have to treat highly infectious sufferers of a novel pathogen, armed with little more than thermometers and aspirin. And it would be a war to the death; either man or virus would win.

Yet that was a preposterous notion. Why was the battle against SARS in Shanxi going to be any more dramatic than in Hong Kong or Singapore or Guangzhou or any of a dozen other afflicted areas? I recalled being in a meeting with Hu Xuli, the brilliant editor of *Caijing*, one of China's leading business journals, which published some groundbreaking SARS stories. I was commenting to her how unsafe I felt in Beijing compared with Hong Kong. She smiled. "Ah, the prejudice of the old infected place versus the new infected place."

Indeed. I was a tourist, in search of a charnel house where Chinese peasants might be dying by the hundreds.

This, I would be able to say, is what an emerging disease outbreak looks like.

We want an apocalypse to look like an apocalypse, not a slow day at the mall.

CHANGZHI NUMBER TWO HOSPITAL HAD ESTABLISHED A MASSIVE fever clinic in the rear of the brown brick-and-ferro-cement six-story building. Signs posted in front of the entrance said, in effect, SARS

CASES IN BACK, which meant that anyone who turned up at the hospital with a fever would report to the quarantine station in the rear. The measure had been imposed by the Shanxi Department of Health at all local hospitals, and Changzhi Number Two had taken it seriously enough to refit a row of storage rooms that lined the dirt parking area to the rear as a quarantine ward. Each concrete-floored room had been swept out and equipped with a wooden cot with no mattress. The rooms looked like prison cells. It was hard to imagine spending the recommended fourteen days locked down in one of these. You did not want to have a fever in Shanxi that spring.

There was a nurses' station in a makeshift office at one end. Twice a day, the nurses would bring around water and congee to those in isolation, sliding the gruel through the window. Generally, I was told, most of the patients were released after their fevers had subsided for seventy-two hours. If they continued to have a fever and developed one or more other SARS symptoms—cough, diarrhea, respiratory distress—they would be sent to People's Hospital Number Two, in Taiyuan. Shanxi had made the decision to centralize its SARS cases

When I inquired at the nurses' station if I could speak to a doctor regarding the hospital's SARS program, I was told to speak to the duty physician. I found him sitting behind a desk in a bungalow that had been converted into a consultation office. As soon as he saw us approach, he quickly slipped on a mask, goggles, plastic cap, and surgical gloves. He stood up and bowed as we came in and quickly took a pencil as if preparing to write down our vital signs. Before him on his desk was a huge mug, like an oversize jelly jar, full of green tea.

I explained to him what I was doing there, that I was curious about how Changzhi was handling the SARS crisis, about how many patients he had treated. He nodded his head and began to flip through a bound casebook on his desk. It was as if he had not spoken to anyone in days and was happy finally to have someone to whom he could justify his efforts.

He told me that at first many patients had been ordered to the rear by the hospital. The nurses and doctors in the regular wards viewed the establishment of the fever clinic and quarantine ward as an excuse to refer almost any patient suffering from a nonchronic ailment.

"I had a dozen patients a day," he said, looking back two weeks in his bound notebook. On each entry, he had written name, gender, age, and then listed symptoms. He said that at first he was referring cases up the bureaucracy to Taiyuan, but then he began to doubt that these could all be SARS cases, because most of them, aside from their fevers, did not deteriorate during their enforced stay. But he could not be sure, because no one ever reported back if the cases he had referred to Taiyuan had turned out to be SARS. But his presumed proximity to the killer virus made him something of a pariah physician around the hospital.

"No one really wants to talk to me anymore," he said, shaking his head. "They are all worried about SARS."

Just then a cleaning lady came in and began dusting the tiny consultation room. She stood near me and kept her dirty rag sweeping in a clockwise motion over the same section of the desk. It was apparent to me she had been sent to find out what was being discussed. After a few minutes, she left.

The young doctor continued. He had volunteered for this duty, he said. He had imagined it would be more interesting than his usual rounds in the general population of the hospital. But instead, he found himself monitoring temperatures and little else. And when he had found suspected SARS cases, he said, he had not been given any medicines with which to treat them. All he did was call an ambulance and wait for it to drive down from Taiyuan. It sometimes took three days, and he would have a patient gasping for breath but would have no treatment for him. He felt terrible, he said. He had begun to note down those cases that he believed were SARS, based on his own observation of patients' symptoms, and he had a chart of those referrals, which he slid across the desk to us. He showed me where he had written a Chinese character to indicate a SARS patient.

Then a remarkable thing happened—the first time I have ever had an interview end this way. Two men in white coats came into the tiny office and seized the doctor, pulled him from his chair, and as he kicked and tried to grab hold of the desk, a chair leg, the frame of the door, they dragged him from the room, kicking but not screaming. The young doctor had apparently not received the memo about "The Three No's."

I stood up and flipped through the doctor's charts, which had been left open on the desk.

I found the same SARS character next to a dozen cases, the first from two weeks before. The frequency of the character declined, and for the past week, there had been no new cases. It looked as though people were beating the virus. This was encouraging, though the circumstances of our collecting this data had certainly been discomfiting.

A nurse came in and interrupted our snooping. She ordered us out. "Dr. Xie is seeing a patient," she said, slamming shut his logbook. "And these are secrets." In a country where even the income tax code is considered an "internal rule" and economic data a "state secret," my flipping through this sort of hospital document could have made this poor doctor the Alger Hiss of Shanxi.

Shanxi had not had an Amoy Gardens–like community explosion or a Beijing-like panic. Instead, there had been a steady accretion of cases, yet with no sudden or ominous spike. If this had been a community outbreak that was truly out of control, then we would have been seeing geometric progressions of cases, entire villages infected, hospitals crashing. So far, Shanxi was surviving.

At this moment, I realized that life might actually get back to normal—or as normal as it ever was in China.

CHAPTER 68

- May 1, 2003
- Jianguomenwai, Beijing, China
- 6,310 Infected, 452 Dead

WHAT WAS GOING ON IN SHANXI WAS A MICROCOSM OF THE FIGHT against SARS across the country. China's mobilization had been to total war footing, and the disease was held in check through measures that might be possible only in a dictatorship. "The fact that China is still an authoritarian state was certainly an important factor in containing SARS," David Heymann would tell me later. "They were able to put resources and assets to work at a level that a democracy might not be able to do. When there is an infectious-disease outbreak, one of the hardest issues for public health officials is imposing quarantine or asking questions that could, in a strict constitutional sense, run up against civil liberties. China did not have these problems."

If you weren't a native son or daughter of a certain village, you would have a hard time even passing through that village. Lower-level officials, especially toward the end of April, seemed to be manning every roadblock and bus station, checking IDs and *Hukous* to ensure the provenance of each voyager. The scenes reminded me of those sheriffs at the California border turning away Okie trucks in *The Grapes of Wrath*. There were many cases of floating people—the migrant workers who crowded the cities—being turned away from hospitals and dying at train stations or bus terminals while other migrants cleared a swath around them.

Chinese train stations, in the best of times, can be packed with sweltering mobs of migrants embarking or returning from long journeys. During the season of SARS, which, I believed, signaled an end to the Era of Wild Flavor, the stations were even more crowded than usual as the entire nation seemed to be on the move at once.

Yet the vast rafts of virus that were supposedly being set adrift with these migrants never appeared.

And when cases did appear, the central government's relentless propaganda campaign made sure every doctor and hospital in the country now knew the protocol: suspected cases had to be isolated, and all their contacts traced and monitored for symptoms; and health care workers were required to maintain the highest standard of infection control. As local hospitals throughout remote provinces such as Shanxi and Hebei began to implement these measures, the greatest threat to the planet, that a newly emerged virus would achieve endemic transmission in one of the poorest regions on earth, appeared to have been curtailed. The strict infection-control measures first advocated by Zhong Nanshan and his fellow clinicians in Guangzhou would prove bluntly effective against the virus. It was a reversion to pre-antibiotic-era hospital protocol; infection-control chiefs—usually senior nurses—enforcing strict protective gear and clothing dress codes. In Singapore, such thorough measures were employed almost from the start, and hence that city-state's early success in ring-fencing the disease. "You had to become religious about it," says Dr. Lim Suet Wun, chief of Tan Tok Seng Hospital. "That meant zero tolerance for any deviating from the protocol."

It has been decades since doctors and nurses practiced this sort of infection control. During the nineteenth century, a routine visit to the hospital could result in grave infection and possibly death. In the maternity hospitals of Paris, for example, the death rate for delivering mothers was a staggering one in nineteen, prompting newspapers to dub a typical lying-in hospital a House of Crime. Women, understandably, were terrified of hospitalization. Louis Pasteur, among others, had insisted that what killed women with so-called child-bed fever was "you doctors, who carry deadly microbes from sick women to healthy ones." Yet Pasteur was always more obsessed with finding microbes than merely interdicting them.

Joseph Lister, a Glasgow surgeon in the 1860s, had been the first to introduce the concept of antisepsis into operating suites. He noticed that if he applied carbolic acid to the wounds and surgical incisions of patients, they would not develop infection. Lister would take this further, having carbolic acid sprayed into the air of operating theaters. Despite the resulting foggy atmosphere and the frequent dressing changes required, antisepsis caught on. Later, this was modified to the concept of asepsis, or sterilizing all instruments and objects that would come into contact with a patient's wound. Asepsis eventually replaced antisepsis, as it was obviously more efficient and less unpleasant for the doctors and nurses. This marked a paradigm shift in the quality of health care. Lister's asepsis was the first successful break in that chain of infection. Hospitals went from being among the deadliest options for someone in need of medical care to actual houses of wellness. In the hundred years from 1800 to 1900, the single greatest advancement in medicine may have been the movement to sterilize operating theaters. In that period immediately prior to the invention of penicillin, the West had the cleanest hospitals in the history of the world.

The advent of penicillin as the first of the "miracle drugs" in the early twentieth century made vigilance against infection less of a priority in hospitals. Why take such aggressive measures against infections and bacteria if most of them could be so easily killed with antibiotics? As a result, infection-control measures at hospitals around the world underwent a gradual deterioration as doctors relied more and more on antibiotics, forcing pharmaceutical companies to invent new drugs faster than the bacteria could mutate into antibiotic strains. (It is in part because of this relaxed vigilance that hospitals in the United States and Canada are plagued by staphylococcus infections.) With the current outbreak, institutions around the world were exposed for having less-than-perfect infection control. "Hospitals emerged as the main amplifiers of the disease," says K. Y. Yuen.

He has a point. In China during the season of SARS, there was only one superspreader outside of a hospital setting: the dialysis patient who lived at Amoy Gardens. Sterilization protocols had been relaxed in part because those respiratory ailments that do show up in modern hospitals are not viewed as serious threats. Measles, for example, is a virus against which almost every health care worker is vaccinated. And common influenza strains will cause at most a few days' discomfort; often, those health care workers who are likely to come into contact with influenza patients will have been vaccinated as well. Most common chest infections can be knocked out with doses of antibiotics. With SARS's arrival, however, hospitals were forced to consider a respiratory ailment that was highly infectious and deadly, an opponent that had not stalked hospitals in decades. "We've never seen a disease that spread like this before," says Aileen Plant of the World Health Organization team in Hanoi. "Millions of people turn up every day in developed countries with infectious but minor illness. You never expect anything like this. We had to think about infection control from the point of view of a respiratory illness that spread this rapidly in hospital settings."

What worked was old-fashioned Florence Nightingale–style proscriptions: protective layers of masks, goggles, gloves, galoshes, and gowns. Sealed wards. Quarantine. Ventilation. This was not Nobel Prize–winning medicine. Yet it was effective. One of the ironies of SARS was that hospitals with less-sophisticated climate-control systems fared better than those with modern air-conditioning. An open window with a fan blowing outward to ensure reverse pressurization was the most effective way to safely disperse the virus. "We noticed that ventilation was essential," says Zhong Nanshan, "that fresh air seemed to curtail the spread. That was before we knew this was a virus." These are methods from a bygone era, yet they were the only procedures that slowed down SARS. Those rural Chinese hospitals with open windows would turn out to be viral dead ends.

And the layering and sealing of the body from patient effluvium was a commonsense approach that turned out to be the only effective course. You protected yourself out of fear, yet that was what worked. It has been half a century since a disease outbreak had caused big-city hospitals to take these sorts of measures; virtually no doctors alive today have had to work in these conditions. Yet it was nineteenth-century medicine that was defeating the twenty-first century's first pandemic. The battle was being fought one hospital ward, one patient at a time. It was the medical equivalent of trench warfare. There were no cures, no magic-bullet treatments. You did what you could for the patient, provided what symptomatic relief you could, and, in a sense, let the disease run its course. Since there was nothing that was proven to be effective at killing the virus itself, the disease made exhausting demands on health care workers. Doctors and nurses were forgoing seeing their families for weeks at a time. "I felt like we went two months without sleeping," said Joseph Sung of Hong Kong's Prince of Wales Hospital.

Yet in a pitched battle between this newly emerged virus and some of the best medical facilities in the world, the virus almost won. This was not supposed to happen. Infectious diseases are supposedly a thing of the past. Since World War II, no modern city had been brought to a halt by an infectious disease, and now, here were a half dozen in Asia and another in North America that had been reduced to panic because of a novel respiratory tract illness. The season of SARS could be viewed as either an anachronism or a harbinger. Unfortunately, scientific evidence suggests that it was an indicator of outbreaks to come and that we had better learn as much as we can from its emergence.

AMONG VIROLOGISTS, THIS LATE-SPRING LULL IN SARS ACTIVITY looked suspiciously like the seasonality that is a trait of most coronaviruses. Scientists are not sure why some viruses become more "active" or "infectious" in colder weather. It might be as mundane a reason as people being huddled together indoors during the winter months, resulting in greater proximity of hosts; or it could have to do with lower relative humidity in cold temperatures. "At the clinical level," says Malik Peiris, "we could see what was working. As hospitals adopted greater infection control, we saw the chains of transmission starting to break. But there was always a nagging question about seasonality. It was getting warmer, and coronaviruses had been shown to react to the climate changes."

Just as influenza had a season, perhaps SARS did as well. Hong Kong, Guangdong, Beijing, and Shanxi were still WHO no-go zones, yet there was a cautious optimism; for the first time, we could see more recovered SARS patients being discharged than new cases being admitted.

It had been only six weeks since the WHO's first Global Advisory. We now knew what this thing was. We knew what it did. We even knew, albeit through prophylactic measures, how to stop it.

We still had no idea where it came from.

CHAPTER 69

■ **May 1, 2003**
■ **Shenzhen Centers for Disease Control, Shenzhen, China**
■ **6,311 Infected, 456 Dead**

I AM A VIRUS, THOUGHT GUAN YI. I RODE BUSES, TRAINS, AND AIR-planes from Guangdong out to infect the world. I am a virus. How did I get to Guangdong? Did I hitch a ride here just as I hitched a ride out?

Guan Yi was obsessed by this question. He strongly suspected that SARS was a zoonotic disease. He had discussed with his Chinese colleagues the coincidence that the earliest clusters of cases—those including Fang Lin and Chou Pei at the Wild Flavor restaurant—had all involved either food handling or food preparation. That meant close contact with animal blood, saliva, and feces. And in Guangdong, that meant *wild* animal blood, saliva, and feces. After Malik Peiris announced that he had found a coronavirus, Guan Yi became obsessed with the idea of finding this virus's host species.

Guan Yi would never admit that he was jealous of Peiris. Yet for reasons he himself may not have understood, he saw Peiris's success as an impugning of his own work. Guan Yi often pointed out to me Peiris's age, as if because Peiris was ten years older and had come later in life to influenza as a specialization, his achievements were somehow less gilded. But as I visited with and interviewed Peiris, I found it hard to see why Guan Yi seemed so resentful of Peiris's success in isolating the SARS coronavirus. Peiris, as K. Y. Yuen explained, had come to represent the Hong Kong virological community to the rest of the world. He was the scientist who had made the big discovery, and accolades were flowing in to him accordingly. How many times in a virologist's career is he likely to be at ground zero of an emerging-disease outbreak, to live in the actual hot zone? Peiris had made the most of it. And technically, the samples he had used to find the virus had not come from Guan Yi's smuggling runs.

Yet Guan Yi was inconsolable. He could outbrood Achilles if it came to that, yet his brooding and his competitive zeal motivated him to try ever harder. He realized that his means to wrest laurels from Peiris was to uncover the great factoid regarding SARS. If he could find where it came from, that would reveal one way to stop it. The most effective measure in combating infectious diseases, ranging from malaria to Lyme disease, is to interrupt or intercept the host and/or vector. SARS had to have a host. Where was it?

I am a virus, Guan reminded himself. It was when he was reading the book *Veterinarian Virology* that he stopped at one particular entry about coronaviruses. There were fourteen known coronaviruses, only two of which infected humans. He noted, however, that genetic sequencing had revealed that all coronaviruses appeared to be descended from a common ancestor. He looked up those genomic sequences and concluded that those two coronaviruses that infected humans, OC43 and 229E, were genetically very similar to the coronaviruses endemic to pigs and mice, respectively, which indicated that they had jumped species at some point in the distant past to cause disease in humans. What if these earlier zoonotic infections represented, in effect, SARS outbreaks that had happened thousands of years ago, when a coronavirus jumped from livestock or pest to human? He paused and thought about that. If SARS was the latest in a line of zoonotic coronavirus jumps, then what was the most likely host species?

The usual suspect in instances like this is a domestic animal. But if it had been from a dog or a cat, then researchers would have found the disease in domestic animals in Hong Kong. Cats had carried the disease in Amoy Gardens, but it was determined that they were what was known as a "mechanical" carrier; a virus could pass through them and remain infectious, but the cats themselves did not develop the illness. Also, any coronavirus in domestic animals that could infect humans would probably have done so thousands of years ago. Humans, cats, and dogs already share most of our microbes. In what species, then, did SARS lurk? Species provenance had never been solved for Ebola, for example, despite thirty years of research and numerous outbreaks.

Most emerging diseases are the result of new or more frequent encounters between human beings and a wider range of animals. Such interactions, as we clear-cut rain forests, mine for gold, or head off on surf trips seeking ever more exotic left-handers, put us in contact with a new range of microbes. Every animal has its own range of viruses. Human beings host a vast and complex array of microbes, among them approximately 150 known viruses, each of these potentially infectious and lethal to an unfamiliar species that comes into contact with us. Likewise, any new animal that we come across in the wild will also have a full deployment of microbes on board, any one of which could induce in an unfamiliar host—you or me—grave illness. Since 1930, there have been fifty emerging virus outbreaks, including Ebola, Rift Valley fever, Sin Nombre, Nipah, and a plethora of others; every one of them, when identified, has been found to be due to an increase or change in the nature of interaction between human and animal. Sin Nombre, for example, was the result of an explosion in the population of field mice carrying a lethal hantavirus in New Mexico and Arizona. Heavy rains had resulted in a surplus of acorns, the preferred meal of deer mice; as a result, the mice reproduced in such numbers that they began to overrun human dwellings, leaving behind virus-laden feces that were churned into dust and inhaled by Navajos, in fatal doses.

All infectious diseases begin as zoonotics, possibly becoming, if we do not find the host and cut off the interaction, endemic. In most of those outbreaks since 1930, the viruses did not achieve widespread human-to-human transmission and therefore could not become endemic to human beings. The notable and terrifying exception, of course, is HIV, which hopped over from a primate sometime in the second half of the century.

Smallpox, measles, meningitis—each at one point in unrecorded history made the jump from a mammal species to infect humans, becoming a notorious and efficient killer in the process. It has been well documented how man's inexorable expansion of his domain has caused us to encounter new viruses with potentially lethal consequences: Machupo and Ebola are two frightening examples. But as Guan Yi looked at Guangdong and the earliest clusters of cases, there didn't seem to be any drastic environmental change that could have triggered the emergence of the virus causing SARS.

But there are several factors at work in a virus's successfully hopping from species to species. Proximity, of course, is the first prerequisite, but even more of a determinant than proximity is frequency of interaction. If man encounters a new mammal species just once, the chances of a sufficient viral load being passed on and infecting the human is highly unlikely. However, if there is a pattern of interaction, of man and host species being in proximity and sharing water, air, space, and, hence, microbes, then the chances of swapping viruses increases geometrically. (Scientists are only just now finding out about how human viruses can affect other animals, and sometimes, as in the case of certain aquatic mammals, fatally.) In recent history, it has been man's invasion of new ecological niches or fundamental alteration of them that has unleashed viral outbreaks. The clearing of savannah in Africa and Asia brought humans into contact with deadly arenaviruses such as Lassa and Hanta. In South America, the plowing of the pampas in preparation for maize growing and cattle ranching caused farmhands and a local species of field mice to swap microbes, unleashing the Junin virus. The increase in agricultural settlement in the Bolivian rain forest had set off Machupo; in Egypt, the building of the Aswan Dam elevated the water table, thus allowing mosquitoes to breed and Rift Valley fever to afflict three hundred thousand people; and in India, forest clear-cutting uncovered Kyasanur Forest disease. What had been the behavioral or environmental change that had created the right circumstances for SARS?

I am a virus, Guan Yi thought. Where did I come from?

■ ■ ■

THE SHENZHEN CDC WAS LIKE ANY OTHER BIG-CITY PUBLIC HEALTH agency during the Era of Wild Flavor. Officials there wrestled with many emerging public health threats while always keeping their gaze firmly fixed on the bottom line. With Shenzhen doubling in population every six years, how was one agency with fewer than a hundred employees to ensure that children were vaccinated, water remained potable, and pest control remained thorough? And in the last few years, the order had come down from the provincial government that the Shenzhen CDC had to become economically self-sufficient. That caused headaches for CDC director Zhuang Zhixiong. He saw his task of policing his city for microbes to be essentially untenable. And a disease outbreak in Shenzhen today, he once explained to me, "will be a disease outbreak in Hong Kong tomorrow."

But SARS had been a shining moment for his office. Shenzhen, despite being one of the points of emergence, had had just 112 cases and only 13 fatalities. Director Zhuang had instituted a massive city-wide cleanup effort that, though it had not really scrubbed out the squalid migrant-worker communities like the Click, had made the city center sparkle. Under the guise of SARS prevention, the city had rounded up cripples and beggars and ordered them to the outskirts; sanitation workers in masks and goggles had sprayed disinfectant on every square inch of government and city property; and private companies had hired their own squads of sprayers. Piled refuse was removed. Public burning of garbage was discouraged. And, best of all, the provincial government had allocated several million *kwai* to pay for it. While it hadn't been good form to banquet during the crisis, Director Zhuang made sure to take the senior staff out for a few evenings of Wild Flavor and *baijiu*.

One night, Guan Yi joined them in a private room of a Shenzhen restaurant and found a receptive audience as he laid out a plan to test his theory of the origins of SARS. For, not surprisingly, many of his Chinese colleagues were also upset that scientists outside China had gained all the glory for isolating and identifying the SARS coronavirus.

If the government had not been initially politically reluctant to admit that this was a Chinese disease, they reckoned it could have been a triumph for Chinese science to have led the way. Indeed, Chinese scientists were still laboring under an inferiority complex caused by the country's tradition of isolation. (It was at this time that I began to hear from almost every Chinese clinician I interviewed, about the central role traditional Chinese medicine had played in treating SARS. Chinese treatments, I would find out, were almost always given in tandem with Western medications and measures, and often after the patient was already over the most critical stages of the disease. I did not find one clinician who had treated a SARS patient exclusively with traditional methods. I have not yet read a peer-reviewed study about the role of traditional Chinese medicine and methods in terms of SARS treatment, and it remains a controversial and politically charged topic, even among clinicians. At any rate, it is not necessarily in the nature of traditional Chinese medicine to address something as specific as a respiratory disease. The practice usually seeks to strengthen and develop the body preventively to ward off such illnesses.)

After having lost out in the SARS search, Chinese scientists desperately wanted to make news. The next prize would be in finding the animal reservoir. Director Zhuang and his staff quickly agreed to assist Guan Yi in his plan to scour the local animal markets for SARS.

THE ERA OF WILD FLAVOR MEANT THAT CHINESE WILD ANIMAL MARkets, including Shenzhen's central Dongmen Market, had created some of the world's most diverse ecosystems in the midst of the world's most crowded cities. The wildlife sold in these markets originated in neighboring regions with varied climates and environments. There were monkeys from India and Bangladesh, musk oxen from Tibet, snakes and lizards from Thailand and Laos, wild dogs from Burma, birds from Vietnam and Indonesia, and bats from Wuhan and Guizhou. The range of species was as diverse as all of Asia, and beyond: occasionally, and for a premium, one might find marsupials from Australia or rare aquatic mammals from the North Pacific. "A significant portion of the wildlife sold in markets in southern China," said a Royal Society paper on emerging disease, "[reaches] China through an expanding regional network of illegal wildlife trade." Each of these imported animals has its own range of endemic viruses. Hemmed in their pens, terrified after days or weeks of travel in darkened cages aboard noisy trucks, these animals are in a state of diminished immunity, and in this panicked condition, they are more likely to shed copious amounts of virus in their feces, urine, and blood. Frank Ryan, the author of *Virus X*, calls these sorts of environments "junctional zones" and believes that "the sheer volume of contacts is enough to ensure that sooner or later a virus will emerge with the formula for success." The overcrowded conditions typical of regional wildlife markets were ideal for cross-infection. The range of reptile, avian, amphibian, and mammalian species exposed to one another and to rodent, avian, and invertebrate pest species moving freely within those markets would lead to what Laurie Garrett called "a microbial swap meet."

Guan Yi had been visiting these markets for years, mainly to inspect poultry populations and look for signs of H5N1 avian influenza infection. Visiting places like the open-air tin-roofed Dongmen Market had given him a firsthand look at the microbial swap meet and had led him to believe that this had been the way station for the SARS virus. He was asking the Shenzhen CDC to join him in testing his hypothesis.

He didn't need the CDC for its expertise or technical skills. He needed Director Zhuang's collaboration for political considerations. The agency, Guan Yi knew, could obtain police cooperation, which meant that the Security Ministry wouldn't interfere with his mission.

His goal was to raid the Dongmen wild animal market, with the Shenzhen CDC providing the official imprimatur that would allow him to wage some guerrilla science. He wanted to take as many fecal, nasal-pharyngeal, and blood samples as possible from as wide a range of mammal species as he could secure and then run PCR tests to see if he could find SARS in the market. But of the hundreds of animals for sale in the market, which should he sample?

Guan Yi considered his options. It had to be an animal that was numerously and frequently sold, so that interaction between handlers and the species passed a certain threshold at which point a microbial swap was likely. SARS, as an RNA virus, was highly mutable, yet that didn't mean each of those mutations would enable it to survive in a new host. It was unlikely that the virus as it was present in its host species could make the jump to humans and thrive. A virus that has coevolved to become symbiotic with a particular species will have a difficult time gaining traction in the cellular machinery of another species. Perhaps SARS had undergone a mutation similar to that which had allowed those two previous coronaviruses—OC43 and 229E—to jump species to humans. It could have been, Guan Yi knew, a minute mutation of one chain of nucleotides at the right moment. Most viral mutations are unsuccessful, but RNA viruses are playing the lottery every time they replicate. Sooner or later, a random mutation will occur that will allow a virus particle to survive where previous iterations could not. Yet probability meant that it would have to be a species that was present in sufficient quantities to allow for that winning—or losing, for the human race—genetic lottery ticket.

That ruled out more obscure species of bears, wild dogs, certain monkeys and mountain cats, and most wild birds, which were less-frequent delicacies. Snakes and lizards, while common in the markets, were so far removed from humans on the evolutionary chain that the chance of their successfully sharing a virus with us was unlikely.

Guan Yi had wandered through the markets and been struck by the prevalence of a few species, especially in winter: hog badgers, ferrets, raccoon dogs, barking deer, wild hares, and palm civets were among the wild animals favored in southern China during the cold months, for their putative immune-enhancing qualities and because they were reputed to "warm" the body. Over the past decade, demand for these species had, according to an official of the Guangdong Department of Forestry, which is entrusted to regulate these markets, "increased ten thousand percent." (This may have been hyperbole. Sometimes Chinese officials are prone to this sort of grandiose pronouncement, such as describing a journey as "a voyage of ten thousand li" or a new measure as "the wellspring of ten thousand smiles.") No matter the exact number, the Era of Wild Flavor had necessitated that merchants and sellers build up supply chains of these animals that extended to the remotest rain forests of Southeast Asia. There were not enough wild palm civets or badger dogs in China to satisfy Guangzhou's appetite for Wild Flavor. Vietnam, in particular, had emerged as a crucial provider of wild animals for the Chinese banquet table. Over the past few years, Vietnamese Forest Protection Department rangers have been seizing truckloads of civet cats, turtles, pangolins, snakes, tigers, and primates bound for China. The Chinese wild animal industry has been valued at five hundred million dollars, and exports from Vietnam amount to about forty million dollars per year. Guan Yi suspected there were truckloads of lethal virus plying the roads between Southeast Asia and Guangdong; he just needed to figure out in what animal it was hitching a ride.

In a zoo, Guan Yi would point out, human beings and animals are kept apart by barriers that also prevent the exchange of any nasty microbes. In wild animal markets such as Dongmen Market, humans and animals sit side by side, the threshold of interaction now high enough to ensure that millions of microbes can pass between them. The sellers themselves often live in or near the market, their children play barefoot in animal feces, and then these animals are sold to Wild Flavor restaurants, where they are skinned, gutted, and prepped—Fang Lin's job—in a process that couldn't be designed better if you wanted to pass a virus between one species and another.

ON MAY 5, ACCOMPANIED BY SEVERAL SHENZHEN CDC OFFICIALS, HIS University of Hong Kong colleague Zheng Bo Jian, and Shenzhen CDC epidemiologist Yi He, Guan Yi arrived at Dongmen Market. All members of the entourage were wearing white lab coats and leather gloves and carried their sample cases in black duffels. It was an overcast day, the sky the muddy gray color of tarnished silver. Beneath the

Quonset roof of the market, it was even darker and damper than outside. No matter how familiar Guan Yi was with the bitter odor of stale animal feces, it would always revolt him anew whenever he encountered it. The shit of a hundred different species combined somehow smelled even worse than any one animal's excrement. As Guan Yi strolled around the market, he focused in on those mammals he wanted to sample. He asked some of the animal traders if he could take blood samples and respiratory swabs from their stock. His imperious manner can be intimidating to those less well educated, but with the animal traders of Dongmen Market, his attempts to browbeat them met with mixed results.

"Where?" one trader asked.

Right here, Guan Yi explained. He would anesthetize the animals and draw blood in the parking lot just outside the market.

"You're going to take blood?" the trader asked.

Guan Yi nodded.

"That's the best part," the trader said.

Guan Yi stepped forward and explained that this was a matter of public health and that no animals would be harmed.

"How do we know that?" asked the trader.

Guan Yi pulled a *Guys and Dolls* wad of Hong Kong dollars out of his pocket, the brown-blue-and-red five-hundred notes suddenly drawing the attention of every trader and seller around him. "If we harm an animal," Guan Yi explained, "we pay for it."

He was soon mobbed by sellers hoping that he and his team would mistakenly kill their animals.

After a veterinary specialist from Shenzhen CDC examined the animals to make sure they were free of obvious disease or illness, Guan Yi and Zheng Bo Jian pointed out which creatures they would be jabbing and prodding and instructed the sellers to take those cages out to the parking lot. The sellers would then pop open the door of the cage. The raccoon dog, a type of canine, would cower in its cage rather than run into the gauntlet of humans wearing leather-and-Kevlar bite gloves and holding prods and Y-shaped animal graspers—long aluminum tubes with handles at one end connected by cables to a jaw that can close around an animal's neck and pin it down. The sellers would goad the animal out, at which point it would make a frenzied charge out of the cage, where it would be backed into a corner and held down at the business end of a grasper. Raccoon dogs, like civets, ferrets, and hog badgers, are carnivores and have the appropriate claws and teeth; the team pinned them, approached them from behind, and then hit them with enough keta-mine to knock them into the K-hole for a few hours. While the ani-mals were down, Guan and Zheng would swab and jab, removing vials of blood and arranging them in their sample cases.

Guan Yi says he thought about Malik Peiris as he hunched over a panicked palm civet, a member of the viverrid species, similar to the mongoose, native to much of western China and Southeast Asia. While Guan and Peiris had made an expedition to Ah Chau Island at the beginning of this outbreak, Guan Yi now wondered how the two of them could ever have worked together. "I won't talk to him for one year," Guan Yi told me. And he took pride in how he would under-take this sort of fieldwork while Peiris stayed in his lab high up in his ivory tower.

"God," Guan Yi said as he inserted a Q-tip into a beaver's anus, "let there be virus in these animals. Let there be SARS in these sam-ples. God, please help me to find this virus. I know you are fair, so be fair to me."

Over the course of two days, Guan Yi and his team took nasal, fecal, and blood samples from twenty-five animals—six palm civets, three hog badgers, three beavers, four Chinese hares, two muntjacs, two Chinese ferret-badgers, one raccoon dog, and four domestic cats. They would also take serum samples from thirty-five animal traders in the market.

UPON HIS RETURN TO HONG KONG, TO HIS LAB ON THE FOURTH FLOOR, Guan Yi put the samples through the usual centrifuging before inoc-ulating fetal monkey kidney cells, the cell line Malik Peiris had used to successfully grow the virus. Guan Yi and Zheng Bo Jian also tested for the presence of the SARS coronavirus using reverse transcriptase polymerase chain reaction (RT-PCR) techniques. The teams at the U.S. CDC in Atlanta, University of Hong Kong, and other labs around the world had created genetic primers, i.e., strings of nucleic acids twenty or so units long that were distinctive to the SARS coronavirus. Guan Yi homogenized the fecal, respiratory, and blood samples and then sought to chemically "pair" those primers with the genetic material present in the samples. Most of the nucleotide sequences were, of course, RNA messenger strands unrelated to any virus, the means of channeling information essential to future generations of civets or beavers. By typing these primers into a computer linked to a nucleotide well, Guan Yi could, in effect, "search" through the molecular matter present in those nasal and blood swabs to find strings of nucleic acids similar to those that had already proven to be in the SARS coronavirus. The presence of these nucleotide chains would indicate the presence of the SARS coronavirus.

Guan Yi watched the sequences of numbers spill across his monitor, vast rows reading like this:

CATTCATATTCAGGGCAAGTGTGTTCCTCGTGCGCGCTGT.

God had been fair to him.

Malik Peiris had underestimated him.

There was virus in these animals, primarily in the masked palm civets. He would verify this by growing the virus in those monkey kidney cells.

Furthermore, he found that of the thirty-five animal traders or slaughterers, no fewer than eleven had been antibody-positive for SARS, meaning that they had been exposed to the virus, despite none of them recalling having been sick. Additionally, there had been a slight mutation between the virus in civet cats and that which had infected humans. The human virus had a twenty-nine-nucleotide deletion, which indicated that the virus had probably gone from civet to human rather than the other way around. It was unlikely that a mutation would have caused the insertion of twenty-nine nucleotides.

The virus had to have jumped species in that market, or in dozens of Wild Flavor markets like it, all over Guangdong.

Guan Yi now knew how to stop the disease.

Close the markets.

Once the human chain of infection was broken, as long as the markets stayed closed, SARS could be contained.

I MET WITH GUAN YI AT HIS OFFICE SHORTLY AFTER HIS ANNOUNCE-ment. He was focused on how *Science* magazine had been skeptical of the evidence supporting his thesis that there had been SARS in Dongmen Market. Guan Yi was, as usual, aggrieved. He had been wronged yet again.

And that would push him even harder.

CHAPTER 70

- **July 1, 2003**
- **Victoria Park, Hong Kong**
- **8,445 Infected, 811 Dead**

MY WIFE AND CHILDREN HAD RETURNED FROM THEIR EXILE, AND ON that sultry Sunday, we all went out to the antigovernment protest together, taking a taxi from The Peak and then waiting near Victoria Station for a few other journalists who were planning to march under a banner from the Foreign Correspondents' Club of Hong Kong. The police had allowed for just two entrances to the park, and the Hong Kong government had cynically divided the park, staging a sparsely attended progovernment rally in a vast stretch of soccer pitches next to the protest rally, which was as crowded as a rush-hour subway. For those of us who were protesting, the wait to get into the park was a few hours, and then the march would take ten hours more. What was amazing was how many hundreds of thousands waited in ninety-degree heat to participate.

Six hundred thousand turned out to oppose the imminent passage of legislation calling for Hong Kong to enact measures against treason, secession, sedition, and subversion, measures that would have allowed law enforcement and security officers to take steps against organizations maintaining links with foreign political bodies. Under the legislation as it was written, a news organization that published state secrets could be charged with treason. What worried journalists, civil libertarians, and vast numbers of Hong Kongers was that almost anything deemed sensitive by the central government in Beijing—information about SARS, for example—could be declared a state secret and thus proscribed, putting Hong Kong's media on the same footing as China's. In the wake of what Hong Kong had discovered through SARS—that the central government in China would indeed lie about a disease that could kill you—it was increasingly apparent that the implementation of this legislation, called for by Article 23 of Hong Kong's Basic Law, could have the worst sort of chilling effect on civil liberties, investigative journalism, and your own immune system. Perhaps in a state where the rule of law did not seem as fragile as in Hong Kong, or as trampled upon as in China, the proscriptions of Article 23 would not have seemed so odious. (There are similar-sounding statutes on the books in the United States, for example.) Yet in the current climate, the article was deemed sufficiently frightening to encourage hundreds of thousands out into the streets in the largest mass protest in Hong Kong history.

Despite the cause for the gathering, the mood was almost festive. Hong Kong had survived the worst, many of us felt. Since 1997, it had had two emerging-virus outbreaks and had suffered more than three hundred fatalities due to these novel diseases. A total of 1,755 Hong Kong residents had been infected with SARS, 1,261 had been put into isolation, 13,300 jobs had been lost, 4,000 businesses had folded, 13,783 flights had been canceled, and a million fewer tourists had come to Hong Kong. The WHO travel advisory had been in effect for fifty-two days. Yet the city had not only survived, it had demonstrated its central role in the globe not only as a financial center but also as an emerging-disease sentinel.

That wasn't, of course, what most of those would say that day. They were here because of concern over basic freedoms, Hong Kong's economic stagnation, high unemployment, and feckless government. But the fact that we could gather en masse meant that Hong Kong's doctors and scientists had done a tremendous job fighting off a killer that had tried to crash our species.

The emergence from the somber season of SARS had been gradual but unmistakable. We watched the numbers of newly infected decline, the ranks of the discharged increase. Those from Amoy Gardens were taken off quarantine. The masks came off. Families returned. The WHO lifted its travel advisory and even scheduled a conference in Kuala Lumpur to discuss how the virus had been beaten in one hundred days.

By the middle of May, we had felt safe enough to hold a fourth-birthday party for my daughter, around the swimming pool of our apartment building. All those precautions of just a few weeks ago now seemed the overly protective measures of hypochondriacs. We still had no idea what to make of SARS, how to put it in context. The world had dodged a viral bullet, but how and why? I was convinced that we had just been lucky. Others, like columnist Nury Vittachi, maintained that we had overreacted all along.

I was chatting with Peter, a former journalist turned corporate communications executive, as we sat around the pool watching our kids dive in and out of water that a few weeks before had been declared off-limits by public health officials. What had happened here was weird, we remarked. This sort of disease scare, this type of panic over an outbreak, was supposed to be reserved for science fiction.

"Maybe this was just some sort of dry run," Peter suggested. "What if these sorts of panics and scares become regular phenomena? New disease, everyone runs for the hills. The markets just take it in stride, it becomes like another factor in business forecasts."

I said I would be surprised if mankind ever became unafraid of new diseases. That fear, I pointed out, had been hardwired into us; indeed, it may even have been why we were here, around this pool, evidence of successful genetic lines.

"But what was that?" He shook his head. "It was like we all lost our minds."

He had a point. Wouldn't we all have been just as safe if we hadn't taken any measures? If we had never worn masks or swabbed our floors with bleach or filled our U-pipes? How many Chicken McNuggets had I eaten? And for what?

But that's how they start. With a few cases in a remote hospital. A doctor treating a patient who just won't get well. And that patient passes the virus to another patient. And another.

CHAPTER 71

- **August 10, 2003**
- **Shantou, Guangdong Province, China**
- **8,445 Infected, 876 Dead**

WILD ANIMAL SALES HAD BEEN BANNED IN CHINA FOR THE LAST FEW months, as per the recommendations of the WHO based on the research done by Guan Yi and subsequently confirmed by a Chinese team. Yet here was a farm composed of several airplane-hangar-size brick buildings, each of which housed about four hundred masked palm civets. There were six of these buildings, at least twenty-four hundred civet cats. Yet the sale of these animals was illegal.

I had taken the train back up to Guangdong and then hired a car out to Shantou. I had been told by Guan Yi that there were still hundreds of thousands of civet cats. "Don't go," he warned. "If you go and come back, stay in a hotel. Don't stay with your family." He had conferred with the government regarding the banning of the sale of the animals. Zhong Nanshan, who had emerged as China's preeminent SARS hero, had been convinced by Guan Yi's data that the wild animal markets were the vector. "I was quite sure that the disease was being passed along there," Zhong Nanshan would tell me. "We needed to do more research to understand the exact mechanism. Until then, I advised the markets to stay closed." Operation Green Sword, launched by Guangdong authorities, confiscated nearly forty thousand wild animals from a thousand markets, hotels, and restaurants. Similar measures had been enacted throughout China. In Shanghai's Fengxian district, for example, more than a million partridges, pheasants, and mallards were slaughtered as part of the public health campaign.

Yet as soon as the worst economic effects of the SARS outbreak had passed, public opinion quickly turned against the ban, and the trade in wild animals continued underground. "Wild game was sold openly; bribes ensured that vendors were alerted to 'surprise' inspections ahead of time," wrote the *Wall Street Journal*. Tens of thousands of wild animals were still being raised and imported. And pressure was mounting from the wild animal traders and sellers throughout China to be allowed to go back to business as usual. In a sense, sitting down to a dinner of pangolin's ears or stewed civet meat in China is something like buying an assault rifle in Alabama. It is viewed as a right by those whose views skew toward a more extreme nationalism. In this era of reforms and excess, Wild Flavor was a means of displaying one's essential "Chineseness" in the face of pernicious and omnipresent "foreign" influences, a concession to questionable Western science. The matter was the subject of a rare publicly televised debate in which environmentalists and scientists argued with farmers and traders, who insisted that the ban was economic cruelty. In the end, the Guangdong authorities announced a toothless compromise, suggesting that Chinese should "give up their habit of eating wild animals," but stopping short of banning the practice outright.

This coincided with a Ministry of Forestry report disputing Guan Yi's findings. Ministry scientists claimed to have tested far more civet cats and badger dogs than Guan Yi's team and not found SARS in any of them. Guan Yi protested that this work had not been peer-reviewed and that the PCR testing was done in Chinese labs that during the SARS outbreak itself had repeatedly been unable to come up with diagnostic criteria. What this ministry report, later discredited, would provide was specious scientific justification for those arguing against the ban.

THIS CIVET FARM WAS ON THE OUTSKIRTS OF SHANTOU, A SEASIDE city of 765,000 in the Chaoshan region of Guangdong. It had been declared one of the original Special Economic Zones during the first post-Mao reforms yet had not boomed as had its neighbors down the coast. But during the Era of Wild Flavor, "not booming" was a relative term, and for the visitor to Shantou, there still seemed to be a dizzying amount of construction in progress. This part of Guangdong looked and felt very different from Guangzhou. In particular, the people here were darker skinned, rounder eyed, and spoke a dialect of Cantonese so thick that my driver from Guangzhou couldn't understand them. They looked almost Indonesian instead of Han Chinese. As we drove through villages on our way into the city, we saw thousands of local farmers on the march, hustling to bring in a crop of local barley they would dry in the sun by the side of the road.

I had called ahead to the owner of the civet plantation. When he indicated he was wary of meeting a foreigner, I had my assistant explain that I was a businessman interested in importing the animals to the United States. I told him that owing to the vast amounts of press these exotic animals had been getting because of their link with SARS, there would be great curiosity to try them in the finer restaurants of New York City.

We drove down a series of dirt roads, past former collective farms now trying to make a go of it as capitalist businesses. At the end of a dusty track, we passed through an arched gate that bore the name Farmers' Unit 209 and found an old woman seated on an overturned beverage crate near the front gate. She waved to us and demanded to know who we were.

When we mentioned civet cats, she pointed to her crate and smiled. Looking closer, I was surprised to see that one of the small black animals with distinctive white and gray facial markings was trapped in it.

The woman hurried inside and found the manager, a fleshy-faced man wearing a green polo shirt, khakis, and black loafers with square gold buckles. He was bemoaning the fact that if the wild animal ban wasn't soon lifted, he wouldn't know what to do with these civet cats. He slid a bar off a saloon-style door, and we walked into a vast, dark warehouse. The smell hit me first. Sour and fecal, like someone was holding a used piece of toilet paper to your nose. It took a moment for my eyes to adjust to the light, but gradually hundreds of scurrying, cowering, and panicking civets came into focus. They stared and fidgeted in cages stacked one atop the other in rows stretching fifty meters to a distant brick wall. I walked down one of the narrow aisles, between the panting, sad-eyed mammals who had no idea that, at least for now, their having been singled out as the bearers of a virus had saved them from becoming dinner. I was about halfway down the aisle when I shouted back to the manager. He was standing in the bright sunlight beyond the dark interior.

"Are you coming inside?" I asked.

He shook his head. "I don't get near those animals."

When I called him a few days later, asking about the price, he told me he was no longer interested. The ban was about to be lifted, he explained. He would be able to sell them for a good price right here in China.

He was right. The Ministry of Forestry lifted the ban on wild animal sales on August 13.

GUAN YI WOULD SHAKE HIS HEAD AND MARVEL AT THE INSANITY OF the lifting of the ban. That fall, he would recommence his shuttling back and forth between Hong Kong, Shenzhen, and Guangzhou, expanding his sample gathering to include wild animal markets throughout Guangdong. He was among the few virologists who had remained focused on SARS after the worst of the crisis had passed. There were teams around the world working on vaccines, yet much of the funding and government support for these projects had withered after the crisis left the front page. And the big pharmaceutical companies were not likely to see a huge profit in coming up with a drug to treat a disease no longer perceived as a global threat.

Yet SARS had been the first disease to go H2H, or human to human, since AIDS. And the reasons it had not wiped the human slate were still unclear. Clinically, the consensus had emerged that the virus had been stopped by better infection control. But if SARS had been as contagious as influenza, "then we could have seen hundreds of millions dead," says Klaus Stöhr of the WHO.

To Guan Yi's mind, this remained a mystery worth studying. A virus crashes through, somehow infects humans, yet the outbreak burns itself out. In recorded history, no disease had previously jumped the species barrier to infect humans, caused an epidemic, and then never threatened us again—not without the discovery of a vaccine or cure to curtail the microbe. Some diseases, such as chicken pox, gradually become endemic to man, resulting, if we are lucky, in nothing more than mild childhood illness. Others, such as Ebola, retreat back to whatever animal reservoir they came from, stalking humanity from their hidden lair and occasionally lashing out to bloody a village or crash a rural hospital. But diseases don't, as a rule, just go away.

Guan Yi couldn't believe our luck.

Throughout that fall, working with the Guangzhou CDC and the Shenzhen CDC, he paid fifty *kwai* for each animal to a trader who supplied Dongmen Market. In Guangzhou's Xinyuan Market, Guan Yi would buy animals and haul them away in cages to the Guangzhou CDC labs, where he would gather samples before sending the creatures to be destroyed. Occasionally, when he was in a hurry, he would sedate the animal right there in the market and draw blood and swab for feces.

When he brought those samples back to Hong Kong, a frightening picture started to emerge. Not only was he again finding the SARS coronavirus in a host of rodent species—in addition to the civet cat, he also detected the virus in hog badgers, Eurasian badgers, raccoon badgers, and ferret badgers—he was astonished, when he did the genomic sequencing, to observe that these coronaviruses had actually mutated to become more similar to the SARS coronavirus samples taken from humans during the first outbreak last spring. Those nucleotide deletions that had previously differentiated the human virus from the civet cat virus were now also present in the virus he was isolating from animals. All this confirmed that the disease that had infected humans was again at large. The animals that showed the highest infection rate by far were the civet cats, with sixteen positive cases out of twenty-one animals tested. These civets had come from several different markets and traders, which meant that the disease was lurking in wild animal markets throughout Guangdong—and maybe the rest of China.

It was only a matter of time, Guan Yi feared.

CHAPTER 72

GUAN YI WAS SITTING IN HIS APARTMENT IN CHI FU, ON HIS LEATHER sofa, watching his big-screen TV, smoking his Mild Seven cigarettes, and wondering about his way forward.

There had been two cases over the summer, laboratory infections in Singapore and Taiwan. Yet there had not been a new civilian infection since July. He believed that the SARS coronavirus's seasonality was the primary reason that there had been no new infections. As he had observed in the markets of Guangdong, all the conditions for the reemergence of SARS were in place. There was once again too much interaction between humans and civets for this virus not to make another jump. But it could take months to get a paper peer-reviewed and published that could affect public health by once again encouraging the Guangdong government to cull the civet cat population or at least limit contact between humans and this animal. In that time, the disease could once again gain a foothold among humans, or a mutated, more infectious strain could pop up in some unfortunate hospital, igniting another viral burn through southern China, Hong Kong, and the world.

But as long as there were no new cases in Guangdong, then perhaps he had time to fast-track his paper and get it published in a few weeks.

His wife called from work. There was a suspected SARS case in Guangdong: a thirty-two-year-old television reporter was in isolation at Number Eight Hospital. Guan knew he had to take action ahead of any publication. "I mean, why do you do science?" he asked me later. "To write papers? Or to make a difference in the real world?"

Despite all that he has seen and learned the hard way through years of virological fieldwork, Guan Yi still believes that China's top officials will always do what is right if they have the relevant information. The problem is getting that data in front of them. To overcome that obstacle, he framed a simple letter to Beijing's Department of Hong Kong and Macao Affairs, which he cc'd to the Ministry of Health and the China CDC. Near the beginning, the letter stated, "With winter coming, the wildlife markets have reopened, providing the perfect conditions for another outbreak of SARS. We started in October systematic sampling in the markets and have discovered these four things." He went on to list his findings: that the civet cat was the major carrier of the SARS coronavirus, that the SARS coronavirus had been found in different civet cats from different regions, that the virus could infect humans, and, most frightening, that the "transmitting mechanism for the resurgence of SARS is in place." He included in this letter four pages of genetic sequences taken from civet cats. The letter was hand-delivered to the recipients on January 2.

Within hours, that letter was passed down the chain of command from the Ministry of Health in Beijing to the Guangdong Department of Health. Guan Yi's reputation as a virologist and his contacts with members of the health establishment were sufficient to earn him an official invitation to Guangdong on January 3, where he would present his case in person to some of Guangdong's highest-ranking public health officials.

At noon, in a hotel conference room near the headquarters of the Guangdong CDC, in the southern part of Guangzhou, Guan Yi appeared as usual with his black satchel and met with representatives from the Department of Health, the Guangdong CDC, and the Ministry of Health and eminent doctors and scientists from other institutions. Every man and woman in this room had lived and worked through the SARS outbreak of 2003; many of them were clinicians in hospital wards and respiratory clinics where they had watched patients wither, suffocate, and eventually die from the disease. Deng Zide of Number Three Hospital was there, as was Tang Xiaping of Number Eight Hospital, Huang Wenjie of Guangzhou General Military Hospital, and of course, Zhong Nanshan, who had emerged from the SARS outbreak as probably the most famous physician in China.

For the team gathered here, this was a solemn reunion. They were pleased, first and foremost, to be alive, for not a few of them had contracted SARS and survived. They mourned fallen colleagues. But most of all, they were taking very seriously the chance that SARS was indeed seasonal and could return that winter. It had been around this time in 2002 that the team, including Deng Zide and Huang Wenjie, had been dispatched from Guangzhou to investigate a curious incident in Heyuan. Last January, the first cases had been streaming from Heyuan and Zhongshan into Guangzhou. The new outbreak was already under way, and citizens were beginning to panic. Each of these doctors most dearly wanted to ward off a viral repeat.

It was remarkable, actually, that this meeting had been convened. And it was a measure of how seriously these men and women took Guan Yi's warning.

As Zhong Nanshan began speaking, it was clear that he had not even seen the genetic sequences that Guan Yi had included in his letters. Another official from the Ministry of Health added that Guan Yi's tone in the letter seemed "a little hysterical." Guan Yi realized that as the letter had been passed down from the Ministry of Health, somehow those four pages of genetic sequences he had included, which provided the evidence backing up his dramatic assertions, had been lost. Guan Yi called his lab back at the University of Hong Kong, where his assistants, who work nearly as hard as he, could e-mail the documents to the CDC headquarters in Guangzhou.

Guan Yi's hosts were skeptical of this notoriously impetuous virologist now sitting in their midst chain-smoking and haranguing them about their duty to avoid another outbreak. Zhong Nanshan pointed out that, during the early days of the epidemic last February, Guan Yi had been among those who kept on insisting, incorrectly, that

SARS was a novel form of avian influenza. And even after the gene sequences arrived from Hong Kong and were run over from the CDC—those long strands reading CATTCATATCCAGGGGGAG— the group of scientists assembled before him remained unconvinced. "When someone is showing you raw data," says Xu Ruiheng, deputy director of the Guangdong CDC, "you have to ask yourself, is this real or is this fabricated? And how relevant is this to what Yi was proposing?"

In turn, Guan Yi asked his counterparts if they had the sequences for the human case now recovering in Guangzhou Number Eight Hospital. They produced their documents. It turned out that while they had done the genetic sequencing, they had not yet analyzed this virus's phylogenetic origins. (It was this sort of analysis—the realization that previous human coronavirus infections had been genetically very similar to animal coronaviruses—that had compelled Guan Yi to begin researching in the wild animal markets in the first place.) Guan Yi suggested they send their sequences to his lab in Hong Kong, where his technicians were standing by. The technicians would use their computer modeling programs to analyze fourteen amino acid sequences to reconstruct, in effect, the evolutionary origins of this year's human virus. That way, they would be able to compare the two viruses and determine, more precisely, the real risk level. If the two phylogenetic sequences were similar, Guan Yi explained, and Zhong Nanshan concurred, then that would confirm that the disease was again afoot and was somehow related to the wild animal markets.

The Guangzhou officials agreed, the new case's sequences were sent, and the men drank green tea and smoked while they waited for Guan Yi's lab to complete the computer modeling.

The data that was sent back revealed that the two viruses were more than similar. They were virtually identical. The fourteen amino acid sequences were the same, which meant that these two viruses not only belonged to the same phylogenetic tree but were also both on the same branch, practically the same leaf. Science doesn't produce many moments like this: good luck coinciding with great research had proven that the same virus that was in those wild animal markets had somehow infected a human being. The data was so compelling that the committee resolved that afternoon to inform the governor of Guangdong and recommend a culling of civet cats.

There was only one man in that room with the clout and reputation to recommend a measure this extreme. Zhong Nanshan was delegated to call Governor Huang Huahua. The argument Zhong could give was simple: the wild animal business in Guangdong was estimated to be worth anywhere from one hundred to two hundred million dollars a year; the economic impact of another SARS outbreak, however, was immeasurable. Zhong made that call on Sunday.

He can be very persuasive: the order was given later that day to the Guangdong Health Department and the Guangdong Forestry Department, among other agencies, to launch a campaign to eradicate civet cats from the province's farms and markets. By Monday morning, said Peng Shangde, deputy director of the Guangdong Forestry Department, "we were staffed and the trucks were rolling."

Chinese journalists invited to a sanitation department crematorium to watch some of the gory spectacle described it as "solemnly patriotic." Dozens of flatbed trucks kept rolling up with their cargo of civet cats in cages. The animals were wrapped in plastic bags and then burned alive. Department of Forestry officials would incinerate some animals, boil others to death, and drown still more in disinfectant. It was as if the Chinese government were sacrificing thousands of the rodentlike creatures to ward off the Fourth Horseman of the Apocalypse.

But what seemed to the world a panicked measure was, Guan Yi believed, the only way to avert another outbreak. Several WHO officials, while displeased at the ruthlessness of the implementation, conceded that the measure made sense. Rob Breiman, the American epidemiologist working in Bangladesh who was part of several WHO missions to China during the SARS scare, told me, "From a political and public health standpoint, it was a reasonable step in response to the reemergence of SARS to act on the most likely source."

Guan Yi's mentor, Rob Webster of St. Jude Children's Hospital in

Memphis and a pioneer in establishing the zoonotic origins of many influenzas, smiled as he told me, "The research is solid, but still, Yi has certainly stuck his neck way out there on this one." Guan Yi, as usual, was dismissive of any doubts. Back in Hong Kong, he repeated to me how the virus found in other rodents such as badgers is genetically less similar to the strains found in humans, before vowing that culling civets would "break the chain of infection."

THE INCUBATION PERIOD FOR SARS IS FOURTEEN DAYS. THE LAST civets were taken from the wild animal markets on January 6, 2004. There has not been another human infection.

CHAPTER 73

- **January 1, 2004**
- **Pok Fu Lam, Hong Kong, China**
- **1 Infected, 884 Dead**

THE EMERGING-VIRUS TRIP WIRE WAS TWITCHING AGAIN. PROMED postings in late December had led with "Avian influenza, human—East Asia (1): Viet Nam," which meant that there had been human cases of avian influenza in Vietnam. SARS had started with fears that it was an avian influenza outbreak, that a strain of killer mutant flu had been unleashed. And now here it was, one year since the initial SARS outbreak, and the world appeared to be staring down another new virus. There had already been fourteen deaths in Vietnam and cases in Thailand and China. And once again, the Chinese government was obfuscating, refusing to admit that there had been any avian influenza in China, even though several tainted chickens in Korea and Japan had been traced back to China. The Vietnamese government, this time, also appeared to have covered up the early phases of disease. There had been massive chicken die-offs in Vinh Phuc province, in northern Vietnam, six months before the government officially acknowledged the emergence of avian flu. Giapfa Comfeed Vietnam Ltd., a poultry company in Vinh Phuc's Tam Duong district, reported that twenty thousand of its chickens had died with symptoms correlating to avian flu. The company says it sent blood samples to the Ministry of Agriculture's Veterinary Department, whose tests revealed that the chickens had been killed by an unknown agent. Van Dang Ky, a veterinarian from the department's epidemiology unit, admitted, "The first signs of an epidemic were found in Tam Duong district in July 2003. At the time, Vietnam was preparing actively for the Twenty-second Southeast Asian Games, and we thought we could control the disease, so we did not announce it for political and economic reasons."

Vietnam and Thailand commenced the slaughter of more than eight million chickens, yet human cases, usually resulting from close contact with infected birds, were by now becoming a regular occurrence. The vast majority of these cases were fatal, unless Tamiflu was quickly administered. So far, however, the virus was not proving as contagious as human influenzas. The great risk was that a carrier infected with a common case of human flu would also become dosed with an H5N1 avian flu. Those two RNA viruses swirling around in a person's respiratory system could swap genes and reassort into a highly contagious, fatal flu. This was the slate-wiper scenario.

No one knew the odds of this type of reassortment happening. It would be random, those genomic tumblers whirling, the future of humanity resting on each result. Guan Yi had done significant work showing that avian flus were already recombining in pigs with porcine influenzas. And pigs were also known as carriers of human flu viruses. "It's only a matter of time," Guan Yi would repeat.

An H5N1 that achieves widespread human-to-human transmission became every virologist's greatest nightmare, and as winter gave way to spring and summer, Guan Yi and Malik Peiris and the World Health Organization and the American CDC would all turn their attention to Vietnam, Thailand, and, once again, China.

A new virus was beginning another roll through humanity. The only question was, how far would it go?

THE REFLECTION ON THE WATER WAS OF SMUDGED NEON SIGNAGE— JUMBO SEAFOOD RESTAURANT, LUCKY PRAWN INN—fluorescent dock strips, and yellow running lights as the steady traffic of junks, yachts, and fishing boats bobbed in and out of crowded Aberdeen Harbor. There is a traffic-control system of sorts, an overlaying skein of logic and order that, superimposed upon this chaotic-seeming nautical highway, would reveal an organizing principle. You just needed to understand how the system worked.

From where we sat on the dock of the Aberdeen Boat Club, sipping drinks in the early evening, Malik Peiris and I struggled to see how the harbor was ordered. If this harbor were a human mind, I already concluded, it would be Guan Yi's mind, rather than Malik Peiris's. Guan Yi's thoughts would emerge in myriad shapes and a random cascade—practical fishing trawlers, fancy luxury yachts, junks—and only after he looked at the sum of them would his arguments and lines of reason be revealed. But glimpsed from too close, Guan Yi could seem mad.

Whereas Peiris was always reasoned.

In the harbor of Malik Peiris's mind, every ship or boat—each idea—was moored perfectly at its corresponding slip. No one unaccounted for, each expressly fitted for the purpose it would serve when it was finally launched. Discussions with Peiris could therefore become frustratingly prosaic affairs, Peiris sticking closely to the matter as it had been laid bare by the provable facts. Emotions and motives were not relevant to his work. Or to his conversation.

I often marveled at how the battle against emerging infectious disease was generally fought so far away from what I considered home. It was waged in African jungles and South American swamps, in Chinese slums and on Vietnamese chicken farms. If I hadn't lived in Hong Kong during the season of SARS, I would never have understood a very basic fact. In the days of slow travel by galley or junk or clipper, a disease outbreak in, say, East Africa would be in Beijing within a year or two. In the era of jet travel, a disease outbreak in Shenzhen is in Toronto just a few days later. Those remote battles fought by men like Malik Peiris and Guan Yi, and even by whistle-blowers like Dr. Jiang Yanyong and journalists like Susan Jakes and Huang Yong, in terms of virology or epidemiology, are battles to save our world. If the virus is there today, it will be here tomorrow. We are as vulnerable as the weakest immune system at the farthest link of the human chain. If the battle is not fought and won in those remote quarters, then it is lost. The trip wires are also our only defense. There can be no clearer warning than SARS.

This was one of those evenings when you are reminded that Hong Kong is only a few degrees removed from the tropics, and its heat and damp and even gauzy light put one more in mind of Southeast Asia than of southern China—the floating restaurant across the harbor, the bow-tie-wearing waiters, the expats and their fish-and-chips and tall lagers at tables all around us. There are vestiges of colonial, tropical decadence and squalor in the air, as well as that smell of fetid fish endemic to most tropical harbors. Behind us were the mountains that divide the south side of Hong Kong Island from central. Go up that hill just a kilometer or two and you would find Peiris's lab and its drums of incubating samples of viruses.

We were discussing a notion that Peiris was a little uncomfortable with—it wasn't quantifiable, nor was it verifiable: that of motive. A murderer's motive, actually. Why did viruses kill? I mean, what was in it for them?

As I traced the course of this disease, from hospital to hospital, country to country, it never really made sense. The virus's goal, surely, was replication. Nothing more. So wouldn't it be in the virus's best interest to find in its human host a comfortable evolutionary niche, to colonize a little, settle down? Causing severe sickness, disease, and death would seem an evolutionary dead end. Yet that is precisely what viruses sometimes do when they encounter a new species. Most frequently, when a virus runs into a new species, it fails to find a way into the new host or to bind with any of that species' cells and passes harmlessly through the respiratory or digestive system. When a virus particle does find a route of trafficking and manages to breach a cell's surface defenses, it faces still more obstacles as it may not be able to subvert that cell's genomic machinery. And even if it manages to convert a host cell into a virus factory, there is still a chance the host immune system will produce sufficient amounts of antibodies to wipe out the new virus particles. "The vast majority of viral attacks must necessarily fail," writes Frank Ryan in *Virus X*, a study of emerging diseases. However, the greater the number of interactions between two species, the more likely that millions of virus particles will be swapped, and then one of those will successfully repli- cate in a new species. The odds of this happening are still unknown—a million to one, a billion to one?—but are certainly shortened by the large numbers involved. Microbes, remember, deal in volume.

The notion I was interested in discussing with Peiris was a concept I had first read about in *Virus X*, that of the aggressive symbiont. The the- ory is that a virus in symbiosis with a host species will not cause disease in that species but will cause severe disease if there is a sufficient amount of interaction with another species that will compete with the virus's host for resources. This serves the interests of both the virus and its host. The virus will continue to enjoy its ecosystem—the host animal—as the host continues to thrive unthreatened in its own ecosystem. Yet when the host species is threatened, its viral symbiont will wipe out its competitors. Nature provides numerous examples of such behavior. There is a herpes virus, for example, that has a coevolutionary relationship with a squirrel monkey that lives in the Amazonian rain forest. When a rival species, say, the marmoset monkey, comes into contact with the squirrel monkey, the virus jumps species and eradicates the intruder monkey. This theory is attractive in that it helps to explain why the virus is unlikely to cause ill- ness in a species vastly different from its symbiotic host—that new species is unlikely to be competing for the same resources. Our viruses, in other words, may be our stoutest defense against those animals that would intrude into our ecological niche. And when we intrude into a rival species' niche . . . look out. Ebola, hantavirus, perhaps SARS—these are viruses programmed to kill.

Peiris agreed that the theory was an attractive one but then quickly pointed out that it was only a theory and had little or no effi- cacy in finding the viral agent behind a disease outbreak.

"Why would it be in SARS's best interest to kill human beings?" Peiris asked.

"Because we are a threat," I said, "to whatever animal is this virus's natural host."

Peiris then pointed out that whoever gets this virus first will be farther along the evolutionary path to developing the sort of herd immunity that would confer a tremendous Darwinian advantage. Think of Cortés and his conquistadors as they lay siege to Montezuma's Tenochtitlán, Pizarro as he ravaged Peru. In both instances, the invaders succeeded primarily because of their symbiotic smallpox viruses, which wiped out the immunologically naïve natives far more efficiently than Spanish steel or gunpowder. Those few Native Americans who survived became hosts to the same viruses as their conquerors, fellow symbionts—but at what terrible cost? Examples abound of such aggressive symbiosis. In the famous instance of the myxomatosis virus, introduced into the Australian rabbit population by scientists seeking to cull the species, only .2 percent of the indigenous rabbit population survived. However, over the following ten years, the remaining rabbit population, through this most brutal natural-selection process, became immunologically tolerant of the new virus.

"Could this be history repeating itself?" I asked. Say the Chinese get a new virus first, and it proves as fatal as smallpox when introduced into a new gene pool. Hundreds of millions die in a horrible repeat of the Black Death. Won't the surviving, immune Chinese then have an evolutionary advantage over the rest of the world? Couldn't a virus be, in effect, a terrible first step toward world domination?

Peiris shook his head at this fantastic scenario. I was in the realm of science fiction, he quickly pointed out. "A new virus," he explained, "will be on a plane and everywhere on earth within a day or two of emerging, especially if it emerges, as this one has appeared to, in southern China."

Cortés and his conquistadors had been reared in a completely different microbial environment from Montezuma and his warriors. Smallpox, measles, and a host of other microbes had never managed to cross great oceans until they could catch a ride in human cells. Now, however, such binary evolutionary paths for human beings had all been merged into one. If I got sick in Hong Kong today, you might catch the same bug in Florida tomorrow, a nasty virus looking to wipe out whatever species was threatening its host. My idea of an immunologically superior Chinese army marching on Washington was preposterous.

Peiris asked for the check.

He had another virus to find.

I WAS BACK IN THE CLICK, STROLLING THROUGH DIRTY ALLEYWAYS, between the hookers calling out for tricks and the video kiosks and the bathtub *baijiu* stalls. The Era of Wild Flavor had been reignited. SARS had been a developmental blip. Chinese economic growth, despite the dire forecasts of 2003, had resumed double-digit expansion and would continue to boom through 2004. The virus had been a bad quarter, nothing more, a production shortfall offset by triple shifts the rest of the year. Those migrants who had streamed out of the cities in the spring had poured back in in the summer, bringing with them another fifty million country cousins looking for piecework in the cities. For neighborhoods like the Click, that meant fresh labor.

For a virus, that meant fresh meat.

That's all I saw as I walked through the damp alleys and inhaled the wet air. Human meat, slabs of protein layered in tiny little apartments, on top of plywood sleeping tablets, curled up on straw mattresses. That's what cities were, and the more crowded, the faster growing, the more efficient they would be as viral amplifiers. And China now had about one hundred cities with populations over one million. SARS, I reckoned, had been a warning. We had gone too far, plundered too many ecosystems, negatively affected the behavior of too many species. And then we had even brought those ecosystems and species into the heart of our cities. Dongmen wild animal market was just a kilometer from here. The government was again talking about lifting the ban on wild animal sales. The trucks from the rain forests of Southeast Asia and the mountain ranges of the Chinese interior were already rolling, bringing the animals, and their viruses, into the city.

I was back looking for Fang Lin, who, I had heard, had returned to Shenzhen after convalescing back in Jiangxi. I didn't find him but was grateful to know that he had been lucky.

In fact, humanity had been lucky. The University of Hong Kong, the WHO, the U.S. CDC, the clinicians, and public health officials had, to varying degrees, done heroic work. But had they actually stopped the virus? Or had the virus, all along, lacked the infectiousness to wipe the slate? Had it broken out before the twentieth century, it would certainly have been a dreadful epidemic, with a 5 to 10 percent mortality rate, causing millions of deaths. Even today, had it been as infectious as influenza, it would have killed millions. Modern medical science and public health hadn't stopped the virus. The virus had stopped itself. The season changed before the bug had mutated into something more infectious. And Guan Yi was proving that it was still out there and mutating, its genetic tumblers clicking away until it might settle on a more deadly combination.

But if we had been lucky, been given, in effect, a warning by the genomic tapestry we had gashed open, had we taken heed? Looking around at slums like the Click, it was apparent that we hadn't. There were teams in Beijing, Washington, and Taiwan working on vaccines for SARS. There were Guan Yi and Malik Peiris and their cohorts in Hong Kong keeping vigilant. But what had really changed in China? The government still expected the local public health agencies to pay their own way. When that TV producer was infected by reemergent SARS and treated at Number Eight Hospital, rather than encourage the media to cover the outbreak, the government had arrested the first reporters to write about the new case. "While the Chinese leadership is serious about fighting SARS," wrote political commentator Baopu Liu, "its No. 1 priority remains the stability of the regime." Any biases in the system that prevented accurate reporting of the initial outbreak remained. There had been no attempts to encourage more openness. One now encountered government officials who believed that the real lesson of SARS was to engage in more effective cover-ups. Even more troubling, Dr. Jiang Yanyong, the whistle-blower who had exposed the Chinese government cover-up, had since been put under house arrest for writing a letter to party leadership criticizing the government's labeling of the student protest at Tiananmen Square as "counterrevolutionary." The takeaway: deny everything.

There would be another outbreak, if not SARS, then another emerging virus. China was transforming so fast that there were bound to be sociological and biological repercussions. With such vast societal shifts, from rural to urban, from agricultural to industrial, from cooperative economy to consumer society, there were dislocations that further tore at nature's fabric. This was common sense: too many people were living too close together with too many wild animals.

Welcome, virus.

NOTES ON SOURCES AND FURTHER READING

The vast majority of the information in this book was gathered through author interviews with the various characters or with surviving family members. In those instances where quotes, data, or description have been taken from another source or publication, I have indicated this in the narrative or in these source notes. I am not a scientist, and so I have had to try the patience of numerous epidemiologists, virologists, physicians, nurses, public health officials, and patients. I may have made some mistakes, either in recounting the scientific processes or understanding the significance of each particular step along the way. I assure you, in all cases, such mistakes are shortcomings of the author rather than of his generous subjects. When I began this book, I knew nothing about viruses. If I learned anything, it is that we should be grateful for the accumulated wisdom of those who make their living finding those viruses, identifying them, and destroying them.

On the advice of numerous physicians and at the request of several patients, I have respected the medical rule of using pseudonyms for patients' names, and for those who requested it—or whose surviving relatives asked—I have altered certain identifying characteristics in ways that I felt did not affect the epidemiological record. In those many cases where a health care worker became a patient, I have used the real name.

The numbers of infected and dead given in each chapter headline are estimates drawn from World Health Organization statistics augmented by local hospital counts, which indicated more cases and fatalities than the WHO knew about at the time of their tally. The figures represent, literally, back-of-the-envelope calculations and, in some instances, educated guesses. By the last chapter, the number of infected is given as zero because the formerly infected either had made full recoveries or had passed away.

It is important to note that the patient called Fang Lin was never tested for seroconversion to the SARS virus.

PROLOGUE

The earliest mentions were in the *Heyuan News* on January 3 and 4, 2003. According to a story by Matthew Forney in *Time Asia* (4/21/2003), provincial party leaders on January 4 barred coverage of what happened in Heyuan and elsewhere until February 11. By January 23, however, Chinese newspapers, in particular *Nanfang Daily, Guangzhou News Express,* and *Yangcheng Evening News,* would all skirt this edict by reporting the panic buying of vinegar. These stories, viewed with the benefit of hindsight, hint at the existence of a new, or at least a very serious, respiratory disease in Guangzhou. Xinhua, the official Chinese government wire service, would also move an item the last week of January. This item would read much more like an account of a curious superstitious practice. These various accounts were picked up and translated into English by the *South China Morning Post,* which was where we first read about the outbreak and which, throughout the SARS crisis, did an admirable job of providing Hong Kong with useful and reliable coverage. In particular, I found the stories of Mary Ann Benitez, Ella Lee, and Leu Siew-Ying to be of great help in researching this book.

CHAPTER 1

My writing about Pearl River development in this chapter and chapter 3 was informed by several books written about China's urbanization, including *Great Leap Forward,* edited by Chuihua Judy Chung, Jeffrey Inaba, Rem Koohaas, and Sze Tsung Leong, in particular the chapters "Ideology" by Mihai Craciun, "Architecture" by Nancy Lin, "Money" by Stephanie Smith, "Landscape" by Kate Orff, and "Infrastructure" by Bernard Chang; *The New Chinese City,* edited by John R. Logan, in particular the chapters "Migrant Enclaves in Large Chinese Cities" by Fan Jie and Wolfgang Taubman, "Social Polarization and Segregation in Beijing," by Chaolin Gu and Haiyong Liu, and "Temporary Migrants in Shanghai: Housing and Settlement Patterns" by Weiping Wu; and *Urbanizing China,* by Gregory Eliyu Guldin, in particular the chapters "Urbanizing the Countryside, Guangzhou, Hong Kong and the Pearl River Delta" by Guldin and "Urbanization under Economic Reform" by R. Yin-Wang Kwok. Many of the economic statistics came from *The State of China Atlas,* by Stephanie Hemelryk Donald and Robert Benewick; *China Statistical Yearbook 2003* and *2004, China: The Investment Agenda for Building an Environmentally Sustainable Economy,* by the Association for Sustainable and Responsible Investment in Asia; and from several articles in the *Asian Wall Street Journal.* Two articles that appeared in *Time Asia* catalyzed some of my thinking about Shenzhen: "Too Much, Too Soon?" (11/24/2003), by Matthew Forney, and, in particular, "Crossing the Line" (5/7/2001), by Alex Perry. Zhou Litai's quote comes from Perry's article.

Conditions in southern China's wild animal markets have been widely documented, but the perils posed by them were made most explicit to me by a presentation by Laurie Garrett at "Globalization's Newest Challenge: SARS," a conference at

Yale University held in September 2003. Four helpful stories on the subject were "Wild Animal Markets in China May Be Breeding SARS," *USA Today* (10/28/2003), David J. Lynch; "Alarm as China's Wild Animal Trade Is Blamed for 'New Case of SARS,'" *Telegraph* (4/1/2004), Cortlan Bennett; "Scouring the Market for SARS," *Time Asia* (6/2/2003), Anthony Spaeth; "Infections Becoming More Widespread," *Washington Post* (6/15/2003), Rob Stein; "Wild-animal Traders Have Antibodies to SARS," *South China Morning Post* (5/26/2003), Ella Lee. "Noxious Nosh," *Time Asia* (6/9/2003), by Hannah Beech, was a good explanation of the putative health benefits of dining on various wild animals.

CHAPTER 2

Information about Penfold Park was provided by the Jockey Club and from "Going the Distance, the 25th Anniversary Celebration of Sha Tin Racecourse," in the *South China Morning Post* (10/7/2005). Sha Tin's history with animal influenzas is recorded in *The Veterinary Record* (5/27/1995), "Outbreak of Equine Influenza Among Horses in Hong Kong During 1992," by D. G. Powell, K. L. Watkins, P. H. Li, and K. F. Shortridge, and in the *Proceedings of the 11th International Conference of Racing Analysts and Veterinarians* (1995), "Endemic and Exotic Equine Infectious Diseases and Their Effect on International Racing," by K. L. Watkins. The chronology of the bird outbreak was taken from an emergency report, "Pathogenic Avian Influenza H5N1 in Waterfowl," prepared by Trevor Ellis and his team at the Hong Kong Department of Agriculture, Fisheries, and Conservation.

CHAPTER 3

Stories about violent crime began to appear regularly in Guangdong newspapers in late 2003. "Police Act on Surging Crime," from *Shenzhen Daily* (9/15/2003), was the first time I had seen the issue discussed in the Chinese press. The quotes from Ren Jintao, from Ma Yong's neighbors, and the account of the serial killers in Buji come from "Predatory Transients," *Time Asia* (12/1/2003), by Matthew Forney. The situation in Buji was further elucidated by an unpublished, extensive, first-person account written by Jodi Xu. I also found these six books about China to be very helpful in framing my own thinking and reporting on economic and cultural development: *River Town*, by Peter Hessler; *The Chinese*, by Jasper Becker; *The New Chinese Empire*, by Ross Terrill; *China Pop*, by Jianying Zha; *Mister China*, by Tom Clissold; and the incomparable *Red Dust*, by Ma Jian.

CHAPTER 4

There have been many excellent accounts of various influenza outbreaks. Among the most recent and eminently readable was *The Great Influenza*, by John M. Barry, which complemented very well the epic account *America's Forgotten Pandemic: The Influenza of 1918*, by Alfred W. Crosby. *The Devil's Flu*, by Pete

Davies, was both a good primer on the history of influenza and an introduction to the risks of a novel zoonotic influenza. One of the best, most concise accounts of the 1997 bird flu outbreak was by Erik Larson: "The Flu Hunters" appeared in *Time* (3/9/1998) and turned out to be wonderfully prescient journalism. There was a *New York Times Magazine* (11/7/2004) cover story by the same name, written by Gretchen Reynolds, that went over much of the same material but did so in the context of 2004's avian influenza outbreak in Vietnam. The article "Well Braced for the SARS Struggle," *Los Angeles Times* (4/13/2003), by Madeline Drexler, was among the earliest to point out how reminiscent the emergence of SARS was with the 1918 flu. Rob Webster contributed a chapter on influenza to *Emerging Viruses*, edited by Stephen S. Morse, that helped me to understand how various influenza strains are related. Among the academic papers that I found helpful were "The Next Influenza Pandemic: Lessons from Hong Kong," *The Society for Applied Microbiology* (2003), K. F. Shortridge, J. S. M. Peiris, and Y. Guan; "Genesis of a Highly Pathogenic and Potentially Pandemic H5N1 Virus in Eastern Asia," *Nature* (7/8/2004), K. S. Li, Y. Guan, et al.; "SARS Exposed, Pandemic Influenza Lurks," *The Lancet* 361 (2003), K. F. Shortridge; "Human Influenza Virus Related to a Highly Pathogenic Avian Influenza Virus," *The Lancet* 351 (1998), E. C. J. Claas, A. D. M. E. Oosterhaus, et al.; "Avian Influenza Viruses of Southern China and Hong Kong," *Bulletin of the World Health Organization* (Vol. 60, 1982), K. F. Shortridge. I also found useful an internal WHO report, "Summary of Urgent Issues Related to H5N1 Cases in Hong Kong, SAR" (3/3/2003), which was a rough draft of a report that was later submitted by the WHO expert team to the Chinese Ministry of Health.

Malik Peiris's career as a virologist started in Sri Lanka; there are two papers that I believe are revealing of his assiduousness and diligence. "Japanese Encephalitis in Sri Lanka: Comparison of Vector and Virus Ecology in Different Agro Climatic Areas," *Transactions of the Royal Society of Tropical Medicine and Hygiene* 87 (1993): 541–48, by J. S. M. Peiris, F. P. Amerasinghe, et al., and "Japanese Encephalitis in Sir Lanka—the Study of an Epidemic: Vector, Incrimination, Porcine Infection and Human Disease," ibid. 86 (1992): 307–14, J. S. M. Peiris, F. P. Amerasinghe, et al. The profile of Peiris that appeared in *Nature Medicine* (9/2004) is the most concise and accurate sketch of him that I have read.

CHAPTERS 5 AND 7

My ideas about the perils of urbanization were shaped by numerous writers who have tackled this subject. When SARS first reached Hong Kong, I became curious about the history of disease and paid a visit to the American Club library. I was looking for the *Hot Zone*, by Richard Preston, as well as *And the Band Played On*, by Randy Shilts. Both books were inspirations and showed different, very useful approaches to writing about emerging viruses for the general reader. I owe a great debt to both works. The American Club library, alas, had neither of them. The only book about epidemics on the shelves was *Plagues and Peoples*, by William H. McNeill, a book I am embarrassed to say I had never read, or heard of, until that morning. I borrowed the book, read it, and still believe it to be among the best books ever written about how the history of man cannot be understood separately from the history of his diseases. I had never seen this idea articulated as clearly as I did in McNeill's book, in particular the observation that modern cities are possible only because of modern medicine. *Guns, Germs, and Steel*, by Jared Diamond, was an equally vivid study of how microbes have influenced civilization, and it set me wondering how the latest microbes to crash the species barrier might influence humanity's going forward. There are other books that cover the confluence of urbanization and epidemics, in particular *Man and Microbes*, by Arno Karlen, to which I am indebted for its poetic evocation of the biological perils of urban life. *Infections and Inequalities*, by Paul Farmer, and *Epidemics and History*, by Sheldon Watts, were also richly informative. Among the fictional accounts that inspired me, *The Plague*, by Albert Camus, *Journal of the Plague Year*, by Daniel Defoe, and the *Decameron*, by Giovanni Boccaccio, were those that I kept going back to. *The Andromeda Strain*, by Michael Crichton, provided a much-needed respite from academic papers during a reporting trip to Geneva—and remains a strikingly original way to write about a disease outbreak.

The specific impact of China's economic reforms and urbanization on public health are best described in "Health and Macroeconomics in China," *The Country Report for the Second World Health Organization Ministerial Consultive Meeting of Macroeconomics and Health* (10/2003); "Public Health Options for China: Using the Lessons Learned from SARS" (7/18/2003), the Office of the WHO Representative in China; "Emerging Epidemics in China: Coinfections and Mobile Populations," *Population Reference Bureau* (5/2003), Drew Thompson; "Ailing Rural Care Raises Risk of Epidemics," *Inter Press Service* (10/15/1999), Antoancta Bezlova; "Development—China: Growing Cities Foul Up the Environment," *Inter Press Service* (5/1999), Antoaneta Bezlova; "Unhappy Returns," *Time Asia* (12/8/2003), Hannah Beech; "China's Failing Health System," *Time Asia* (5/19/2003), Matthew Forney. The ethical dilemmas that this system imposes on its doctors are well described by Matthew Forney in "Heal Thyself," *Time Asia* (6/16/2003).

CHAPTER 6

That various species were already swapping influenzas and that genetic reassortment was already happening was made clear to me in several influenza- and SARS-related papers for which Guan Yi had either the first, second, or last author's line, in particular "Emergence of Avian H1N1 Influenza Virus in Pigs in China," *Journal of Virology* 70 (1996); "Characterization of H5N1 Viruses That Continue to Circulate in Geese in Southeastern China," *Journal of Virology* 76 (2002);

"H9N2 Influenza Possessing H5N1-like Internal Genomes Continue to Circulate in Poultry in Southeastern China," *Journal of Virology* 73 (2000); "Isolation and Characterization of Viruses Related to the SARS Coronavirus from Animals in Southern China," *Science* 302 (2003); "Genesis of a Highly Pathogenic and Potentially Pandemic H5N1 Influenza Virus in Eastern Asia," *Nature* (7/8/2004).

CHAPTER 8

In trying to understand viruses, I turned to several sources. The starting points for me were the books *Viruses*, by Arnold J. Levine, and *The Coming Plague*, by Laurie Garrett, the latter of which became so dog-eared by the time I finished my first reading that rereading just the highlighted passages took as long as rereading most books. My education continued with several parts of *Field's Virology: Volume 1*, David M. Knipe and Peter M. Howley (editors in chief). Of particular help: "The Origins of Virology," by Arnold J. Levine; "Principles of Virology," by Richard C. Condit; "Principles of Virus Structure," by Stephen C. Harrison; "Virus–Host Cell Interactions," by David M. Knipe, Charles E. Samuel, and Peter Palese; "Pathogenesis of Viral Infections," by Kenneth L. Tyler and Neal Nathanson; "Cell Transformation by Viruses," by Joseph Nevins; and "Virus Evolution," by Victor R. DeFilippis and Luis P. Villarreal. *Notes on Medical Virology*, by Morag C. Timbury, was an early gift from K. Y. Yuen that kept on giving. *Emerging Viruses*, edited by Stephen Morse, was helpful throughout the writing of this book. Morse's own chapter, "Examining the Origins of Emerging Viruses," has provided the blueprint for so much writing on viruses that it was a pleasure for me finally to read the source. For explanations of the mechanics of virology, I drew from two other chapters of that same book: "Virus Detection Systems," by Douglas D. Richman, and "New Technologies for Virus Detection," by David C. Ward. *Viruses, Plagues, and History*, by Michael B. A. Oldstone, also provided a very good introduction to the principles of virology.

CHAPTERS 9 AND 10

Material in these chapters was drawn almost entirely from interviews with the principals involved. I have reconstructed the conversation and meeting dialogue based on the accounts of different participants and have provided the version that seemed the most logical. I have also reviewed a copy of "The Preliminary Report on the Investigation and Consultation at Heyuan People's Hospital" (1/2/2003). The contents of the fax sent by the director of Heyuan Number One Hospital were re-created for me by several of the doctors and officials present at the meeting. None of them had retained a copy. The article "Epidemic Is Only a Rumor" is from the *Heyuan Daily* (1/4/2003), Li Jianhua. Guangzhou's *News Express* would have two stories, "No Reason to Worry" (1/10/2003) and "Heyuan: Back to Normal" (1/13/2003), Xiao Ping.

CHAPTER 11

Censorship and freedom of information in China are frequent topics of conversation and editorializing in the international press. Among the best at illuminating the issues has been Perry Link in articles such as "China: Wiping Out the Truth," *New York Review of Books* (2/24/2005), with He Qinglian; "China: The Anaconda in the Chandelier," *New York Review of Books* (4/11/2002); and "Will SARS Transform China's Chiefs?" *Time Asia* (5/5/2003). The quote from Link in the paragraph beginning "In the other, parallel, system . . ." comes from the third story. *The Tiananmen Papers*, compiled by Zhang Liang and edited by Andrew J. Nathan and Perry Link, provides a look into the Chinese information apparatus, a look that has never been equaled. In "Arrested in China," *New York Review of Books* (9/20/2001), Kang Zhengguo reveals how the party operates in the twenty-first century. "SARS Lays Bare a Contradiction," *Los Angeles Times* (5/1/2003), by Ross Terrill, and "Tight Rein Stays on Mainland Media," *South China Morning Post* (6/23/2003), by Leu Siew Ying, illuminate nicely the specific information issues raised by SARS.

The Zhongshan report referred to in the paragraph beginning "The document the team produced . . ." is entitled "Investigation of Unexplained Pneumonia in Zhongshan City." Later accounts, including "China's Slow Reaction to Fast-Moving Illness," *Washington Post* (4/2/2003), John Pomfret, and "On the Trail of an Asian Contagion," *Time Asia* (3/31/2003), Simon Elegant, et al., would first raise the issue of why this document had not been more widely disseminated. Though *Time* reported the top-secret document first, it was John Pomfret who recognized the importance of the narrow circulation of the document and the pattern of secrecy it represented.

The *People's Daily*, the official Communist Party newspaper, would itself, belatedly, lament the burying of information related to SARS in "SARS, a Valuable Lesson for the Chinese Government to Learn" (June 9, 2003).

CHAPTER 12

The clinical progression of these early cases was reconstructed from physicians' descriptions, based on their recollections of the cases, and from patient charts, where they were available. Among useful accounts of early SARS cases, clinical progression, and lung pathology were "Case Report of the First Severe Acute Respiratory Syndrome Patient in China," *Artificial Organs* (9/2003), Hong-Tao Luo, Min Wu, and Min-Min Wang; "Lung Pathology of Fatal Severe Acute Respiratory Syndrome," *The Lancet* (5/24/2003), John M. Nicholls, Leo L. M. Poon, et al.; "Clinical Progression and Viral Load in Community Outbreak of Coronavirus-associated SARS Pneumonia: A Prospective Study," *The Lancet* (5/24/2003), J. S. M. Peiris, C. M. Chu, et al.; "Description and Clinical Treatment of an Early Outbreak of Severe Acute Respiratory Syndrome in Guangzhou," *Journal of Medical Microbiology* 52 (2003), Z. Zhao, F. Zhang, et al.; "Development of a Standard Treatment Protocol for Severe Acute Respiratory Syndrome," *The Lancet* (5/10/2003), Loletta K.-Y. So, Arthur C. W. Lau, et al.; "Severe Acute Respiratory Syndrome in Guangdong Province of China," *Chinese Journal of Preventive Medicine* (7/2003), He Jian-feng, Xu Ruiheng, et al. There is also a vivid account of the efforts to save early patients in the Chinese book *Heroes of the Anti-SARS Battle*, by Zheng Wenqiong.

I also had access to an internal WHO report on Guangdong.

Details on the history of intubation were drawn from *Fighting Infection*, by Harry F. Dowling. I first read about Avicenna's introducing tubes into the breathing passage from "Anesthetic and Analgesic Practices in Avicenna's Canon of Medicine," *American Journal of Chinese Medicine* (Winter 2000), E. Aziz. Avicenna's most notable work was *The Canon of Medicine*, only a portion of which has been translated into English. When I had trouble visualizing the descriptions given to me by physicians, I consulted with *The Textbook of Surgery*, edited by David C. Sabiston, in particular the chapter "Tracheostomy and Its Complications," by Hermes C. Grillo, and *ACS Surgery*, by Douglas W. Wilmore, Laurence Y. Cheung, et al., in particular the chapter "Initial Emergency Management of Noninjured Patients," by Steven R. Lowenstein and Alden H. Harken.

I also found helpful four very good stories on the early epidemiology of SARS, "Stalking a Killer," *Time Asia* (4/21/2003), Matthew Forney; "Tracking the Roots of a Killer," *Science* (7/18/2003), Dennis Normile and Martin Enserink; "In Chinese Village, Few Clues to Illness," *Washington Post* (4/9/2003), John Pomfret; "Tracing the Path of SARS: A Tale of Deadly Infection," *Los Angeles Times* (4/22/2003), Ching-Ching Ni.

I should add that the book *Twenty-First Century Plague: The Story of SARS*, by Thomas Abraham, does an excellent job describing the outbreak.

CHAPTER 13 AND 14
Drawn from author interviews.

CHAPTER 15
The content of the paragraph beginning "By early February . . ." comes from several members of both bureaucracies in a position to observe, though not from Chou Yenfou and Huang Qingdao themselves. (Neither would comment on the subject.) That these men were in communication with the Ministry of Health in Beijing came from sources in the Ministry of Health.

CHAPTER 16
There were several official Hong Kong inquiries that delved into the flawed lines of communication with the mainland. The two that I drew from were the *Report of the SARS Expert Committee*, chaired by Cyril Chandler and Sian

Griffiths, and the *Report of the Legislative Council Select Committee to Inquire into the Handling of the Severe Acute Respiratory Syndrome Outbreak by the Government and the Hospital Authority* (July 2004). The book *At the Epicenter: Hong Kong and the SARS Outbreak*, by Christine Loh and Civic Exchange, is an excellent account of SARS in Hong Kong, and Loh's chapter "The Politics of SARS: The WHO, Hong Kong and Mainland China" helped me frame my thinking about these issues.

CHAPTER 17

The account of Hong Kong's 1894 plague outbreak is drawn primarily from *The Plague Race*, by Edward Marriott. Two other informative books on the subject are *The Barbary Plague*, by Marilyn Chase, and *Bubonic Plague in Nineteenth-Century China*, by Carol Benedict. I also found helpful "Manchurian Plague: Medicine and Politics, East and West," *Harvard Asia-Pacific Review* 6, no. 2 (2002), William C. Summers, and "China and the Origins of Immunology," a monograph by Joseph Needham.

CHAPTER 18

The quote from the doctor in Renai Hospital comes from ProMed (2/10/2003).

CHAPTER 19

The account of the February 10 press conference comes from several reporters who were there, as well as from two of the participants.

The account of the Roche press conference comes from several reporters who were present.

The quote from Cai Lihui comes from "Stalking a Killer," *Time Asia* (4/21/2003), Matthew Forney.

The quote from Zhang Dejiang comes from a source present at the meeting.

CHAPTER 20

Zhong Nanshan's treatment strategies are laid out in "Diagnosis and Treatment of SARS," *Chinese General Clinical Journal* 19 (2003), N. S. Zhong; "Our Strategies for Fighting Severe Acute Respiratory Syndrome (SARS)," *American Journal of Respiratory and Critical Care Medicine* 168 (2003), N. S. Zhong and G. Q. Zeng; and "Chinese Medical Association and Chinese Association of Chinese Traditional Medicine: Consensus for the Management of Severe Acute Respiratory Syndrome," *Chinese Medical Journal* 116 (2003), N. S. Zhong (editor in chief).

CHAPTER 21

The quote in the first paragraph comes from the *Constitution of the World Health Organization*.

Two papers that describe the evolution of the WHO's mission are "Ebola

Hemorrhagic Fever: Lessons from Kikwit, Democratic Republic of the Congo," *Journal of Infectious Diseases* 179, suppl. 1 (1999), David L. Heymann, Deo Barakamfitiye, Guénaël Rodier, et al.; "Hot Spots in a Wired World: WHO Surveillance of Emerging and Reemerging Infectious Diseases," *The Lancet* 1 (December 2001), David L Heymann, Guénaël R. Rodier, and the WHO Operational Support Team to the Global Outbreak Alert and Response Network. The quote in the paragraph beginning "It was Heymann . . ." comes from the latter paper. The WHO's strategy for combating infectious disease is put forth in "Global Defense Against the Infectious Disease Threat," *Communicable Diseases 2003*, World Health Organization. "On the front line of an epidemic," *South China Morning Post* (6/19/2003), Ella Lee, is a good profile of David Heymann.

The fax referred to in the paragraph beginning "That evening, Henk Bekedam sent . . ." is headed "Reported Outbreak in Guangdong Province" and was sent on February 12 (February 11, Geneva time) from the office of the WHO representative in China.

Stöhr's quote in the last paragraph comes from "A Multicentre Collaboration to Investigate the Cause of Severe Acute Respiratory Syndrome," *The Lancet* 361 (2003), World Health Organization Multicentre Collaborative Network for Severe Acute Respiratory Syndrome (SARS) Diagnosis.

Also informative was the account of the early history of the WHO given in *Scourge: The Once and Future Threat of Smallpox*, by Jonathan B. Tucker.

CHAPTER 22

Drawn from author interviews.

CHAPTER 23

The faxes on the seventeenth and nineteenth were headed "Reported Outbreak in Guangdong Province" and were sent to Liu Peilong, director-general, Department of International Cooperation, Ministry of Health. The fax referred to in the paragraph beginning "Bekedam faxed Liu Peilong a few days later . . ." was sent on February 21, marked "very urgent," and headed "Disease Outbreak in Guangdong and H5N1 Cases in Hong Kong."

CHAPTER 24

The visit of Liu Jianlun was reconstructed from relatives, fellow physicians, and staff at the Metropole Hotel. To get some sense of Dr. Liu's experience, I also rode a bus from Shenzhen and stayed in room 911 at the Metropole Hotel. I cross-referenced my own findings with the WHO internal reports, "Cluster of Pneumonic Illness Linked to Metropole Hotel, Hong Kong" and "Hotel M Cohort Study," both by Denise Werker and Chris Braden. An excellent account of Liu Jianlun is given in "SARS Signals Missed in Hong Kong; Physician's Visit May Have Led to Most Known Cases," *Washington Post* (5/20/2003), Ellen Nakashima.

My recounting of Dr. Liu's treatment has been drawn from author interviews and the *Report of the Hospital Authority Review Panel* (October 2003) and the *Report of the Legislative Council Select Committee* (July 2004).

CHAPTER 25

The claims of the various Chinese laboratories and the initial announcement of *Chlamydia* are best covered in the fine articles "China's Missed Chance," *Science* (7/18/2003), Martin Enserink, with reporting by Ding Yimin and Xiong Lei; and "Beijing Missteps on Virus Show Research Failings," *Asian Wall Street Journal* (6/4/2003), Charles Hutzler. Hong Tao's claims are laid out in "Chlamydia-like and Coronavirus-like Agents Found in Dead Cases of Atypical Pneumonia by Electron Microscopy," *National Medical Journal of China* 83 no. 8 (2003), Hong Tao, et al.

The history of virology was taken from the inimitable *Microbe Hunters*, by Paul de Kruif; from the chapter "The Origins of Virology," by Arnold J. Levine, in *Field's Virology*; from *Viruses*, by Arnold J. Levine; and from *Viruses, Plagues and History*, by Michael D. A. Oldstone.

CHAPTER 26

Danny Yang Chin's account is reconstructed from his co-workers and from the physicians at Hanoi French Hospital.

The descriptions of Carlo Urbani come from his colleagues and friends in Hanoi, Manila, and Geneva. Also useful in reconstructing Urbani's experience and the situation at Hanoi French Hospital were the daily e-mails sent by Urbani to Hitoshi Oshitani at the Western Pacific regional office of the WHO. These were sent on: March 4, "Severe Pneumonia Case in Hanoi"; March 5, "Urgent: Severe Pneumonia Case in Hanoi"; March 6, "Pneumonia in Hanoi: Situation Report"; March 7, "Pneumonia in Hanoi: Situation Report"; and March 8, "Outbreak in Hanoi: Situation Report." There was also a fax from the WHO to Tinh Quan Huan, director of the Department of Preventive Medicine at the Vietnamese Ministry of Health, on March 7, 2003, which revealed the mind-set of Urbani, Brudon, and others in Hanoi at that time.

The Vietnamese Ministry of Health sent a formal request for help via fax on March 12, 2003, to Brian Doberstyn of the World Health Organization in Manila.

An excellent account of the WHO's early efforts is given in "Inside the WHO as It Mobilized to Fight Battle to Control SARS," *Wall Street Journal* (5/2/2003), Margot Cohen, Guatam Nai, and Matt Pottinger.

CHAPTER 27

Drawn from author interviews.

CHAPTER 28

Luong Nguyen's account came from her family.

The Global Outbreak Alert and Reponse Network request was headed "Alert/Request for Assistance: Acute Respiratory Illness, Hanoi, Vietnam" and was sent on March 7, 2003.

CHAPTER 29

Du Ping's account is based on author interviews. I also found very helpful several accounts in *Caijing* magazine. "Fight to Win . . . or Die," *Caijing* (April 23, 2003), and Fight to Win . . . or Die" (part 2), *Caijing* (May 4, 2003), were among the best accounts of the outbreak in Shanxi and Beijing. In English, the first story that pieced together the outbreak in Shanxi was "A 'Superspreader' of SARS," *Washington Post* (May 29, 2003), Philip P. Pan.

CHAPTER 30

The account of the national CDC meeting was reconstructed from interviews with four WHO and two Chinese participants, from two attendees' handwritten notes, from a WHO meeting record filed by Hitoshi Oshitani, and from the "Summary of Urgent Issues Related to H5N1 Cases in Hong Kong SAR," submitted by the WHO expert team.

CHAPTER 31

Drawn from author interviews.

CHAPTER 32

David Heymann first suggested to me that there were three phases of response to an infectious-disease outbreak: denial, panic, and rationale. I also read about this idea in *And the Band Played On*, by Randy Shilts, where the consequences of denial were made dreadfully clear. The article "What Is the Next Plague?" *New York Times* (11/11/2003), Lawrence Altman, was helpful in putting SARS into the context of historical epidemics.

CHAPTER 33

There have been several published accounts of Prince of Wales Hospital's clinical experience combating SARS. Mine was drawn primarily from physicians, nurses, and Hong Kong Hospital Authority officials and was cross-referenced with the *Report of the Legislative Council Selection Committee;* the *Report of the Hospital Authority Review Panel;* "A Major Outbreak of Severe Acute Respiratory Syndrome in Hong Kong," *New England Journal of Medicine* 348, no. 1986–94 (2003), Nelson Lee, David Hui, et al.; "SARS: Experience at the Prince of Wales Hospital, Hong Kong," *The Lancet* 361, no. 1486–87 (2003), Brian Tomlinson and Clive Cockram; "Hospital Vetoed SARS Ward Closure," *South China Morning Post* (7/31/2003), Ella Lee; and the chapter "Healing Myself: Diary of a SARS Patient and Doctor," by Gregory Cheng, from *At the Epicenter*.

CHAPTER 34

The document referred to in the paragraph beginning "This was something new . . ." is the "EPI Update of Acute Respiratory Syndrome in Hanoi," based on data provided by the WHO office in Hanoi.

Much has been written about the 1993 Sin Nombre hantavirus outbreak in the American Southwest, which, as a fatal respiratory virus, provided a frightening precursor to SARS. *Virus X*, by Frank Ryan, and *The Coming Plague*, by Laurie Garrett, both eloquently recount that outbreak, which killed twenty-five. "Hantavirus Pulmonary Syndrome: An Emerging Infectious Disease," *Science* (11/5/1993), James M. Hughes, C. J. Peters, et al., is a very good primer on the subject.

My calculation of how long it took the virus to infect three hundred people is based on the number of AIDS cases in the United States in April 1982, according to Randy Shilts, and on when Dr. Grethe Rask returned to Denmark from Zaire in 1976. While it is unlikely that Rask was the index patient for AIDS, the other alternatives would be even earlier, making it a longer burn than six years. However, it is also very likely there were more than three hundred cases outside the United States, so the comparison could be extended in duration as well as number of cases, still making the same point, I believe.

The lab network received much attention during the outbreak and was covered in both the academic and the mainstream press. Among the best accounts, "A Multicentre Collaboration to Investigate the Cause of Severe Acute Respiratory Syndrome," *The Lancet* 361 (2003), World Health Organization Multicentre Collaborative Network for Severe Acute Respiratory Syndrome (SARS) Diagnosis; "How Global Effort Found SARS Virus in Matter of Weeks," *Asian Wall Street Journal* (April 17–20, 2003), Matt Pottinger, Elena Cherney, Gautam Naik, and Michael Waldholz; "Deferring Competition, Global Net Closes in on SARS," *Science* (4/11/2003), Martin Enserink and Gretchen Vogel; and "Step by Step, Scientists Track Mystery Ailment," *New York Times* (4/1/2003), Lawrence K. Altman.

CHAPTER 35

The history of the origins of the University of Hong Kong came partially from *The Life and Work of Sir Patrick Manson*, by Philip H. Manson-Bahr and A. Alcock.

CHAPTER 36

The quote from Jim LeDuc was lifted, with much gratitude, from *Virus X*, by Frank Ryan.

Ksiazek and/or Zaki have appeared in *Virus X*, *The Coming Plague*, *The Hot Zone*, and *Beating Back the Devil*, by Marilyn McKenna, among other texts.

CHAPTER 37

Three good accounts of Taiwan's late and brutal SARS outbreaks are "Fever Pitch," *Time Asia* (5/26/2003), Andrew Perrin; "Living on a Prayer," ibid.

(6/2/2003), Andrew Perrin; "Public Alarm in Taipei over SARS Measures," *South China Morning Post* (5/11/2003), Joe Tang.

The Global Alert was announced by the press release "WHO Issues a Global Alert about Cases of Atypical Pneumonia," of March 12, 2003.

CHAPTER 38

Bryan Walsh's story was "Outbreak in Asia," *Time Asia* (3/24/2003).

The Shanghai doctor anecdote in the paragraph beginning "Other medical officials . . ." comes from "How Bad Is It?" *Time Asia* (5/5/2003), Hannah Beech.

CHAPTER 39

The e-mail in the paragraph beginning "Peiris agreed that it was time . . ." was sent on March 3, 2003, under the subject line "Family contacts of index patients (Liu)."

CHAPTER 40

The exchange among the University of Hong Kong virologists is well illustrated in "Facing the Unknowns of SARS" monograph by K. Y. Yuen.

CHAPTER 41

Drawn from author interviews and from *A Defining Moment: How Singapore Beat SARS*, by Chua Mui Hoong; and *The Silent War*, by Ng Wan Ching, edited by Lee Jin Jin, Zane Chan, and Olivia Branson; and "Inquiry Clears SARS-hit Doctor Taken off Plane," *Straits Times* (9/2/2003), Salma Khalik.

CHAPTER 42

Drawn from author interviews.

The Global Alert appeared as "World Health Organization Issues Emergency Travel Advisory," on March 15, 2003.

CHAPTER 43

John Tam sent an e-mail on his preliminary findings to ProMed, which appeared under the heading "Severe Acute Respiratory Syndrome—Worldwide (9)," on March 19, 2003. The WHO's announcement was called "Severe Acute Respiratory Syndrome (SARS) Multi-country Outbreak—Update 4," and was posted on March 19, 2003.

David Heymann's quote in the paragraph beginning "Chinese University had the bug . . ." is from a World Health Organization SARS press briefing made on March 21, 2003.

My explanation of the relationship between the human immune system and viruses is drawn from several sources, including *Virus*, by Arnold J. Levine; *The Coming Plague*; and *Field's Virology*.

The description of the grid in the paragraph beginning "Meanwhile, downstairs in the . . ." was inspired—if not outright lifted—from a passage in *The Hot Zone*, by Richard Preston.

"Coronavirus as a Possible Cause of Severe Acute Respiratory Syndrome," *The Lancet* (4/8/2003), J. S. M. Peiris, S. T. Lai, K. Y. Yuen, et al., was the paper that announced the breakthrough.

Twenty-First-Century Plague, by Thomas Abraham, and "The Struggle to Profile a Killer," *South China Morning Post* (6/25/2003), Mary Ann Benitez, both provided very good accounts of the University of Hong Kong team.

The e-mail from Peiris was headed "Re: Teleconference" and was sent on March 21 at 4:22 P.M. Hong Kong time.

CHAPTER 44

I was fortunate enough to have access to several files prepared for *Time* magazine by Kay Johnson on the situation in Vietnam. Additionally, *Beating Back the Devil*, by Maryn McKenna, does an excellent job of describing the Hanoi outbreak and Urbani's hospitalization in Bangkok. "Disease's Pioneer Is Mourned as a Victim," *New York Times* (4/8/2003), Donald G. McNeil Jr., is a lovely, and very helpful, eulogy of Carlo Urbani.

The fax from Doberstyn was "Critical Care/Infectious Disease Specialist" and was sent on March 20, 2003.

CHAPTER 45

Several accounts crediting CDC efforts appeared, including: "NCID-led Laboratory Teams Discover Novel Coronavirus, Sequence Entire Viral Genome," *National Center for Infectious Diseases Focus* (Spring 2003); "CDC Lab Analysis Suggests New Coronavirus May Cause SARS," CDC Press Release (3/24/2003); and "Global Collaboration on SARS Bears Fruit," *New York Times* (5/26/2003), Denise Grady and Lawrence K. Altman. The CDC team published its findings in "A Novel Coronavirus Associated with Severe Acute Respiratory Syndrome," *New England Journal of Medicine* (5/15/2003), Thomas G. Ksiazek, Dean Erdman, Cynthia S. Goldsmith, et al. "Stopping a Scourge," *Smithsonian* (9/2003), David Brown, is a thorough narrative account of CDC efforts. There is also a more general profile of Julie Gerberding, which appeared as "Fear Factor," *Vogue* (July 2003), Robert Sullivan.

The e-mail mentioned in the paragraph beginning "This sample had come . . ." was sent on Friday, March 21, 2003, at 12:11 P.M.

CHAPTER 46

This chapter is drawn primarily from author interviews and observations.

The quote from Edward Kass comes from "Legionnaire's Disease," *New England Journal of Medicine* (12/1/1977), E. H. Kass.

The quote from E. K. Yeoh comes from "Outbreak in Asia," *Time Asia* (3/24/2003), Bryan Walsh.

The quote from Laurie Garrett comes from page 569 of *The Coming Plague*, as does the concept of expressing the surface area of the human lung.

The material on immunology was informed by *Viruses*, by Arnold Levine, and by the relevant chapters in *Field's Virology*.

Information on Amoy Gardens comes from author interviews as well as from "Outbreak of Severe Acute Respiratory Syndrome (SARS) at Amoy Gardens, Kowloon Bay, Hong Kong," Hong Kong Department of Health (April 2003); "Report of the Legislative Council Select Committee to Inquire into the Handling of the Severe Acute Respiratory Syndrome Outbreak by the Government and Hospital Authority," Legislative Council of the Hong Kong Special Administrative Region (July, 2004); "Report of the SARS Expert Committee," SARS Expert Committee (2003).

CHAPTER 47

The quote in the first paragraph comes from *Virus Hunter*, by C. J. Peters and Mark Olshaker.

The material on the role of the Hospital Authority is drawn from author interviews as well as from the July 2004 "Report of the Legislative Council Select Committee" cited in chapter 46; "Stinging Rebuke for SARS Chiefs," *South China Morning Post* (8/22/2003), Mary Ann Benitez; "SARS Nurses May Sue Hospital Authority," *South China Morning Post* (10/1/2003), Mary Ann Benitez; "War on SARS Had No Command Chain," *South China Morning Post* (10/8/2003), Ella Lee.

The rats-as-vector theory appeared as "Possible Role of an Animal Vector in the SARS Outbreak at Amoy Gardens," *The Lancet* (8/16/2003), Stephen K. C. Ng.

CHAPTER 48

In the first paragraph, Margaret Chan's explanation would be given in her testimony to the Legislative Council Select Committee.

The quote in the paragraph beginning "Air traffic into and out . . ." comes from "Living in a Hot Zone," *Time Asia* (4/7/2003), Jim Erickson.

Details of the prank come, in part, from "Year of Guidance for SARS Site Prankster," *Standard* (8/5/2003), staff reporter.

Indira A. R. Lakshmanan's account appeared as "Exploring China's Silence on SARS—New Details Surface on Initial Cover-up," *Boston Globe* (5/23/2003). Two other helpful stories on SARS and air travel are "Carrier of New Virus Made 7 Flights Before Treatment," *New York Times* (4/11/2003), Keith Bradsher, and "A Germ Has a Ticket to Ride, and Airlines Can't Stop It," *Hartford Courant* (4/28/2003), Garret Condon.

CHAPTER 49

The account of the barefoot doctor comes from "Regional Affair," *Time Asia* (4/28/2003), Hannah Beech.

The Julie Gerberding editorial was published as "Faster . . . but fast enough?" *New England Journal of Medicine* (4/2/2003), Julie Louise Gerberding.

The quote from C. J. Peters comes from page 94 of *Virus Hunter*.

The quote from Julie Gerberding beginning "The biggest unknown . . ." comes from a March 29, 2003, CDC press conference.

The quote from Gerberding beginning "We don't know . . ." comes from an April 22, 2003, CDC press conference.

Much of the economic data comes from "The Surprising Impact of SARS," *Time* (5/5/2003), Michelle Orecklin, and from "The Truth About SARS," *Time* (5/5/2003), Michael D. Lemonick and Alice Park.

The WHO team gave an account of SARS and China in "Role of China in the Quest to Define and Control Severe Acute Respiratory Syndrome," *Emerging Infectious Diseases* (9/2003), Robert F. Breiman, Meirion R. Evans, Wolfgang Preiser, et al.

My thinking on the influence of epidemics on society and commerce was shaped in part by two wonderful books on the subject: *In the Wake of the Plague: The Black Death and the World It Made*, by Norman Cantor, and *The Great Plague: The Story of London's Most Deadly Year*, by A. Lloyd Moote and Dorothy C. Moote.

CHAPTER 50

Drawn primarily from author interviews and from: Dr. Jiang's letter to the chairman and vice-chairman of the National People's Congress, sent on March 11; "An Honest Doctor," *Beijing Review* (2/11-2/17, 1991), Zhou Chunnong; "Postmortem," *Time Asia* (6/11/2004), Perry Link.

The material regarding the events of June 3–4 comes from *The Tiananmen Papers*, compiled by Zhang Liang, edited by Andrew J. Nathan and Perry Link. The quote comes from page 374 of the hardcover edition.

The quote beginning "Lying before me were . . ." comes from Dr. Jiang's letter to the National People's Congress.

I also found very helpful "Jiang Yanyong: The People's Interests Matter Most," *Sanlian Life Weekly* (6/9/2003), Li Jing.

CHAPTER 51

The paper in which the University of Hong Kong team announced its breakthrough was "Coronavirus as a Possible Cause of Severe Acute Respiratory Syndrome," *The Lancet* (4/8/2003), J. S. M. Peiris, S. T. Lai, K. Y. Yuen, et al.

CHAPTER 52 THROUGH 53

Drawn from author interviews.

CHAPTER 54

The quote from Li Liming in the paragraph beginning "The WHO team, from the . . ." comes from other participants in the meetings.

The quote from Gro Harlem Brundtland in the paragraph beginning "For the WHO team—all scientists . . ." comes from page 101 of *Twenty-First Century Plague*, by Thomas Abraham.

CHAPTER 55

The quote from Huang Huahua comes from "SARS Struggle," *Nanfang Daily* (4/8/2003), staff reporter.

The quote from the Beijing newspaper in that same paragraph comes from "Outbreak Gave China's Hu an Opening," *Washington Post* (5/13/2003), John Pomfret.

The quote from Kong Quan in the same paragraph comes from a Foreign Ministry press conference on March 26, 2003.

The quotes in the paragraph beginning "A physician at Beijing's . . ." come from the story posted as "Beijing's SARS Attack," TIMEasia.com (4/8/2003), Susan Jakes.

The number of text messages and the second text message given as an example come from "Outbreak Gave China's Hu an Opening," cited above.

The quote from the blog *Peking Duck* is from April 9, 2003, http://peking duck.org/archives/00152.php. In general, the blog provided a fascinating commentary on life in Beijing during SARS and was a great help to me as I reconstructed that period.

CHAPTER 56

Drawn from author interviews, from Huang Yong's notes, and from "Unmasking a Crisis," *Time Asia* (4/21/2003), Hannah Beech, with reporting by Susan Jakes, Huang Yong, and Bu Hua.

CHAPTER 57

The WHO discussed its findings in "Visit of WHO Expert Team to Review the Outbreak of Atypical Pneumonia in Guangdong Province, 24 March–9 April 2003," Wolfgang Preiser, Meirion Evans, James Harvey Maguire, et al.

The account of Wen Jiabao's visit to the national CDC comes from "Outbreak Gave China's Hu an Opening," cited above.

The account of Bi Shengli comes from "China's Crisis Has a Political Edge," *Washington Post* (4/27/2003), John Pomfret.

CHAPTER 58

The case fatality ratio and quotes from the WHO come from "Update 49—SARS Case Fatality Ratio, Incubation Period" (May 7, 2003) http://www.who.int/csr/sarsarchive.

The Department of Health press conference was held on April 17, 2003. The information about rat droppings comes from the *Report of the Legislative Council Select Committee.*

The quotes in the paragraph beginning "The prevailing hypothesis . . ." come from "Outbreak of Severe Acute Respiratory Syndrome at Amoy Gardens, Kowloon Bay, Hong Kong: Main Findings of the Investigation," by the Hong Kong Department of Health and eight other agencies.

E. K. Yeoh's quote is from the Department of Health press conference of April 17.

My doubts were greatly informed by the paper "Evidence of Airborne Transmission of the Severe Acute Respiratory Syndrome Virus," *New England Journal of Medicine* (4/22/2004), Ignatius T. S., Yuguo Li, Tze Wai Wong, et al.

The genome sequence was published in "The Genome Sequence of the SARS-Associated Coronavirus," *Science* (May 1, 2003), Marco A. Marra, Steven J. M. Jones, Caroline Astell, et al.

The study of the SARS virus evolution was published in "Molecular Evolution of the SARS Coronavirus During the Course of the SARS Epidemic in China," *Science* (1/29/2004), Chinese SARS Molecular Epidemiology Consortium.

CHAPTERS 59-62

Drawn from author interviews.

CHAPTER 63

The quote in the paragraph "The doctor then recounted . . ." comes from "Regional Affair," *Time Asia* (4/28/2003), Hannah Beech, with reporting by Bu Hua, Matthew Forney, Huang Yong, Susan Jakes, and Kaiser Kuo.

Susan Jakes's story was posted as "Beijing Hoodwinks WHO Inspectors," TIMEasia.com (4/18/2003).

CHAPTER 64

The question in the first paragraph was posed to Alan Schnur at a press conference given by the WHO team in China on April 16, 2003.

"Health Officials Tell China Virus Data Aren't Trusted." *Asian Wall Street Journal* (4/17/20), Peter Wonacott, Susan V. Lawrence, and Matt Pottinger, outlines the WHO's position at this time.

CHAPTER 65

Two very good stories on the situation in Beijing at this time are "SARS Panic Growing," *Newsday* (4/25/2003), Laurie Garrett, and "How Bad Is It?" *Time Asia* (5/5/2003), Hannah Beech.

Wen Jiabao's quote in the paragraph beginning "Now, Hu Jintao embarked . . ." comes from a Xinhua news release (3/14/2003).

The quote in the paragraph beginning "In Beijing, it was as if . . ." comes from "How Bad Is It?" cited above.

The WHO travel advisories were published as "Update 37—WHO extends Its SARS-related Travel Advice to Beijing and Shanxi Province in China and to Toronto Canada" (4/23/2003).

The quote relating to India comes from "Update 33—On Hong Kong and China, First SARS Case Reported in India" (4/18/2003).

CHAPTER 66

Dr. Zhang Hanwei's quotes come from "Regional Affair," cited above.

CHAPTER 67

Drawn from author interviews.

CHAPTER 68

My explanations of the history and origins of asepsis were informed by *Fighting Infection*, by Harry F. Dowling, and *Framing Disease*, edited by Charles E. Rosenberg and Janet Golden, in particular the chapter "Parasites and the Germ Theory of Disease," by John Farley.

The quote from Louis Pasteur comes from *Microbe Hunters*, by Paul de Kruif.

CHAPTER 69

Virus X, by Frank Ryan; *The Coming Plague*, by Laurie Garrett; *The Hot Zone*, by Richard Preston; *Six Modern Plagues*, by Mark Jerome Walters; and *The New Killer Diseases*, by Elinor Levy and Mark Fischetti, are among the many recent books that delve into the subject of zoonotic diseases and emerging infections.

Microbial Threats to Health: Emergence, Detection, and Response, edited by Mark S. Smolinski, Margaret A. Hamburg, and Joshua Lederberg, was a constant source and companion whenever my writing turned to emerging diseases.

The quote in the paragraph beginning "The Era of Wild Flavor . . ." comes from "Animal Origins of SARS Coronavirus: Possible Links with the International Trade in Small Carnivores," *Philosophical Transactions—Royal Society of London Series* (7/24/2004), Diana Bell, Scott Roberton, and Paul R. Hunter.

The breakthrough paper that emerged from this work is "Isolation and

Characterization of Viruses Related to the SARS Coronavirus from Animals in Southern China," *Science* (9/4/2003), Guan Yi, B. J. Zheng, Y. Q. He, et al.

CHAPTER 70

The statistics in the paragraph beginning "Despite the cause . . ." come from "WHO Gives the All-Clear," *South China Morning Post* (6/24/2003), Jimmy Cheung, Chow Chung-yan, and Benjamin Wong.

The information about Operation Green Sword comes from "China Lifts Wild Animal Ban Despite Risk of Link to SARS," *Wall Street Journal* (8/14/2003), Matt Pottinger and Ben Dolven.

CHAPTERS 71 AND 72

Drawn from author interviews and from "Averting an Outbreak," *Time Asia* (1/11/2004), Karl Taro Greenfeld.

"New Clues Zero In on Killer Virus," *South China Morning Post* (1/21/2004), Ella Lee, was also a useful story on the reemergence

CHAPTER 73

This account of the H5N1 outbreak in Vietnam, as well as the quote from Van Dang Ky, comes from "On High Alert," *Time Asia* (1/24/2004), Karl Taro Greenfeld, with reporting by Do Minh Thuy and Sam Taylor.

The quote in the paragraph "But if we had been lucky . . ." comes from "Five Myths About China and SARS," *Asian Wall Street Journal* (5/13/2003), Baopu Liu.

An account of the arrests is given in "7 at Chinese Paper Held After Report of New SARS Case," *New York Times* (1/7/2004), Joseph Kahn.

Dr. Jiang Yanyong's arrest was written about in "Prisoner of Conscience," *Time Asia* (7/19/2004), Susan Jakes.

ACKNOWLEDGMENTS

This book could not have been written without the generous help and support of Jodi Xu, who helped me research, report and, most important, to make sense of what we were discovering.

Susan Jakes and Huang Yong, two of my colleagues in Beijing, generously shared their stories with me.

I also owe a great debt to Flora Yi, Davena Mok, Alice Fung, and Kim Wu. They were all my collaborators.

There were so many scientists, doctors, public health officials, nurses, and patients who shared their time and memories with me that merely listing them seems insufficient recompense for what they have given me, but this is all I have to offer at the moment.

In Hong Kong: Guan Yi, Malik Peiris, K. Y. Yuen, Zheng Bo Jian, John Nichols, Joseph Sung, Wilina Lim, K. H. Chan, Henry Chan, E. K. Yeoh, Margaret Chan, Stephen Tsoi, John Tam, Dennis Lo, Ken Shortridge, Rob Webster, Lo Wiong Lok, Danny Yip, C. H. Leong, Trevor Ellis, Keith Watkins, Lois Yip, Michael Young, K. K. Liu, Tam Lai-shan, Wang Shouje, Sam Cheung, Raymond Wong, Chui See-to, Gavin Joynt, and Peter Kunick.

In China: Jiang Yanyong, Zhong Nanshan, Xu Ruiheng, Deng Zide, Xiao Zhenglun, Luo Huiming, Du Lin, Zhuang Zhixiong, Huang Wenji, Lian Junhao, He Huiyu, Deng Tei Tao, Liu Erming, Zhong Xue Bing, Huang Datian, Ma Xiaowei, Tang Xiaoping, Li Jinguo, Xie Jinkei, Cao Hong, Xiao Ping, Song Zuing, Zhou Fu Yao, Luo Hong Meisu, Liu Pilong, Tang Toh Mah, Su Qinshan, Jingping Zheng, and Peng Shangde.

In Singapore: Lim Suet Wun, Olivia Branson, Balaji Sadasivan, Dessmon Tai, and Genevieve Wilkinson.

In Hanoi: Kay Johnson and Do Minh Thuy.

At the WHO: David Heymann, Klaus Stöhr, Dick Thompson, Hitoshi Oshitani, Brian Dobertsyn, Eva Cristofle, Pascale Brudon, Mike Ryan, Guénaël Rodier, Henk Bekedam, Alan Schnur, Julie Hall, John McKenzie, Pierre Fermenty, Maria Cheng, Meirion Evans, Wolfgang Preiser, Rob Breiman, Aileen Plant, Hume Field, Rob Condon, and Richard Nesbit.

At the United States CDC: Keiji Fukuda, James Maguire, Thomas Ksiazek, Sherif Zaki, Cynthia Goldsmith, Charles Humphrey, Julie Gerberding, and James Hughes.

At Erasmus: A. B. Oosterhaus, Thijs Kuiken, and Ron A. M. Fouchier.

My colleagues at TIME Asia who were generous and supportive when I needed them most: Hannah Beech, Matthew Forney, Bu Hua, Alex Perry, Dennis Wong, Cecelia Wong, Bryan Walsh, Jim Erickson, Anthony Spaeth, Aryn Baker, Zoher Abdoolcarim, William Green, Neil Gough, Ilya Garger, Virginia Lau, Yuman Wong, Kate Drake, Chaim Estulin, Austin Ramzy, Leslie Leung, Queenie Chung, Chi Cham, K. K. Kwok, and Wei Leng Tay.

I have to extend a big *doh je* to the reporters and editors of *Southern Metropolis Daily* who provided help and support throughout my research, despite their own repeated run-ins with the authorities.

I owe special debts of gratitude to Walter Isaacson, who hired me at TIME, Adi Ignatius, who moved me to TIME Asia, and to Terry Mcdonell, who brought me over to *Sports Illustrated*. Norm Pearlstine, John Huey, and Jim Kelly at TIME Inc. gave me space and time when I needed them.

Adi Ignatius and Peter Palese thoughtfully read and commented on this manuscript.

William Areson in Hong Kong and Herb and Gail Crews in Pacific Palisades, California, generously provided me with space to write.

The fellowship up on Borrett Road and a very different fellowship at the Aspen Institute were great sources of inspiration.

Daniel Eisenberg came up with the title. Nick Pachetti, Ptolemy Tompkins, Evan Wright, William Gibson, and William T. Vollman were friends of this book.

The wise Mark Bryant at HarperCollins got this book started, and the sage Gail Winston blessed me with her intelligence and support. Also at HarperCollins, Katherine Hill was a steady hand throughout.

Gail Ross, as always, was on my side.

To Josh Greenfeld and Foumiko Kometani, thank you, thank you, thank you.

To Silka, Esmee, and Lola, all my love and gratitude.

On November 18, 2003, after I had been working on this book for five months, my briefcase was snatched in Hong Kong's Hung Hom station. Inside was my laptop computer on which I had written the first 40,000 words of this book and many of my working notes and interviews. I had been backing up my work on an iPod, which I had also unwisely shoved into that briefcase just moments before it was swiped.

I felt as if all my dreams had been destroyed.

On December 19, 2003, I began to rewrite what has become this book. I relate this only because some reader might one day suffer a similar loss. It is possible to start again.

INDEX